Lecture N

Springer-Verlag Berlin Heidelberg GmbH

The Editorial Policy for Proceedings

The series Lecture Notes in Physics reports new developments in physical research and teaching – quickly, informally, and at a high level. The proceedings to be considered for publication in this series should be limited to only a few areas of research, and these should be closely related to each other. The contributions should be of a high standard and should avoid lengthy redraftings of papers already published or about to be published elsewhere. As a whole, the proceedings should aim for a balanced presentation of the theme of the conference including a description of the techniques used and enough motivation for a broad readership. It should not be assumed that the published proceedings must reflect the conference in its entirety. (A listing or abstracts of papers presented at the meeting but not included in the proceedings could be added as an appendix.)
When applying for publication in the series Lecture Notes in Physics the volume's editor(s) should submit sufficient material to enable the series editors and their referees to make a fairly accurate evaluation (e.g. a complete list of speakers and titles of papers to be presented and abstracts). If, based on this information, the proceedings are (tentatively) accepted, the volume's editor(s), whose name(s) will appear on the title pages, should select the papers suitable for publication and have them refereed (as for a journal) when appropriate. As a rule discussions will not be accepted. The series editors and Springer-Verlag will normally not interfere with the detailed editing except in fairly obvious cases or on technical matters.
Final acceptance is expressed by the series editor in charge, in consultation with Springer-Verlag only after receiving the complete manuscript. It might help to send a copy of the authors' manuscripts in advance to the editor in charge to discuss possible revisions with him. As a general rule, the series editor will confirm his tentative acceptance if the final manuscript corresponds to the original concept discussed, if the quality of the contribution meets the requirements of the series, and if the final size of the manuscript does not greatly exceed the number of pages originally agreed upon. The manuscript should be forwarded to Springer-Verlag shortly after the meeting. In cases of extreme delay (more than six months after the conference) the series editors will check once more the timeliness of the papers. Therefore, the volume's editor(s) should establish strict deadlines, or collect the articles during the conference and have them revised on the spot. If a delay is unavoidable, one should encourage the authors to update their contributions if appropriate. The editors of proceedings are strongly advised to inform contributors about these points at an early stage.
The final manuscript should contain a table of contents and an informative introduction accessible also to readers not particularly familiar with the topic of the conference. The contributions should be in English. The volume's editor(s) should check the contributions for the correct use of language. At Springer-Verlag only the prefaces will be checked by a copy-editor for language and style. Grave linguistic or technical shortcomings may lead to the rejection of contributions by the series editors. A conference report should not exceed a total of 500 pages. Keeping the size within this bound should be achieved by a stricter selection of articles and not by imposing an upper limit to the length of the individual papers. Editors receive jointly 30 complimentary copies of their book. They are entitled to purchase further copies of their book at a reduced rate. As a rule no reprints of individual contributions can be supplied. No royalty is paid on Lecture Notes in Physics volumes. Commitment to publish is made by letter of interest rather than by signing a formal contract. Springer-Verlag secures the copyright for each volume.

The Production Process

The books are hardbound, and the publisher will select quality paper appropriate to the needs of the author(s). Publication time is about ten weeks. More than twenty years of experience guarantee authors the best possible service. To reach the goal of rapid publication at a low price the technique of photographic reproduction from a camera-ready manuscript was chosen. This process shifts the main responsibility for the technical quality considerably from the publisher to the authors. We therefore urge all authors and editors of proceedings to observe very carefully the essentials for the preparation of camera-ready manuscripts, which we will supply on request. This applies especially to the quality of figures and halftones submitted for publication. In addition, it might be useful to look at some of the volumes already published. As a special service, we offer free of charge LaTeX and TeX macro packages to format the text according to Springer-Verlag's quality requirements. We strongly recommend that you make use of this offer, since the result will be a book of considerably improved technical quality. To avoid mistakes and time-consuming correspondence during the production period the conference editors should request special instructions from the publisher well before the beginning of the conference. Manuscripts not meeting the technical standard of the series will have to be returned for improvement.

For further information please contact Springer-Verlag, Physics Editorial Department II, Tiergartenstrasse 17, D-69121 Heidelberg, Germany

Oluş Boratav Alp Eden Ayşe Erzan (Eds.)

Turbulence Modeling and Vortex Dynamics

Proceedings of a Workshop
Held at Istanbul, Turkey,
2–6 September 1996

Springer

Editors

Oluş Boratav
Department of Mechanical and Aerospace Engineering
University of California
Irvine, CA 92697, USA
e-mail: oboratav@maemaster.eng.uci.edu

Alp Eden
Department of Mathematics, Boğaziçi University
80815 Bebek, Istanbul, Turkey
e-mail: eden@boun.edu.tr

Ayşe Erzan
Department of Physics, Istanbul Technical University
80626 Maslak, Istanbul, Turkey
e-mail: erzan@sariyer.cc.itu.edu.tr

Cataloging-in-Publication Data applied for.

Die Deutsche Bibliothek - CIP-Einheitsaufnahme

Turbulence modeling and vortex dynamics : proceedings of a workshop, held at Istanbul, Turkey, 2 - 6 September 1996 / Oluş Boratav ... (ed.).
(Lecture notes in physics ; Vol. 491)
ISBN 978-3-662-14140-3 ISBN 978-3-540-69119-8 (eBook)
DOI 10.1007/978-3-540-69119-8

ISSN 0075-8450
ISBN 978-3-662-14140-3

Originally published by Springer-Verlag Berlin Heidelberg New York in 1997
Softcover reprint of the hardcover 1st edition 1997

Typesetting: Camera-ready by the authors/editors
Cover design: *design & production* GmbH, Heidelberg
SPIN: 10550780 55/3144-543210 - Printed on acid-free paper

Preface

It is a pleasure to present to the scientific community this volume consisting of the proceedings of the "Workshop on Turbulence Modeling and Vortex Dynamics" held at the Istanbul Technical University, Istanbul, Turkey, from September 2 to 6, 1996. It is an even greater pleasure to begin by thanking the lecturers and invited speakers of this workshop for their selfless collaboration, which made the meeting possible, their outstanding contributions, which led to its being a scientific success, and the extra burden that they shouldered in agreeing to submit their reports for publication in these proceedings.

Bringing together physicists, mathematicians, and engineers, the workshop aimed at achieving a balance between the universal and non-universal or more practical aspects of the problem, as well as giving a flavor of mathematical rigour. The significant advances achieved in building a statistical description of turbulence, which can be found in these reports, attests to the timeliness of this workshop.

I. Procaccia, P. Constantin, Zhen-Su She, J.-P. Eckmann, A. Fursikov, and S. Biringen were the main speakers who gave 2-3 hours of lectures. With the exception of the elegant series of lectures by J.-P. Eckmann on "Evolution equations of the hydrodynamic type in infinite domains," their contributions have been brought together in this volume. In addition, one-hour talks were given by L. Biferale, O. Boratav, Y. Couder, H. Dekker, U. Frisch, O.Y. Imanuvilov, V. Kalantorov, R. Kaykayoğlu, A. Libchaber, and A. Wirth. In arranging the contributions, we have tried to achieve some degree of coherence by putting related talks next to each other, while trying to be as faithful to their order of presentation as possible.

Most of the participants were young researchers who were new to this field. Their enthusiasm, which carried over to discussions by the poolside late into the night, made this workshop a worthwhile exercise for all of us. We hope that the list of participants included at the end will aid in the continued exchange of recent work and ideas.

We would like to take this opportunity to express our gratitude to Prof. E. S. Şuhubi and Dr. Haluk Örs, co-organizers of the workshop, for their invaluable collaboration. It is a pleasant duty to thank our young friends Yıldız Silier and İpek Erkin for expert assistance during the workshop and İnanç Birol, Ziya Perdahçı and Kayhan Ülker who were always ready with a helping hand. We also thank Mrs. Gülizar İplikçi for her warm hospitality at the Istanbul Technical University guesthouse. Thanks are also due to the Economic and Social History Foundation of Turkey for bringing Istanbul and history closer to our reach. Last but not least, we would like to thank Sabine Lehr and Brigitte Reichel-Mayer from Springer-Verlag for their precious help in the preparation of these proceedings for publication.

The Workshop on Turbulence Modeling and Vortex Dynamics was made possible by financial support from the Scientific and Technical Research Council of Turkey (TUBITAK), under the International Symposia Program, the United Nations UNISTAR program, Çelebi Holding, the Boğaziçi University Foundation, and the Istanbul Technical University Foundation.

Irvine – İstanbul
April 1997

Oluş Boratav
Alp Eden
Ayşe Erzan

Table of Contents

List of Participants

E. Abit abit@sariyer.cc.itu.edu.tr
Department of Engineering Sciences, Istanbul Technical University
80626 Maslak, Istanbul, Turkey

A. Ak
Middle East Technical University
Department of Mechanical Engineering, Ankara, Turkey

B. Andreotti andreott@physique.ens.fr
45 Rue d'Ulm, 75005 Paris, France

K. Atalık atalik@hamlin.boun.edu.tr
Department of Mechanical Engineering, Boğaziçi University
80815 Bebek, Istanbul, Turkey

L. Biferale biferale@roma2.infn.it
Universita di Roma "Tor Vergata", Dip. di Fisica, 00133 Roma, Italy.

S. Biringen biringen@spot.colorado.edu
Department of Aerospace Engineering Sciences, University of Colorado,
Campus Box 429, Boulder, Colorado 80309-0429, USA

İ. Birol inanc@caju.phys.boun.edu.tr
Department of Physics, Boğaziçi University
80815 Bebek, Istanbul, Turkey

O. Boratav oboratav@maeroot.eng.uci.edu
University of California, Mechanical and Aerospace Engineering Dept.,
Irvine, CA 92697-3975, USA

E. Can
Yeşil Köşk sokak, No: 34/4, 81040 Kadıköy, Istanbul, Turkey

M. Can mcan@sariyer.cc.itu.edu.tr
Department of Mathematics, Istanbul Technical University
80626 Maslak, Istanbul, Turkey

P. Constantin const@cs.uchicago.edu
Department of Mathematics, University of Chicago
Chicago, Illinois 60637, USA

Y. Couder couder@peterpanf.ens.fr
Laboratoire de Physique Statistique, Ecole Normale Superieure
24 Rue Lhomond, 75231 Paris Cedex 05, France

A. Çetin
Department of Physics, Trakya University, Edirne, Turkey

H. Dekker gkun2@fel.tno.nl
TNO Physics and Electronics Laboratory & Institute for Theoretical Physics, University of Amsterdam, P.O.Box 96864, 2509 JG Den Haag, The Netherlands

J-P. Eckmann eckmann@sc2a.unige.ch
Department de Physique Theorique, Universite de Geneve, 24 Ernest Ansermet, 1211 Geneva 4, Switzerland

A. Eden eden@boun.edu.tr
Department of Mathematics, Boğaziçi University
80815 Bebek, Istanbul, Turkey

H. Erbay feherbay@tritu.bitnet
Department of Mathematics, Istanbul Technical University
80626 Maslak, Istanbul, Turkey

S. Erbay feherbay@tritu.bitnet
Department of Mathematics, Istanbul Technical University
80626 Maslak, Istanbul, Turkey

Ş. Ertürk
Ship Construction and Marine Sciences Department
Istanbul Technical University, Ayazağa, Istanbul, Turkey

A. Erzan erzan@sariyer.cc.itu.edu.tr
Department of Physics, Faculty of Sciences and Letters
Istanbul Technical University, 80626 Maslak, Istanbul, Turkey

U. Frisch uriel@obs-nice.fr
CNRS, Observatoire de Nice, B.P. 4229, 06304 Nice Cedex 4, France

A. Fursikov fursikov@dial01.msu.ru
Department of Mechanics and Mathematics, Moscow State University
119899 Moscow, Russia

A. Gel
Middle East Technical University
Department of Mechanical Engineering, Ankara, Turkey

A. Gorbon gorbon@cc.itu.edu.tr
Department of Physics, Istanbul Technical University
80626 Maslak, Istanbul, Turkey

H. Gümral hasan@yunus.mam.tubitak.gov.tr
Physics Section, TUBITAK MAM, Gebze, Istanbul, Turkey

A. Hacınlıyan avadis@boun.edu.tr
Department of Physics, Boğaziçi University
80815 Bebek, Istanbul, Turkey

M. Hortaçsu hortacsu@yunus.mam.tubitak.gov.tr
Department of Physics, Istanbul Technical University
80626 Maslak, Istanbul, Turkey

O. Imanuvilov oleg@math.snu.ac.kr
Korea Institute for Advanced Study, 207-43 Chungryangri-dong
Dongdaemoon-ku, Seoul, Korea 130-012

V. Kalantarov varga@eti.cc.hun.edu.tr
Hacettepe University, Faculty of Science, Department of Mathematics, 06532 Beytepe, Ankara, Turkey

M. Kaya
İstasyon Caddesi, No: 30/13, Feneryolu, Istanbul, Turkey

R. Kaykayoğlu gokcol@sariyer.cc.itu.edu.tr
Istanbul University, Faculty of Engineering, Avcılar 34850, İstanbul, Turkey

Z. Kovacs kz@poe.elte.hun
Institute of Theoretical Physics. Eötvös University
Puskin u. 5-7, H-1088 Budapest, Hungary

Y. Lenger lenger@ana.cc.yildiz.edu.tr
Department of Physics, Yıldız Technical University
Şişli, Istanbul, Turkey

A. Libchaber
Center for Studies in Physics and Biology, Rockfeller University
1230 York Avenue, Box 265, New York, N. Y. 10021-6399, USA

B. Lütfüoğlu
Moda Caddesi, Ağabey Apt., No: 23/12, 81300 Kadıköy
Istanbul, Turkey

Z. Neufeld neufeld@hercules.elte.hu
Department of Atomic Physics, Eötvös University
Puskin u. 5-7, H-1088 Budapest, Hungary

H. Örs ors@boun.edu.tr
Department of Mechanical Engineering, Boğaziçi University
80815 Bebek, Istanbul, Turkey

S. Özeren ozeren@geol.itu.edu.tr
Department of Earth Sciences, Istanbul Technical University
80626 Maslak, Istanbul, Turkey

Z. Perdahçı ziya@caju.phys.boun.edu.tr
Department of Physics, Boğaziçi University
80815 Bebek, Istanbul, Turkey

D. Pierotti pierotti@vaxrom.roma1.infn.it
Dipartimento di Fisica, Universita'Dell'Aquila
Via Vetoio 1, Coppito, L'Aquila, Italy

I. Procaccia cfprocac@weizmann.weizmann.ac.il
Department of Chemical Physics, The Weizmann Institute of Science
Rehovot 76100, Israel

F. Savacı savaci@tritu.bitnet
Electrical Engineering Department, Istanbul Technical University
80626 Maslak, Istanbul, Turkey

Z-S. She she@math.ucla.edu
UCLA, Department of Mathematics, Los Angeles, CA 90095, USA
Ü. Sönmezler sonmezler@hamlin.boun.edu.tr
Department of Mechanical Engineering, Boğaziçi University
80815 Bebek, Istanbul, Turkey
R. Şahin A11318@rorqual.cc.metu.edu.tr
Middle East Technical University
Department of Mechanical Engineering, Ankara, Turkey
E. Şuhubi
Department of Engineering Sciences, Istanbul Technical University
80626 Maslak, Istanbul, Turkey
U. Tırnaklı tirnakli@fenfak.ege.edu.tr
Ege University, Department of Physics, 35100 Bornova, İzmir, Turkey
F. Toschi toschi@cibs.sns.it
Dipatimento di Fisica, Piazza Toricelli 2, I-56126 Pisa, Italy
A. Umur fmv@boun.edu.tr
1.C. 160/12 Arnavutköy, 80820 Istanbul, Turkey
K. Ülker
Department of Physics, Istanbul Technical University
80626 Maslak, Istanbul, Turkey
G. Ünal feunal@tritu.edu.tr
Department of Engineering Sciences, Istanbul Technical University
80626 Maslak, Istanbul, Turkey
A. Wirth wirth@obs-nice.fr
CNRS, Observatoire de Nice, B.P. 4229, 06304 Nice Cedex 4, France

Hydrodynamic Turbulence: a 19th Century Problem with aChallenge for the 21st Century

Victor L'vov and Itamar Procaccia

Department of Chemical Physics,
The Weizmann Institute of Science, Rehovot 76100, Israel

Abstract. The theoretical calculation of the scaling exponents that characterize the statistics of fully developed turbulence is one of the major open problems of statistical physics. We review the subject, explain some of the recent developments, and point out the road ahead.

1 Turbulence for the Physicist and for the Engineer

Sir Horace Lamb once said "I am an old man now, and when I die and go to Heaven there are two matters on which I hope enlightenment. One is quantum electro-dynamics and the other is turbulence of fluids. About the former, I am really rather optimistic" (Goldstein 1969). Possibly Lamb's pessimism about turbulence was short-sighted. There exist signs that the two issues that concerned Lamb are not disconnected. The connections have been brewing for some while, and began to take clearer form recently. They promise renewed vigor in the intellectual endeavour to understand this long-standing problem. This is not due to some outstanding development of new tools in the theoretical or experimental study of turbulence *per se*, but rather due to developments in neighboring fields. The great successes of the theory of critical and chaotic phenomena and the popularity of nonlinear physics of classical systems attracted efforts that combined the strength of fields like quantum field theory and condensed matter physics leading to renewed optimism about the solubility of the problem of turbulence. It seems that this area of research will have a renaissance of rapid growth that promises excitement well into the next century.

It is possibly a happy coincidence that hydrodynamic turbulence is considered a problem of immense interest by both physicists and engineers. The physicist tends to appreciate phenomena that display universal characteristics; the engineer may find such characteristics irrelevant since they cannot be manipulated. The engineer seeks control, and control means a ready response to perturbations. Universal phenomena are immune to perturbations. The point is of course that "turbulence" means different things to different researchers. All agree that hydrodynamic turbulence arises in fluids that are highly stressed, or stirred, such that there exist significant fluid velocities (or winds) on the largest scales of motion. The engineer is typically interested in the flow characteristics near the boundaries of the fluid (boundary layers, airplane wings, pipes, turbines etc). By understanding how to manipulate the boundary region one may reduce drag and improve the performance of technological devices. The physicist is interested in the small scale structure of turbulence away from any boundary, where the action of fluid mechanics effectively homogenizes the flow characteristics and where universal phenomena may be sought. In this context "universal" means those phenomena that are independent of the nature of the fluid (water, oil, honey etc), independent of the mechanism of stirring the flow, and independent of the form of the container of the fluid. They are inherent features of fluid mechanics as a classical field theory. Understanding these universal features may permit only marginal technological improvements. But this is the theoretical challenge that excites the physicist.

To many people turbulence research seems orthogonal to the two main lines of progress in modern physics. On the one hand, tremendous effort has been invested in understanding the structure of matter, with later developments concentrating on ever-diminishing scales of constituent particles using the ever-increasing energies of particle accelerators. On the other hand, astronomy and cosmology have exploded with a rich tableau of discoveries at ever-increasing distances from our galaxy. The physics of phenomena on the human scale, phenomena that are of acute interest to the scientist and layman alike, were relegated into a secondary position in the course of the development of the first half of 20th century physics. Of course, problems related to the health and well-being of humans are deservedly being studied in biology and medicine. But physical phenomena that can be observed by simply looking out the window are considered by many as "non-fundamental" and belonging to 19th century research. It is the conviction of the present writers as well as of a growing number of researchers that physics on the human scale offers tremendously rewarding intellectual challenges, some of which were at the core of the recent interest in chaotic phenomena and in the area which is vaguely termed "physics of complex systems". Fluid turbulence, which is the highly complex, chaotic and vortical flow that is characteristic of all fluids under large stresses, is a paramount example of these phenomena that are immensely challenging to the physicist and the mathematician alike. The aim of this paper is to explain why this problem is exciting, why it is difficult, and what are the possible routes that one can traverse in finding the solution. The point of view described here is that of the physicist whose interest is biased in favour of universal phenomena.

2 Some History

The mathematical history of fluid mechanics begins with Leonhard Euler who was invited by Frederick the Great to Potsdam in 1741. According to a popular story (which we have not been able to corroborate) one of his tasks was to engineer a water fountain. As a true theorist, he began by trying to understand the laws of motion of fluids. In 1755 he wrote Newton's laws for a fluid which in modern notation reads (for the case of constant density) (Lamb 1945)

$$\frac{\partial \boldsymbol{u}(\boldsymbol{r},t)}{\partial t} + \boldsymbol{u}(\boldsymbol{r},t) \cdot \boldsymbol{\nabla} \boldsymbol{u}(\boldsymbol{r},t) = -\boldsymbol{\nabla} p(\boldsymbol{r},t) \ . \tag{1}$$

Here $\boldsymbol{u}(\boldsymbol{r},t)$ and $p(\boldsymbol{r},t)$ are the fluid velocity and pressure at the spatial point $\boldsymbol{r}$ at time t. The LHS of this "Euler equation" for $u(r,t)$ is just the material time derivative of the momentum, and the RHS is the force, which is represented as the gradient of the pressure imposed on the fluid. In fact, trying to build a fountain on the basis of this equation was bound to fail. This equation predicts, for a given gradient of pressure, velocities that are much higher than anything observed. One missing idea was that of the viscous dissipation that is due to the friction of one parcel of fluid against neighboring ones. The appropriate term was added to (1) by Navier in 1827 and by Stokes in 1845 (Lamb 1945). The result is known as the "Navier-Stokes equations" :

$$\frac{\partial \boldsymbol{u}(\boldsymbol{r},t)}{\partial t} + \boldsymbol{u}(\boldsymbol{r},t) \cdot \boldsymbol{\nabla} \boldsymbol{u}(\boldsymbol{r},t) = -\boldsymbol{\nabla} p(\boldsymbol{r},t) + \nu \nabla^2 \boldsymbol{u}(\boldsymbol{r},t) \ . \tag{2}$$

Here ν is the kinematic viscosity, which is about 10^{-2} and 0.15 cm^2/sec for water and air at room temperature respectively. Without the term $\nu\nabla^2\boldsymbol{u}(\boldsymbol{r},t)$ the kinetic energy $u^2/2$ is conserved; with this term kinetic energy is dissipated and turned into heat. The effect of this term is to stabilize and control the nonlinear energy conserving Euler equation (1).

Straightforward attempts to assess the solutions of this equation may still be very non-realistic. For example, we could estimate the velocity of water flow in any one of the mighty rivers like the Nile or the Volga which drop hundreds of meters in a course of about a thousand kilometers. The typical angle of inclination α is about 10^{-4} radians, and the typical river depth L is about 10 meters. Equating the gravity force αg ($g \simeq 10^3 cm/sec^2$) and the viscous drag $\nu d^2u/dz^2 \sim \nu u/L^2$ we find u to be of the order of 10^7 cm/sec instead of the observed value of about 10^2 cm/sec. This is of course absurd, perhaps to the regret of the white water rafting industry. This estimate contradicts even simple energy conservation arguments. After all, we cannot gain in kinetic energy more than the stored potential energy which is of the order of $\rho g H$ where H is the drop in elevation of the river bed from its source. For the Volga or the Nile H is about 5×10^4 cm, and equating the potential energy drop with the kinetic energy we estimate $u \sim \sqrt{2gH} \simeq 10^4$ cm/sec. This is still off the mark by two orders of magnitude. The resolution of this discrepancy was suggested by Reynolds (Reynolds 1894) who stressed the importance of a dimensionless ratio of the nonlinear term to the viscous term in (2). With a velocity drop of the order of U on a scale L the nonlinear term is estimated as U^2/L. The viscous term is about $\nu U/L^2$. The ratio of the two, known as the Reynolds number Re, is UL/ν. The magnitude of Re measures how large is the nonlinearity compared to the effect of the viscous dissipation in a particular fluid flow. For Re$\ll 1$ one can neglect the nonlinearity and the solutions of the Navier-Stokes equations can be found in closed-form in many instances (Lamb 1945). In many natural circumstances Re is very large. For example, in the rivers discussed above Re$\simeq 10^7$. Reynolds understood that for $Re \gg 1$ there is no stable stationary solution for the equations of motion. The solutions are strongly affected by the nonlinearity, and the actual flow pattern is complicated, convoluted and vortical. Such flows are called turbulent.

Modern concepts about high Re number turbulence started to evolve with Richardson's insightful contributions (Richardson 1922) which contained the famous "poem" that paraphrased J. Swift: "*Big whirls have little whirls that feed on their velocity, and little whirls have lesser whirls and so on to viscosity -in the molecular sense*". In this way Richardson conveyed an image of the creation of turbulence by large scale forcing, setting up a cascade of energy transfers to smaller and smaller scales by the nonlinearities of fluid motion, until the energy dissipates at small scales by viscosity, turning into heat. This picture led in time to innumerable "cascade models" that tried to capture the statistical physics of turbulence by assuming something or other about the cascade process. Indeed, no one in their right mind is interested in the full solution of the turbulent velocity field at all points in space-time. The interest is in the statistical properties of the turbulent flow. Moreover the statistics of the velocity field itself is too heavily dependent on the particular boundary conditions of the flow. Richardson understood that universal properties may be found in the statistics of velocity *differences* $\delta\boldsymbol{u}(\boldsymbol{r}_1,\boldsymbol{r}_2) \equiv \boldsymbol{u}(\boldsymbol{r}_2) - \boldsymbol{u}(\boldsymbol{r}_1)$ across a separation $\boldsymbol{R} = \boldsymbol{r}_2 - \boldsymbol{r}_1$. In taking such a difference we subtract the non-universal large scale motions (known as the "wind" in atmospheric flows). In experiments (see for example (Monin and Yaglom 1973, Anselmet et al. 1984, Sreenivasan and Kailasnath 1993, Benzi et al. 1993, Praskovskii and Oncley 1994, Frisch 1995)) it is common to consider one dimensional cuts of the velocity field, $\delta u_\ell(R) \equiv \delta\boldsymbol{u}(\boldsymbol{r}_1,\boldsymbol{r}_2)\cdot\boldsymbol{R}/R$. The interest is in the probability distribution function of $\delta u_\ell(R)$ and its moments. These moments are known as the "structure functions"

$$S_n(R) \equiv \langle \delta u_\ell(R)^n \rangle \ , \tag{3}$$

where $\langle \dots \rangle$ stands for a suitably defined ensemble average. For Gaussian statistics the whole distribution function is determined by the second moment $S_2(R)$, and there is no information to be gained from higher order moments. In contrast, hydrodynamic experiments indicate that turbulent statistics are extremely non-Gaussian, and the higher order moments contain important new information about the distribution functions.

Possibly the most ingenious attempt to understand the statistics of turbulence is due to Kolmogorov who in 1941 (Kolmogorov 1941) proposed the idea of universality (turning the study of smallscale turbulence from mechanics to fundamental physics) based on the notion of the "inertial range". The idea is that for very large values of Re there is a wide separation between the "scale of energy input" L and the typical "viscous dissipation scale" η at which viscous friction become important and dumps the energy into heat. In the stationary situation, when the statistical characteristics of the turbulent flow are time independent, the rate of energy input at large scales (L) is balanced by the rate of energy dissipation at small scales (η), and must be also the same as the flux of energy from larger to smaller scales (denoted $\bar{\epsilon}$) as it is measured at any scale R in the so-called "inertial" interval $\eta \ll R \ll L$. Kolmogorov proposed that the only relevant parameter in the inertial interval is $\bar{\epsilon}$, and that L and η are irrelevant for the statistical characteristics of motions on the scale of R. This assumption means that R is the only available length for the development of dimensional analysis. In addition we have the dimensional parameters $\bar{\epsilon}$ and the mass density of the fluid ρ. From these three parameters we can form combinations $\rho^x \bar{\epsilon}^y R^z$ such that with a proper choice of the exponents x, y, z we form any dimensionality that we want. This leads to detailed predictions about the statistical physics of turbulence. For example, to predict $S_n(R)$ we note that the only combination of $\bar{\epsilon}$ and R that gives the right dimension for S_n is $(\bar{\epsilon}R)^{n/3}$. In particular for $n = 2$ this is the famous Kolmogorov "2/3" law which in Fourier representation is also known as the "-5/3" law. The idea that one extracts universal properties by focusing on statistical quantities can be applied also to the correlations of gradients of the velocity field. An important example is the rate $\epsilon(r,t)$ at which energy is dissipated into heat due to viscous damping. This rate is roughly $\nu|\nabla u(r,t)|^2$. One is interested in the fluctuations of the energy dissipation $\epsilon(\boldsymbol{r},t)$ about their mean $\bar{\epsilon}$, $\hat{\epsilon}(r,t) = \epsilon(r,t) - \bar{\epsilon}$, and how these fluctuations are correlated in space. The answer is given by the often-studied correlation function

$$K_{\epsilon\epsilon}(R) = \langle \hat{\epsilon}(r+R,t)\hat{\epsilon}(r,t)\rangle \ . \tag{4}$$

If the fluctuations at different points were uncorrelated, this function would vanish for all $R \neq 0$. Using Kolmogorov's dimensional reasoning one estimates $K_{\epsilon\epsilon}(R) \simeq \nu^2 \bar{\epsilon}^{4/3} R^{-8/3}$, which means that the correlation decays as a power, like $1/R^{8/3}$.

Experimental measurements show that Kolmogorov was remarkably close to the truth. The major aspect of his predictions, i.e. that the statistical quantities depend on the length scale R as power laws is corroborated by experiments. On the other hand, the predicted exponents seem not to be exactly realized. For example, the experimental correlation $K_{\epsilon\epsilon}(R)$ decays according to a power law,

$$K_{\epsilon\epsilon}(R) \sim R^{-\mu} \quad for\ \eta \ll R \ll L \ , \tag{5}$$

with μ having a numerical value of $0.2 - 0.3$ which is in large discrepancy compared to the expected value of 8/3 (Sreenivasan and Kailasnath 1993). The structure functions also behave as power laws,

$$S_n(R) \simeq R^{\zeta_n} \ , \tag{6}$$

but the numerical values of ζ_n deviate progressively from $n/3$ when n increases (Anselmet et al. 1984, Benzi et al. 1993). Something fundamental seems to be missing. The uninitiated reader might think that the numerical value of this exponent or another is not a fundamental issue. However one needs to understand that the Kolmogorov theory exhausts the dimensions of the statistical quantities under the assumption that $\bar{\epsilon}$ is the only relevant parameter. Therefore a deviation in the numerical value of an exponent from the prediction of dimensional analysis requires the appearance of another dimensional parameter. Of course there exist two dimensional parameters, i.e. L and η, which may turn out to be relevant. Experiments indicate that for the statistical quantities mentioned above the energy-input scale L is indeed relevant and it appears as a normalization scale for the deviations from Kolmogorov's predictions: $S_n(R) \simeq (\bar{\epsilon}R)^{n/3}(L/R)^{\delta_n}$ where $\zeta_n = n/3 - \delta_n$. Such forms of scaling, which deviate from the predictions of dimensional analysis, are referred to as "anomalous scaling". The realization that the experimental results for the structure functions were consistent with L rather than η as the normalization scale developed over a long time and involved a large number of experiments; recently the accuracy of determination of the exponents has increased appreciably as a result of a clever method of data analysis by Benzi, Ciliberto and coworkers (Benzi et al. 1993). Similarly a careful demonstration of the appearance of L in the dissipation correlation was achieved by Sreenivasan and coworkers (Sreenivasan and Kailasnath 1993). A direct analysis of scaling exponents ζ_n and μ in a high Reynolds number flow was presented by Praskovskii and Oncley, leading to the same conclusions (Praskovskii and Oncley 1994).

3 Turbulence as a Field Theory

Theoretical studies of the universal small scale structure of turbulence can be classified broadly into two main classes. Firstly there is a large collection of phenomenological models that by attempting to achieve agreement with experiments have given important insights into the nature of the cascade or the statistics of the turbulent fields (Frisch 1995). In particular there appeared influential ideas, following Mandelbrot (Mandelbrot 1974), about the fractal geometry of highly turbulent fields which allow scaling properties that are sufficiently complicated to include non-Kolmogorov scaling. Parisi and Frisch showed that by introducing multifractals one can accommodate the nonlinear dependence of ζ_n on n (Parisi and Frisch 1985). However these models are not derived from the equations of fluid mechanics; one is always left with uncertainties about the validity or relevance of these models. The second class of approaches is based on the equations of fluid mechanics. Typically one acknowledges the fact that fluid mechanics is a (classical) field theory and resorts to field theoretic methods in order to compute statistical quantities. Even though there has been a continuous effort for almost 50 years in this direction, the analytic derivation of the scaling laws for $K_{\epsilon\epsilon}(R)$ and $S_n(R)$ from the Navier-Stokes equations (2) and the calculation of the numerical value of the scaling exponents μ and ζ_n have been among the most elusive goals of theoretical research. Why did it turn out to be so difficult?

To understand the difficulties, we need to elaborate a little on the nature of the field theoretic approach. Suppose that we want to calculate the average response of a turbulent fluid at some point r_0 to forcing at point r_1. The field theoretic approach allows us to consider this response as an infinite sum of all the following processes: firstly there is the direct response at point r_0 due to the forcing at r_1. This response is caused by *linear* processes in the fluid, and is instantaneous if we assume that the fluid is incompressible (and therefore the speed of sound is infinite). Then there are processes which are inherently nonlinear. Nonlinear procesesses are mediated by inermediate points, but take time. Forcing at r_1 causes a response at an intermediate point r_2, which then acts as a forcing for the response at r_0. Since this intermediate process can take time, we need to integrate over all the possible positions of point r_2 and all times. This is the second-order term in perturbation theory. Then we can force at r_1, the response at r_2 acting as a forcing for r_3 and the response at r_3 forces a response at r_0. We need to integrate over all possible intermediate positions r_2 and r_3 and all the intermediate times. This is the third-order term in perturbation theory. And so on. The actual response is the infinite sum of all these contributions. In applying this field theoretical method one encounters three main difficulties:

(A) The theory has no small parameter. The usual procedure is to develop the theory perturbatively around the linear part of the equation of motion. In other words, the zeroth order solution of Eq.(2) is obtained by discarding the terms which are quadratic in the velocity field. The expansion parameter is then obtained from the ratio of the quadratic to the linear terms; this ratio is of the order of the Reynolds number Re which was defined above. Since we are interested in Re$\gg 1$, naive perturbation expansions are badly divergent. In other words the contribution of the various processes described above increases as $(Re)^n$ with the number n of intermediate points in space-time.

(B) The theory exhibits two types of nonlinear interactions. Both are hidden in the nonlinear term $\boldsymbol{u}\cdot\nabla\boldsymbol{u}$ in Eq. (2). The larger of the two is known to any person who has watched how a small floating object is entrained in the eddies of a river and swept along a complicated path with the turbulent flow. In a similar way any fluctuation of small scale is swept along by all the larger eddies. Physically this sweeping couples any given scale of motion to all the larger scales. Unfortunately the largest scales contain most of the energy of the flow; these large scale motions are what is experienced as gusts of wind in the atmosphere or the swell in the ocean. In the perturbation theory for $S_n(R)$ one has the consequences of the sweeping effect from all the scales larger than R, with the main contribution coming from the largest, most intensive gusts on the scale of L. As a result these contributions diverge when $L \rightarrow$

∞. In the theoretical jargon this is known as "infrared divergences". Such divergences are common in other field theories, with the best known example being quantum electrodynamics. In that theory the divergences are of similar strength in higher order terms in the series, and they can be removed by introducing finite constants to the theory, like the charge and the mass of the electron. In the hydrodynamic theory the divergences become stronger with the order of the contribution, and to eliminate them in this manner one needs an infinite number of constants. In the jargon such a theory is called "not renormalizable". However, sweeping is just a kinematic effect that does not lead to energy redistribution between scales, and one may hope that if the effect of sweeping is taken care of in a consistent fashion a renormalizable theory might emerge. This redistribution of energy results from the second type of interaction, that stems from the shear and torsion effects that are sizable only if they couple fluid motions of comparable scales. The second type of nonlinearity is smaller in size but crucial in consequence, and it may certainly lead to a scale-invariant theory.

(C) Nonlocality of interaction in r space. One recognizes that the gradient of the pressure is dimensionally the same as $[(\boldsymbol{u}\cdot\boldsymbol{\nabla})]\boldsymbol{u}$, and the fluctuations in the pressure are quadratic in the fluctuations of the velocity. This means that the pressure term is also nonlinear in the velocity. However, the pressure at any given point is determined by the velocity field everywhere. Theoretically one sees this effect by taking the divergence of Eq.(2). This leads to the equation $\nabla^2 p = \boldsymbol{\nabla}\cdot[(\boldsymbol{u}\cdot\boldsymbol{\nabla})\boldsymbol{u}]$. The inversion of the Laplacian operator involves an integral over all space. Physically this stems from the fact that in the incompressible limit of the Navier-Stokes equations sound speed is infinite and velocity fluctuations at all distant points are instantaneously coupled.

Indeed, these difficulties seemed to complicate the application of field theoretic methods to such a degree that a wide-spread feeling appeared to the effect that it is impossible to gain valuable insight into the universal properties of turbulence along these lines, even though they proved so fruitful in other field theories. The present authors (as well as other researchers starting with Kraichnan (Kraichnan 1965) and recently Migdal (Migdal 1994), Polyakov (Polyakov 1993), Eyink (Eyink 1993) etc.) think differently, and in the rest of this paper we will explain why.

The first task of a successful theory of turbulence is to overcome the existence of the interwoven nonlinear effects that were explained in difficulty (B). This is not achieved by directly applying a formal field-theoretical tool to the Navier-Stokes equations. It does not matter whether one uses standard field theoretic perturbation theory (Wyld 1961), path integral formulation, renormalization group (Yakhot and Orszag 1986), ϵ-expansion, large N-limit (Mou and Weichman 1995) or one's formal method of choice. One needs to take care of the particular nature of hydrodynamic turbulence as embodied in difficulty (B) *first*, and *then* proceed using formal tools.

The removal of the effects of sweeping is based on Richardson's remark that universality in turbulence is expected for the statistics of velocity *differences across a length scale R* rather than for the statistics of the velocity field itself. The velocity fields are dominated by the large scale motions that are not universal since they are produced directly by the agent that forces the flow. This forcing agent differs in different flow realizations (atmosphere, wind tunnels, channel flow etc.). Richardson's insight was developed by Kraichnan who attempted to cast the field theoretic approach in terms of Lagrangian paths, meaning a description of the fluid flow which follows the path of every individual fluid particle. Such a description automatically removes the large scale contributions (Kraichnan 1965). Kraichnan's approach was fundamentally correct, and gave rise to important and influential insights in the description of turbulence, but did not provide a convenient technical way to consider all the orders of perturbation theory. The theory did not provide transparent rules on how to consider an arbitrarily high term in the perturbation theory. Only low order truncations were considered.

A way to overcome difficulty (B) was suggested by Belinicher and L'vov (Belinicher and L'vov 1987) who introduced a novel transformation that allowed on one hand the elimination of the sweeping that leads to infrared divergences, and on the other hand allowed the development of simple rules for writing down any arbitrary order in the perturbation theory for the statistical quantities. The essential idea in this transformation is the use of a coordinate frame in which velocities are measured relative to the velocity of *one* fluid particle. The use of this transformation allowed the examination of the structure functions of velocity differences $S_n(R)$ to all orders in perturbation theory. Of course, difficulty (A) remains; the perturbation series still diverges rapidly for large values of Re, but now standard field theoretic methods can be used to reformulate the perturbation expansion such that the viscosity is changed by an effective "eddy viscosity". The theoretical tool that achieves this exchange is known in quantum field theory as the Dyson line resummation (L'vov and Procaccia 1994). The result of this procedure is that the effective expansion parameter is no longer Re but an expansion parameter of the order of unity. Of course, such a perturbation series may still diverge as a whole. Nonetheless it is crucial to examine first the order-by-order properties of series of this type.

Such an examination leads to a major surprise: every term in this perturbation theory remains finite when the energy-input scale L goes to ∞ and the viscous-dissipation scale η goes to 0 (L'vov and Procaccia 1995a). The meaning of this is that the perturbative theory for S_n does not indicate the existence of any typical length-scale. Such a length is needed in order to represent deviations in the scaling exponents from the predictions of Kolmogorov's dimensional analysis in which both scales L and η are assumed irrelevant. In other areas of theoretical physics in which anomalous scaling has been found it is common that the perturbative series already indicates this phenomenon. In many cases this is seen in the appearance of logarithmic divergences that must be tamed by truncating the integrals at some renormalization length. Hydrodynamic turbulence seems at this point different. The nonlinear Belinicher-L'vov transformation changes the underlying linear theory such that the resulting perturbative scheme for the structure functions is finite order by order (Belinicher and L'vov 1987, L'vov and Procaccia 1995a). The physical meaning of this result is that as much as can be seen from this perturbative series the main effects on the statistical quantities for velocity differences across a scale R come from activities on scales comparable to R. This is the perturbative justification of the Richardson-Kolmogorov cascade picture in which widely separated scales do not interact.

Consequently the main question still remains: how does a renormalization scale appear in the statistical theory of turbulence?

It turns out that there are two different mechanisms that furnish a renormalization scale, and that finally *both* L *and* η appear in the theory. The viscous scale η appears via a rather standard mechanism that can be seen in perturbation theory as logarithmic divergences, but in order to see it one needs to consider the statistics of gradient fields rather than the velocity differences themselves (L'vov and Lebedev 1995, L'vov and Procaccia 1995b). For example, considering the perturbative series for $K_{\epsilon\epsilon}(R)$, which is the correlation function of the rate of energy dissipation $\nu|\nabla \boldsymbol{u}|^2$, leads immediately to the discovery of logarithmic ultraviolet divergences in every order of the perturbation theory. These divergences are controlled by an ultraviolet cutoff scale which is identified as the viscous-dissipation scale η acting here as the renormalization scale. The summation of the infinite series results in a factor $(R/\eta)^{2\Delta}$ with some anomalous exponent Δ which is, generally speaking, of the order of unity. The appearance of such a factor means that the actual correlation of two R-separated dissipation fields is much larger, when R is much larger than η, than the naive prediction of dimensional analysis. The physical explanation of this renormalization (L'vov and Lebedev 1995, L'vov and Procaccia 1995c) is the effect of the multi-step interaction of two R-separated small eddies of scale η with a large eddy of scale R via an infinite set of eddies of intermediate scales. The net result on the scaling exponent is that the exponent μ changes from $8/3$ as expected in the Kolmogorov theory to $8/3 - 2\Delta$.

At this point it is important to understand what is the numerical value of the anomalous exponent Δ. In (L'vov and Procaccia 1995b) there was found an exact sum rule that forces a relation between the numerical value of Δ and the numerical value of the exponent ζ_2 of $S_2(R)$, $\Delta = 2 - \zeta_2$. Such a relation between different exponents is known in the jargon as a "scaling relation" or a "bridge relation". Physically this relation is a consequence of the existence of a universal nonequilibrium stationary state that supports an energy flux from large to small scales (L'vov and Procaccia 1995b, L'vov and Procaccia 1996a). The scaling relation for Δ has far-reaching implications for the theory of the structure functions. It was explained that with this value of Δ the series for the structure functions $S_n(R)$ diverge when the energy-input scale L approaches ∞ as powers of L, like $(L/R)^{\delta_n}$. The anomalous exponents δ_n are the deviations of the exponents of $S_n(R)$ from their Kolmogorov value. This is a very delicate and important point, and we therefore expand on it. Think about the series representation of $S_n(R)$ in terms of lower order quantities, and imagine that one succeeded to resum it into an operator equation for $S_n(R)$. Typically such a resummed equation may look like $[1-\hat{O}]S_n(R)]$ =RHS, where $\hat{O}$ is some integro-differential operator which is not small compared to unity. If we expand this equation in powers of $\hat{O}$ around the RHS we regain the infinite perturbative series that we started with. However, now we realize that the equation possesses also homogeneous solutions, solutions of $[1-\hat{O}]S_n(R)] = 0$ which are inherently nonperturbative since they can no longer be expanded around a RHS. These homogeneous solutions may be much larger than the inhomogeneous perturbative solutions. Of course, homogeneous solutions must be matched with the boundary conditions at $R = L$, and this is the way that the energy input scale L appears in the theory. This is particularly important when the homogeneous solutions diverge in size when $L \to \infty$ as is indeed the case for the problem at hand.

The next step in the theoretical development is to understand how to compute the anomalous exponents δ_n. The divergence of the perturbation theory for $S_n(R)$ with $L \to \infty$ forces us to seek a nonperturbative handle on the theory. In the rest of this article we describe briefly how this is done.

Firstly one needs to understand that the natural statistical objects that appear in the field theoretic approach are not the structure functions (3), but rather statistical quantities that depend on many spatial and temporal coordinates simultaneously. Defining the velocity difference $\boldsymbol{w}(\boldsymbol{r},\boldsymbol{r}',t)$ according to

$$\boldsymbol{w}(\boldsymbol{r},\boldsymbol{r}',t) \equiv \boldsymbol{u}(\boldsymbol{r}',t) - \boldsymbol{u}(\boldsymbol{r},t) \ , \tag{7}$$

one considers the n-rank tensor space-time correlation function

$$\begin{aligned} &\boldsymbol{F}_n(\boldsymbol{r}_1,\boldsymbol{r}'_1,t_1;\boldsymbol{r}_2,\boldsymbol{r}'_2,t_2;\ldots;\boldsymbol{r}_n,\boldsymbol{r}'_n,t_n) \\ &= \langle \boldsymbol{w}(\boldsymbol{r}_1,\boldsymbol{r}'_1,t_1)\boldsymbol{w}(\boldsymbol{r}_2,\boldsymbol{r}'_2,t_2)\ldots\boldsymbol{w}(\boldsymbol{r}_n,\boldsymbol{r}'_n,t_n)\rangle \ . \end{aligned} \tag{8}$$

The equal time correlation function is obtained when $t_1 = t_2 \ldots = t_n$. In stationary turbulence the equal time correlation function is time independent, and we denote it as

$$\begin{aligned} &\boldsymbol{F}_n(\boldsymbol{r}_1,\boldsymbol{r}'_1;\boldsymbol{r}_2,\boldsymbol{r}'_2;\ldots;\boldsymbol{r}_n,\boldsymbol{r}'_n) \\ &= \langle \boldsymbol{w}(\boldsymbol{r}_1,\boldsymbol{r}'_1)\boldsymbol{w}(\boldsymbol{r}_2,\boldsymbol{r}'_2)\ldots\boldsymbol{w}(\boldsymbol{r}_n,\boldsymbol{r}'_n)\rangle \ . \end{aligned} \tag{9}$$

One expects that when all the separations $R_i \equiv |\boldsymbol{r}_i - \boldsymbol{r}'_i|$ are in the inertial range, $\eta \ll R_i \ll L$, the same time correlation function is scale invariant in the sense that

$$\begin{aligned} &\boldsymbol{F}_n(\lambda\boldsymbol{r}_1,\lambda\boldsymbol{r}'_1;\lambda\boldsymbol{r}_2,\lambda\boldsymbol{r}'_2;\dots;\lambda\boldsymbol{r}_n,\lambda\boldsymbol{r}'_n)\\ &=\lambda^{\zeta_n}\boldsymbol{F}_n(\boldsymbol{r}_1,\boldsymbol{r}'_1;\boldsymbol{r}_2,\boldsymbol{r}'_2;\dots;\boldsymbol{r}_n,\boldsymbol{r}'_n)\ , \end{aligned} \tag{10}$$

and the exponent ζ_n is numerically the same as the one appearing in Eq.(6).

One of the major difference between the study of statistical turbulence and other examples of anomalous scaling in physics (like critical phenomena) is that there is no theory for the same time correlation functions (9) that does not involve the many time correlation functions (8). *Turbulence is a truly dynamical problem*, and there is no free energy functional or a Boltzmann factor to provide a time-independent theory of the statistical weights. A point of difficulty of the field theoretic approach is that such a theory for the same time quantities (9) involves *integrals over the time variables of the many time quantities (8)*. One must therefore learn how to perform these time integrations properly.

The first step in this direction is obtained from the analysis of the asymptotic properties of the correlation functions (9). Obviously the structure functions (3) are obtained from (9) by "fusing" all the unprimed coordinates into one position, and all the primed coordinates into a second position which is displaced by R from the first. In this process of fusion the dissipative scale η is unavoidably crossed, and this has many consequences for the scaling theory. Such "fusions" cannot be done blindly; one needs to study the asymptotics of (9) when two or more coordinates are brought together (L'vov and Procaccia 1996b, L'vov and Procaccia 1996c). These asymptotics are summarized by the "fusion rules" which address the scaling behavior of $\mathcal{F}_n$ when a group of p points, $p < 2n-1$, coalesce together within a ball of radius ρ while all the other coordinates remain separated from each other and from this group by a larger distance R. In particular under the two general assumptions of scale invariance and universality of the scaling exponents the fusion rules state that to leading order in ρ/R

$$\boldsymbol{F}_n(\lambda\boldsymbol{r}_1,\lambda\boldsymbol{r}'_1;\lambda\boldsymbol{r}_2,\lambda\boldsymbol{r}'_2;\dots;\lambda\boldsymbol{r}_n,\lambda\boldsymbol{r}'_n)\sim\frac{S_p(\rho)}{S_p(R)}S_{2n}(R). \tag{11}$$

This forms holds as long as ρ is in the inertial range.

The availability of the fusion rules opens up the possibility of studying the scaling exponents of correlations of gradients fields. Assuming that below the viscous-dissipation scale η derivatives exist and the fields are smooth, one can estimate gradients at the end of the smooth range by dividing differences across η by η. The question is, what is the appropriate cross-over scale to smooth behaviour? Is there just one cross-over scale η, or is there a multiplicity of such scales, depending on the function one is studying? For example, when does the above n-point correlator become differentiable as a function of ρ when p of its coordinates come together? Is that typical scale the same as the one exhibited by $S_p(\rho)$ itself, or does it depend on p and n and on the remaining distances of the remaining $n-p$ coordinates that are still separated by a large distance R?

The answer is that there is a multiplicity of cross-over scales. For the n-point correlators discussed above we denote the dissipative scale as $\eta(p, n, R)$, and it depends on each of its arguments (L'vov and Procaccia 1996c, L'vov and Procaccia 1996e). In particular it depends on the inertial range variables R and this dependence must be known when one attempts to determine the scaling exponents ζ_n of the structure functions. In brief, this line of thought leads to a set of non-trivial scaling relations between the scaling exponents characterizing correlations of gradient fields and the exponents ζ_n of the structure functions. For example we derive the phenomenologically conjectured (Frisch 1995) "bridge relation" $\mu = 2 - \zeta_6$ for the exponent appearing in (5). This predicted numerical value of μ is in close agreement with the experimental values.

The most important and unexpected consequence of the fusion rules was the understanding of the temporal properties of the time correlation functions. Without going into details here, we can state that the main finding is (L'vov et al. 1996a) that the time correlation functions *are not scale invariant in their time arguments.* Naively one could assume that the scale invariance property (10) extends to the time correlation functions in the sense that

$$\begin{aligned} &\boldsymbol{F}_n(\lambda \boldsymbol{r}_1, \lambda \boldsymbol{r}'_1, \lambda^{z_n} t_1; \dots; \lambda \boldsymbol{r}_n, \lambda \boldsymbol{r}'_n, \lambda^{z_n} t_n) \\ &= \lambda^{\zeta_n} \boldsymbol{F}_n(\boldsymbol{r}_1, \boldsymbol{r}'_1; \dots; \boldsymbol{r}_n, \boldsymbol{r}'_n) \,, \quad (\textit{not true!}) \end{aligned} \tag{12}$$

We have shown that this is not the case. Nevertheless, it is possible to present a set of rules that allow one to evaluate time-integrals of many point correlation functions. Such rules are essential in the context of the field theoretic approach, since every term in this kind of theory involves integrals over space and time. At the time of writing of this paper the authors are involved in setting up a scheme of calculation of the scaling exponents ζ_n from first principles, based on the building blocks that were briefly mentioned here. It is our hope that this scheme will provide the long awaited theory of anomalous scaling in turbulence.

4 Summary

It appears that there are five conceptual steps in the construction of a theory of the universal anomalous statistics of turbulence on the basis of the Navier-Stokes equations. First one needs to take care of the sweeping interactions that mask the scale invariant theory (Belinicher and L'vov 1987, L'vov and Procaccia 1995a). After doing so the perturbation expansion converges order by order, and the Kolmogorov scaling of the velocity structure functions is found as a perturbative solution. Secondly one understands the appearance of the viscous-dissipation scale η as the natural normalization scale in the theory of the correlation functions of the gradient fields (L'vov and Lebedev 1995, L'vov and Procaccia 1995b). This step is similar to critical phenomena and it leads to a similarly rich theory of anomalous behaviour of the gradient fields. Only the tip of the iceberg was considered above. In fact when one considers the correlations of tensor fields which are constructed from $\partial u_\alpha/\partial r_\beta$ (rather than the scalar field ϵ) one finds that every field with a different transformation property under the rotation of the coordinates has its own independent scaling exponent which is the analog of Δ above (L'vov et al. 1996b). The third step is the understanding of the divergence of the diagrammatic series for the structure functions as a whole (L'vov and Procaccia 1996a). This sheds light on the emergence of the energy-input scale L as a normalization length in the theory of turbulence. This means that the Kolmogorov basic assertion that there is no typical scale in the expressions for statistical quantities involving correlations across a scale R when $\eta \ll R \ll L$ is doubly wrong. In general both lengths appear in dimensionless combinations and change the exponents from the predictions of dimensional analysis. Examples of correlation functions in which both normalization scales L and η appear simultaneously were given explicitly (L'vov et al. 1996b). Next is the formulation of the fusion rules and the exposition of the multiplicity of the dissipative scales. Last, but not least, is the elucidation of the temporal properties of the time correlation functions. This last step is crucial in our opinion, and with it we believe that one can attempt to set up a calculation of the scaling exponents from first principles.

The road ahead is not fully charted, but it seems that some of the conceptual difficulties have been surmounted. We believe that the crucial building blocks of the theory are now available, and they begin to delineate the structure of the theory. We hope that the remaining 4 years of this century will suffice to achieve a proper understanding of the anomalous scaling exponents in turbulence. Considerable work, however, is still needed in order to fully clarify many aspects of the problem, and most of them are as exciting and important as the scaling properties. There are universal aspects that go beyond exponents, such as distribution functions and the eddy viscosity, and there are important non-universal aspects like the role of inhomogeneities, the effect of boundaries and so on. Progress on these issues will bring the theory closer to the concern of the engineers. The marriage of physics and engineering will be the challenge of the 21st century.

Acknowledgements. Our thinking about these issues were influenced by discussions with V. Belinicher, R. Benzi, P. Constantin, G. Falkovich, U. Frisch, K. Gawedzki, S. Grossmann, L.P. Kadanoff, R.H. Kraichnan, V.V. Lebedev, M. Nelkin, E. Podivilov, A. Praskovskii, K.R. Sreenivasan, P. Tabeling, S. Thomae and V.E Zakharov. We thank them all. Our work has been supported in part by the Minerva Center for Nonlinear Physics, by the Minerva Foundation, Munich, Germany, the German-Israeli Foundation, the US-Israel Binational Science Foundation and the the Naftali and Anna Backenroth-Bronicki Fund for Research in Chaos and Complexity.

References

F. Anselmet, Y. Gagne, E.J. Hopfinger, and R.A. Antonia, J. Fluid Mech. **140**, 63 (1984).

V. I. Belinicher and V. S. L'vov. Sov. Phys. JETP, **66**, 303 (1987).

R. Benzi, S. Ciliberto, R. Tripiccione, C. Baudet, F. Massaioli and S. Succi, Phys. Rev. E **48**, R29 (1993).

G.L. Eyink, Phys.Lett. A **172**, 355 (1993)

Uriel Frisch. *Turbulence: The Legacy of A.N. Kolmogorov.* Cambridge University Press, Cambridge, 1995.

S. Goldstein, Ann. Rev. Fluid Mech, **1**, 23 (1969)

A.N. Kolmogorov, Dokl. Acad. Nauk SSSR, **30**, 9 (1941).

R.H. Kraichnan, Phys. Fluids, **8**, 575 (1965)

H. Lamb *Hydrodynamics*, (Dover publications, 1945)

V.S. L'vov and I. Procaccia "Exact Resummation in the Theory of Hydrodynamic Turbulence": 0. Line Resummed Diagrammatic Perturbation Approach" in F. David P. Ginsparg, and J. Zinn-Justin, eds. Les Houches session LXII, 1994, "Fluctuating Geometries in Statistical Mechanics and Field Theory" (Elsevier, 1995).

V.S. L'vov and I. Procaccia "Exact Resummation in the Theory of Hydrodynamic Turbulence": I. The Ball of Locality and Normal Scaling",Phys. Rev. E, **52**, 3840 (1995)

V.S. L'vov and V. Lebedev, Europhys. Lett. **29**, 681 (1995)

V.S. L'vov and I. Procaccia "Exact Resummation in the Theory of Hydrodynamic Turbulence": II. The Ladder to Anomalous Scaling",Phys. Rev. E, **52**, 3858 (1995)

V.S. L'vov and I. Procaccia Phys.Rev. Lett. **74**, 2690 (1995) .

V.S. L'vov and I. Procaccia "Exact Resummation in the Theory of Hydrodynamic Turbulence": III. Scenarios for Multiscaling and Intermittency". Phys.Rev. E, **53**, 3468 (1996)

V.S. L'vov and I. Procaccia, "Fusion Rules in Turbulent Systems with Flux Equilibrium", Phys. Rev. Lett. 76, 2896 (1996).

V.S. L'vov and I. Procaccia "Towards a Nonperturbative Theory of Hydrodynamic Turbulence: Fusion Rules, Exact Birdge Relations and Anomalous Viscous Scaling Functions", Phys.Rev E, **54**, 6268 (1996).

V.S. L'vov and I. Procaccia "The Viscous Lengths in Hydrodynamic Turbulence are Anomalous Scaling Functions", Phys.Rev. Lett, **77**,3541 (1996).

V.S. L'vov E. Podivilov and I. Procaccia, "Hydrodynamic Turbulence has Infinitely Many Anomalous Dynamical Exponents, Phys.Rev. Lett, Rejected. chao-dyn/9607011.

V.S. L'vov, E. Podivilov and I. Procaccia, "Scaling Behaviour in Turbulence is Doubly Anomalous", Phys. Rev. Lett. **76**, 3963 (1996).

B. Mandelbrot, J. Fluid Mech. **62**,331 (1974)

A.A. Migdal, Int. J. of Mod. Phys. A **9**, 1197 (1994)

A. S. Monin and A. M. Yaglom. *Statistical Fluid Mechanics: Mechanics of Turbulence*, volume II. The MIT Press, Cambridge, Mass., 1973.

C-Y. Mou and P. B. Weichman. Phys.Rev E **52**, 3738 (1995)

G. Parisi and U. Frisch in"Turbulence and Predictability in Geophysical Fluid Dynamics, eds. M. Ghil, R. Benzi and G. Parisi, (North Holland, Amsterdam 1985)

A. M. Polyakov, Nucl. Phys. B **396**, 367 (1993)
A. Praskovskii and S. Oncley, Phys. Fluids, **6**,2886 (1994)
O. Reynolds, Phil. Trans. Roy. Soc., London **A186**, 123 (1894).
L.F. Richardson, *Weather Prediction by Numerical Process*, (Cambridge, 1922)
K.R. Sreenivasan and P. Kailasnath. Phys. Fluids A **5**, 512, (1993).
H. W. Wyld. Ann. Phys., **14**, 143 (1961).
V. Yakhot and S.A. Orszag, Phys.Rev. Lett **57**,1722 (1986)

Exponents of bulk heat transfer in convective turbulence

Peter Constantin*

Department of Mathematics
University of Chicago
Chicago, Illinois 60637, USA

Abstract. We discuss rigorous results regarding the dependence of the Nusselt number upon Rayleigh number. We find that the exponent 2/7 is obtained if a certain quantitative condition holds; this condition can be interpreted as one of relative horizontal smoothnes of time averaged temperature profiles or as a separation between the thermal and viscous boundary layers.

1 Introduction

In this lecture I would like to describe somewhat informally recent work on the problem of estimating bulk heat transport in Rayleigh-Bénard convection; my own work in this area is done in collaboration with Charles R. Doering. The classical approach on this problem was initiated by Malkus in the fifties, realized in the basic work of Howard and continued by Busse and collaborators from the sixties to the present (Malkus 1954 - Busse 1970). The main mathematical problem is to bound the long time average of the total dissipation in solutions of the Boussinesq equations governing Rayleigh - Bénard convection. Because this system is driven at the boundaries the most straightforward energy estimate may fail to show even that the long time dissipation of energy is bounded. A mathematical observation of E. Hopf (Hopf 1941), allows nevertheless a finite upper bound to be derived. The question then arises of the optimal bound and its dependence on control parameters, and this leads naturally to the formulation of a variational approach associated with the upper bound method. These techniques were previously applied to boundary driven shear layers (Doering and Constantin 1992, Doering and Constantin 1994) and channel flow (Constantin and Doering 1995a), and the general conceptual framework was presented both abstractly and on a specific example in (Constantin and Doering 1995b). An extensive paper concerning the application of this approach and the associated variational problem to convective turbulence, and its possibly deep but yet poorly understood connection to the analysis of Howard and Busse, is (Doering and Constantin 1996). A more recent work (Constantin and Doering 1996) discusses dynamical issues and exponents at finite but large Rayleigh numbers. In this talk I

* Partially supported by NSF/DMS 9207080 and DOE/DE-FG02-92ER-25119.

would like to take stock of what has been achieved so far, to explain what are the mathematical and physical conditions that produce different exponents and to venture some guesses as to what can be done further.

Let me start by describing briefly the method. The systems that can be treated with this method share the following features: they are dissipative (Hale 1988, Témam 1988, Constantin et al. 1988), the transport quantities of physical interest are balanced by the total dissipation, and the driving is at the boundary. The method can be described concisely as follows: there exists a closed convex set $\mathcal{B}$ in phase space (a function space) such that every member of this set (let us call them *admissible backgrounds*) has norm (dissipation) larger or equal than the sought-after bound. The set of admissible backgrounds is defined in terms of certain constraints; they must solve certain differential equations, satisfy the driving boundary conditions and, in addition, a certain bilinear form in which they enter as coefficients should be non-negative. The bilinear forms are similar to those which appear in investigations of nonlinear energy stability of stationary solutions of the equations of motion (Joseph 1976, Straughan 1992), and so this final constraint may be interpreted as a "stability" constraint on the admissible backgrounds. In the paper (Constantin and Doering 1996), instead of admissible backgrounds we considered a larger set, that of *sufficient* backgrounds. For these backgrounds the bilinear form is required to be positive only in the temporal average along true solutions of the system. This allowed us to obtain, *under suitable technical assumptions*, rigorous bounds on the Nusselt number (Nu) with exponents that are physically relevant at high Rayleigh numbers (Ra). In particular, not only the scaling $Nu \sim Ra^{1/2}$, but also the scalings $Nu \sim Ra^{1/3}$ and $Nu \sim Ra^{2/7}$, well documented in the experimental and numerical literature (Heslot et al. 1987, DeLuca et al. 1990, Siggia 1994), were derived from the Boussinesq equations under these natural relative regularity assumptions. This talk will be based mainly on (Constantin and Doering 1996); the last two sections are new and deal with the physical conditions for the realization of different exponents.

2 The problem

We consider convection in a finite container using the Boussinesq equations

$$\frac{\partial u}{\partial t} + u \cdot \nabla u + \nabla p = \sigma \Delta u + \sigma Ra \hat{e} T,$$

$$\nabla \cdot u = 0$$

$$\frac{\partial T}{\partial t} + u \cdot \nabla T = \Delta T.$$

The velocity vector $u = u(x,t)$ and the temperature scalar $T(x,t)$ are functions of four variables, x representing position and t representing time. The parameters $\sigma > 0$ and $Ra > 0$ are respectively Prandtl number and Rayleigh number. The first is the ratio $\sigma = \frac{\nu}{\kappa}$ of kinematic viscosity and thermal diffusivity. The second, $Ra = \frac{g\alpha h^3 \Gamma}{\kappa\nu}$ is our control parameter. Here α is the thermal expansivity constant, g is the acceleration of gravity, h is the height of the container and Γ is the difference between the temperatures at the bottom and at the top of the container. We are interested in the large Rayleigh number regime . The unit vector $\hat{e}$ points up in the vertical direction. We measure lengths in units of h, the vertical height. We measure time in units of $\frac{h^2}{\kappa}$ (this has been already reflected by coefficients in the Boussinesq equations). With this non-dimensionalization we will consider functions that are periodic in the horizontal variables with period L (taken to be the same in both directions, for simplicity of exposition). So we are dealing with a basic box of height 1 and lateral side L. We will denote by $z = x_3$ the vertical coordinate and by u_3 or w the corresponding velocity component, $u_3 = u \cdot \hat{e}$. We will denote by Q the square of side L. The boundary conditions are periodic in the horizontal directions for both velocity and temperature. The velocity vanishes at the top and bottom boundaries. The temperature is held constant at the vertical boundaries. The inhomogeneous boundary conditions are $T = 1$ at the bottom boundary and $T = 0$ at the top. Because of the imposed boundary conditions the conductive heat flux in the vertical direction,

$$a(z,t) = -\frac{1}{L^2}\int_Q \hat{e}\cdot\nabla T(x_1,x_2,z,t)dx_1dx_2$$

has constant average

$$\int_0^1 a(z,t)dz = 1.$$

The non-dimensional convective vertical heat flux is

$$b(z,t) = \frac{1}{L^2}\int_Q u_3(x_1,x_2,z,t)T(x_1,x_2,z,t)dx_1dx_2.$$

We use $\overline{\cdots}$ to denote horizontal average over Q, for instance

$$\overline{T}(z,t) = \frac{1}{L^2}\int_Q T(x_1,x_2,z,t)dx_1dx_2.$$

The vertical heat flux is

$$j(z,t) = a(z,t) + b(z,t).$$

Because

$$a(z,t) = -\frac{\partial}{\partial z}\overline{T}(z,t),$$

the equation obeyed by $\overline{T}$ is

$$\left(\partial_t - (\partial_z)^2\right)\overline{T}(z,t) + \partial_z b(z,t) = 0,$$

where we use $\partial_z = \frac{\partial}{\partial z}$. The long time limit(superior) of $N(t)$ where

$$N(t) = \left[1 + \frac{1}{t}\int_0^t \int_0^1 b(z,t')dzdt'\right] = \frac{1}{t}\int_0^t \int_0^1 j(z,t')dzdt'.$$

is a non-dimensional measure of bulk heat transfer, the Nusselt number Nu:

$$Nu = \limsup_{t\to\infty} \frac{1}{t}\int_0^t \left(\int_0^1 j(z,t')dz\right) dt'.$$

We use $< \cdots >_t$ to denote time average,

$$\langle F(t)\rangle_t = \frac{1}{t}\int_0^t F(t')dt',$$

and we use $< \cdots >$ to denote long time averages in the sense

$$\langle F(t)\rangle = \limsup_{t\to\infty} \frac{1}{t}\int_0^t F(t')dt'.$$

Thus, the Nusselt number is

$$Nu = \left\langle \int_0^1 b(z,t)dz \right\rangle + 1 = \limsup_{t\to\infty} N(t)$$

and one can easily verify that

$$\left\langle \int_0^1 \overline{|\nabla T|^2}(z,t')dz \right\rangle = Nu. \tag{1}$$

Multiplying the momentum equation by u and integrating we also obtain

$$\left\langle \int_0^1 \overline{|\nabla u|^2}(z,t')dz \right\rangle = Ra(Nu - 1). \tag{2}$$

Thus the Nusselt number is a measure of the dissipation (mean-square gradients) of velocity and of temperature.

3 Backgrounds

We say that the pair $\Phi = (U, \tau)$ is a *background* if U and τ satisfy the "viscous" boundary conditions described in the first section and, in addition they solve the the inviscid equations:

$$U \cdot \nabla U + \nabla P = \sigma Ra \tau \hat{e}$$

$$\nabla \cdot U = 0$$

and

$$U \cdot \nabla \tau = 0.$$

The simplest nontrivial examples are

$$U = U(z)\hat{e}_{horiz}, \quad \tau = \tau(z),$$

where $\hat{e}_{horiz}$ is the unit vector in a horizontal direction.

Translating the solutions by a background, i.e., writing

$$u = v + U, \quad T = \theta + \tau,$$

one obtains from the Boussinesq equations

$$\frac{1}{2Ra}\left\{\langle\|\nabla u\|^2\rangle_t - \|\nabla U\|^2\right\} = \frac{1}{t\sigma Ra}\left(\|v_0\|^2 - \|v(\cdot,t)\|^2\right) - \langle A_\Phi\rangle_t \tag{3}$$

and

$$\frac{1}{2}\left\{\langle\|\nabla T\|^2\rangle_t - \|\nabla\tau\|^2\right\} = \frac{1}{t}\left(\|\theta_0\|^2 - \|\theta(\cdot,t)\|^2\right) - \langle B_\Phi\rangle_t \tag{4}$$

where

$$\|f\|^2 = \int_0^1 \overline{|f(\cdot,z)|^2}dz,$$

$$A_\Phi = \frac{1}{2Ra}\|\nabla v\|^2 + \int_0^1 dz \overline{\frac{1}{\sigma Ra}[(v\cdot\nabla U)\cdot v] - \theta v\cdot e}, \tag{5}$$

and

$$B_\Phi = \frac{1}{2}\|\nabla\theta\|^2 + \int_0^1 dz\overline{(v\cdot\nabla\tau)\theta}. \tag{6}$$

Adding the equations 3 and 4 we obtain using 1 and 2

$$2Nu - 1 \le \frac{1}{Ra}\|\nabla U\|^2 + \|\nabla\tau\|^2 \tag{7}$$

if

$$\liminf_{t\to\infty}\langle H_\Phi\rangle_t \ge 0 \tag{8}$$

where

$$H_\Phi = A_\Phi + B_\Phi.$$

We say that the background $\Phi = (U,\tau)$ is *admissible* if $H_\Phi(v,\theta) \ge 0,\ \forall\theta \in H_0^1, v \in H_0^1, \nabla\cdot v = 0$, and that the background $\Phi = (U,\tau)$ is *sufficient* if the condition (8) is satisfied if for all initial data for the Boussinesq equations. One can prove, thus

Theorem 1 *Let $\Phi = (U,\tau)$ be a sufficient background . Then*

$$Nu \le \frac{1}{2} + \frac{1}{2}\left\{\frac{1}{Ra}\|\nabla U\|^2 + \|\nabla\tau\|^2\right\}.$$

4 A general result: the 1/2 law

Let us consider the case of backgrounds of the form $\Phi = (0, \tau(z))$ satisfying the boundary conditions $\tau(0) = 1, \tau(1) = 0$. Let us note that in this case

$$H_{\phi} = \frac{1}{2Ra}\|\nabla u\|^2 + \frac{1}{2}\|\nabla\theta\|^2 + \int_0^1 (\tau'(z) - 1)\,\beta(z)dz$$

where

$$\beta(z) = \overline{(u \cdot e)\theta}.$$

Let us seek τ of the form

$$\tau'(z) - 1 = -2\frac{1}{\delta}\Psi\left(\frac{z}{\delta}\right)$$

where Ψ is a smooth function of one variable, vanishing outside the interval $[0, 1]$ and satisfying

$$\int_0^1 \Psi(\zeta)d\zeta = 1;$$

we will adjust δ to ensure that Φ is a sufficient background. Let us denote by

$$E_{\Phi}^2 = \frac{1}{2Ra}\|\nabla u\|^2 + \frac{1}{2}\|\nabla\theta\|^2.$$

In view of the fact that both u and θ vanish at $z = 0$, obviously

$$|\beta(z)| \leq z\sqrt{Ra}E_{\Phi}^2$$

holds. Therefore

$$\left|\int_0^1 (\tau'(z) - 1)\,\beta(z)dz\right| \leq C_{\Psi}\delta\sqrt{Ra}E_{\Phi}^2$$

holds and a choice of $\delta \sim Ra^{-\frac{1}{2}}$ guarantees that Φ is sufficient (actually, even admissible).

Theorem 2 *There exists an absolute constant C, independent of Rayleigh number Ra, aspect ratio L and Prandtl number σ such that*

$$Nu \leq C\sqrt{Ra}$$

holds for all solutions of the Boussinesq equations.

This result is developed in greater detail in (Doering and Constantin 1996) where the issue of the best prefactor C is also addressed. The 1/2 law is of course derived in the classical Howard, Busse and Malkus approach.

5 Relative horizontal regularity

There are no rigorous results for other exponents than 1/2. Moreover some people (including myself) believe that the asymptotic exponent (meaning as Rayleigh number tends to infinity) is 1/2. However the numerical and experimental data are obtained at large but finite Rayleigh numbers and the observed behavior differs from the conjectured asymptotic one (Heslot et al. 1987, DeLuca et al. 1990, Siggia 1994). One way to make contact with the observed behavior without leaving the realm of rigorous analysis of the Boussinesq equations is to make judicious mathematical assumptions regarding the nature of the solutions and assess their implications. I give below two examples that are similar in nature but slightly different.

In the first example we consider the horizontal Fourier coefficients of the temperature. We write

$$T(x_1, x_2, z, t) = \sum_{k \in \frac{2\pi}{L}\mathbf{Z}^2} T_k(z,t) e^{i(x\cdot k)}.$$

One can prove (Constantin and Doering 1996)

Theorem 3 *Assume that there exists a number $m \in \boldsymbol{N}$ and a function $\Gamma_m(z)$ belonging to $L^p([0,1])$ for some $1 \le p \le \infty$, such that*

$$\left\langle |T_k(z,\cdot)|^2 \right\rangle^{\frac{1}{2}} \le \Gamma_m(z)|k|^{-m} \left\langle G_k^2 \right\rangle^{\frac{1}{2}}$$

holds for all $k \in \frac{2\pi}{L}\boldsymbol{Z}$, $k \neq 0$, where

$$G_k^2 = \frac{1}{2}\int_0^1 \left(|k|^2 |T_k(z,t)|^2 + |T_k'(z,t)|^2\right) dz.$$

Then there exists a constant C such that

$$Nu \le C Ra^{q_{m,p}}$$

where

$$q_{m,p} = \frac{mp}{4mp - p - 2}.$$

The minimal power p is $p = 2$ (i.e., $\Gamma_m(z)$ square-integrable). Indeed, the case $p = 2, m = 1$ requires almost no assumption at all. For $p = 2$ the sequence of exponents is

$$q_{m,2} = \frac{m}{4m-2}$$

i.e., $q_{m,2} = 1/2, 1/3, \cdots 1/4$. At $m = 1$ we recover the 1/2 law (which is true without assumptions), and at $m = 2$ the 1/3 law. The most stringent requirement is $p = \infty$ i.e., $\Gamma_m(z)$ bounded for $z \in [0,1]$. For $p = \infty$ the sequence of exponents is

$$q_{m,\infty} = \frac{m}{4m-1}$$

i.e., $q_{m,\infty} = 1/3, 2/7, \cdots, 1/4$. Thus, at $m = 1$ we recover the 1/3 law and at $m = 2$ we find the exponent **2/7**.

The second result, similar to the first, is (Constantin and Doering 1996):

Theorem 4 *Assume that there exists $r \in \boldsymbol{R}$ and $K > 0$ such that*

$$\left\langle \int_0^1 \overline{|\nabla_h T(\cdot, z, t)|^2} z^{-1-r} dz \right\rangle \leq K \left\langle \int_0^1 \overline{|\nabla_h T(\cdot, z, t)|^2} dz \right\rangle$$

holds. Then there exists a constant C such that

$$Nu \leq C Ra^{s(r)}$$

with

$$s(r) = \frac{1}{3+r}.$$

holds.

Here ∇_h is horizontal gradient. The hypothesis is automatically verified for $r = -1$, and in that case we recover the 1/2 law. If $r = 0$ we obtain the 1/3 law, if $r = 1/2$ we obtain the 2/7 law and if $r = 1$ we get the exponent of 1/4. Note that the technical relative regularity hypotheses used are essentially assumptions concerning the magnitude of horizontal gradients of the temperature field in the boundary layer relative to their vertical average.

6 Connection to velocity

The last result involved the horizontal gradients of temperature. These are in turn connected to the velocity. Because these results have not yet appeared anywhere else I will give some details of their derivation.

Let us consider the time evolution of

$$\overline{|\nabla_h T(\cdot, z, t)|^2}.$$

After a few integrations by parts one can write:

$$\frac{1}{2}(\partial_t - \partial_{zz}) \|\nabla_h T\|^2_{L^2(dx)} + \|\nabla \nabla_h T\|^2_{L^2(dx)} =$$

$$\frac{1}{2}\partial_z \left\{ \overline{T(w \Delta_h T - \nabla_h T \cdot \nabla_h w)} \right\} + \overline{T(\nabla_h u \cdot \nabla \nabla_h T)} \tag{9}$$

Δ_h represents the horizontal Laplacian. Now we multiply this expression by z^{1-r} and integrate in space and time. We use the fact that $0 \leq T \leq 1$, the vanishing of the gradient at the boundaries and elementary inequalities. We also use the fact that

$$\int_0^1 \overline{|w(\cdot, z, t)|^2} z^{-1-r} dz \le C \int_0^1 \overline{|\nabla_h u(\cdot, z, t)|^2} z^{1-r} dz$$

that follows from writing $z^{-1-r} = -\frac{1}{r}(z^{-r})'$, integrating by parts, using the incompressibility and the Schwartz inequality. In the end we obtain

$$\left\langle \int_0^1 \overline{|\nabla_h T(\cdot, z, t)|^2} z^{-1-r} dz \right\rangle$$

$$\le C \left\langle \int_0^1 \overline{|\nabla_h u(\cdot, z, t)|^2} z^{1-r} dz \right\rangle . \quad (10)$$

n

This is the main result of this section.

7 Discussion

One can obtain without difficulties completely rigorous upper bounds of the type

$$Nu \le C Ra^{\frac{1}{2}}.$$

The constant C is explicit. Obtaining a closer estimate at finite Rayleigh numbers involves assumptions. Mathematically these assumptions represent certain relative horizontal regularity assumptions about the temperature field: horizontal gradients of the temperature are assumed to be smaller in the boundary layer than their vertical average. For instance one can prove that if

$$\left\langle \int_0^1 \overline{|\nabla_h T(\cdot, z, t)|^2} z^{-1-r} dz \right\rangle \le K Nu$$

with a Rayleigh independent constant K then

$$Nu \le C Ra^{\frac{1}{3+r}}$$

with another Rayleigh independent constant C. Note that the horizontal gradient of temperature squared and space-time averaged is less than the Nusselt number. In particular, if the assumption is satisfied with $r = \frac{1}{2}$ then

$$Nu \le C Ra^{\frac{2}{7}}$$

follows. The mathematical interpretation of this assumption is the following: at moderate Rayleigh numbers the attractor lies in certain conical sets in function space representing the relative horizontal regularity of the temperature fields.

Now (10) allows one to obtain a sufficient condition expressed in terms of the velocity that implies the assumption on the relative horizontal regularity of the temperature. This condition is

$$\left\langle \int_0^1 \overline{|\nabla_h u(\cdot, z, t)|^2} z^{1-r} dz \right\rangle \le C Nu$$

with a constant C not depending on the Rayleigh number. Note that the vertical integral (without the weight z^{1-r}) of the long time average of the whole gradient of velocity is of the order $RaNu$. The physical interpretation of the condition above emerges when one assumes that the velocity gradients near the boundary are large enough to capture a fixed fraction of the whole gradient. Then one can write the integral as

$$\left\langle \int_0^1 \overline{|\nabla_h u(\cdot, z, t)|^2} z^{1-r} dz \right\rangle \sim c\ell^{4-r} NuRa$$

where ℓ is the width of the velocity (viscous) boundary layer (Recall that $\nabla_h u$ vanishes at the boundary). Therefore the condition is fulfilled if

$$\ell \le C Ra^{-\frac{1}{4-r}}.$$

If this happens then it follows that the width of the thermal boundary layer δ obeys

$$\delta \le C Ra^{-\frac{1}{3+r}}.$$

Now the reason why $r = \frac{1}{2}$ is selected becomes clear: it is precisely the scaling exponent at which the two boundary layers become asymptotically equal.

The mathematical conditions either in terms of the velocity or in terms of the horizontal gradients of temperature are in principle accessible for verification in two dimensional numerical models. The possibility of proving them from first principles seems remote at this time. It is however quite clear that one can produce passive scalars with velocities satisfying the inequality above and whose dissipation is of the required magnitude. It is less clear but perhaps whithin reach to prove that certain physically significant active scalar models can exhibit the required properties.

References

F.H. Busse, J. Fluid Mech. **37** (1969), 457.

F. H. Busse, J. Fluid Mech **41** (1970), 219.

P. Constantin, C.R. Doering, Phys. Rev E **51** (1995), 3192.

P. Constantin, C.R. Doering, Physica D **82** (1995), 221.

P. Constantin, C.R. Doering, Nonlinearity **9**, (1996), 1049-1060.

P. Constantin and I. Procaccia, Nonlinearity **7** (1994), 1045.

C.R. Doering, P. Constantin, Phys. Rev. Lett. **69** (1992), 1648.

C.R. Doering, P. Constantin, Phys. Rev. E **49**, (1994), 4087;

C.R. Doering, P. Constantin, Phys. Rev. E **53** (1996) 5957-5981.

L.C. Evans, R. F. Gariepy, *Measure theory and fine properties of functions*, (Studies in Adv. Math, CRC Press, Boca Raton, 1992).

J. Hale, *Asymptotic Behavior of Dissipative Systems* (American Mathematical Society, Providence, 1988); R. Témam, *Infinite-Dimensional Dynamical Systems in Mechanics and Physics* (Springer, New York, 1988); P. Constantin, C. Foias, B. Nicolaenko and R. Témam, *Integral and Inertial Manifolds for Dissipative Partial Differential Equations* (Springer, New York, 1988).

F. Heslot, B. Castaing and A. Libchaber, Phys. Rev. A **36** (1987), 5870; E. DeLuca, J. Werne and R. Rosner, Phys. Rev. Lett. **64** (1990), 2370; for a review, see: E. Siggia, Ann. Rev. Fluid Mech. **26** (1994), 137.

E. Hopf, Math. Annalen **117** (1941), 764.

L.N. Howard, J. Fluid. Mech. **17** (1963), 406.

L.N. Howard, Annu. Rev. Fluid Dyn. **4** (1972), 473.

D. Joseph, *Stability of Fluid Motions* (Springer, Berlin, 1976); B. Straughan, *The Energy Method, Stability and Nonlinear Convection* (Springer, Berlin, 1992).

W. V. R. Malkus, Proc. R. Soc. London Ser. A **225** (1954), 185.

Hierarchical Structures and Scalings in Turbulence

Zhen-Su She

Department of Mathematics, UCLA
Los Angeles, CA 90095, USA

Abstract. Turbulence is a mixture of hierarchical structures (eddies) of different sizes, different amplitudes and different degree of coherence. The size distributes continuously from the integral scale ℓ_0 to dissipation scale η. At each scale ℓ ($\ell_0 < \ell < \eta$), the most intermittent structures have the highest amplitude with the highest degree of coherence. Structures of lower amplitudes are related to the most intermittent structures according to a symmetry relation which defines the hierarchy. This is the Hierarchical Structure model (She & Leveque, 1994) which links fluctuation structures of various sizes and amplitude all together.

Moments of the velocity increments across distance ℓ vary as a power law with ℓ in the inertial range. The whole set of scaling exponents ζ_p of the p-th order moments carry a rich set of statistical information about the fully developed turbulent field. It is now widely recognized that ζ_p vary nonlinearly with p, which is often described as the multi-scaling or anomalous scaling problem. The Hierarchical Structure model describes ζ_p in teams of parameters which characterize the most intermittent structures. In isotropic turbulence, they are conjectured to be filaments, and the resulting expression for ζ_p reads $p/9 + 2(1 - (2/3)^{p/3})$.

The Hierarchical Structure model can be derived by a random cascade model of the log-Poisson type. A more elegant derivation calls for an invariance principle and the plausible assumptions that eddies have no characteristic size other than ℓ_0 and no characteristic amplitude other than that of the most intermittent structures. The experimental validation of the model and its implications are discussed in great detail. It is concluded that the description of turbulence in terms of hierarchical structures is physically sound and promising.

1 Turbulence: An Old Problem

Turbulence is a subject of long history. Over the past century, a huge amount of experimental and theoretical investigation has been devoted to its study, yet general consensus have not been reached regarding what is *the solution* of turbulence. It is perhaps naive to try to look for *the solution* of turbulence, it is nevertheless possible to identify a few basic features which the community generally agree that turbulence possesses. First, turbulence (velocity, temperature, etc..) is a highly irregular vector field with excitations spreading *continuously* over many scales both in space and time. The wide range of excited scales in both space and time imply the existence of a tremendous amount of disorder and complexity, or, in theoretical terms, of a huge number of degrees of freedom. Second, strong nonlinear interaction is primarily responsible for spreading excitations over many scales. Consequently, turbulent fluctuations, however random they may be, contain certain structures of various degree of coherence. Experimentalists and computing scientists have carried out measurements of various physical quantities, struggling to define this mixed correlation. Theoreticians have developed many mathematical tools (singular perturbation, renormalized perturbation, etc.), fighting for describing such a mixture of orders and disorders. Although the success has been limited, the impact of a breakthrough will undoubtedly go beyond the area of fluid mechanics; it will increase our analytical ability to characterize the complexity generated by nonlinear processes in general.

1.1 Statistical Description of Turbulence

In the development of the theory of turbulence, several major trends may be noticed. Early in 20th century, turbulence was characterized by the second order correlation functions whose Fourier transform is related to the energy spectrum. It is a natural tool to use to describe a random field which was strongly inspired by the development of statistical mechanics and of the theory of random functions. Indeed, it gives the first approximation to a near Gaussian stochastic field which carries minimal internal correlation, as turbulence appears to be (see further discussions later). Work in this direction has led to the paradigm now described as the statistical description of turbulence. While many pioneers have contributed to the development of the ideas and tools, the 1941 work of Kolmogorov (1941) has made a particular impact. The reason seems to be that Kolmogorov (1941) made a specific prediction about the energy spectrum of turbulence at very large Reynolds number, which, whether right or wrong, has made one to consider more seriously some ideas such as the local cascade and the dimensional argument. These ideas have since been instrumental in the study of the dynamics of turbulence in many areas, particularly in the geophysical context.

Starting in the late fifties, there have been important developments in the analytical description of turbulence stimulated by the progress in quantum field theory. The pioneer work of Kraichnan (1959) and many others aimed at rationalizing the statistical description of turbulence with field-theoretical analysis of stochastic fields constrained by the dynamics of the Navier-Stokes equation (for more discussions, see Proccacia & L'vov in this volume). Difficulties appeared immediately because an infinite series resummation is necessary. Closure assumptions must be introduced to regulate this resummation of divergent series, and it is now realized that except for a few model problems, the regulated problems could be arbitrarily different from the original problem. There is no known regulatory procedure (including the renormalization group analysis) which is free of uncontrolled deviation from the original problem. There is one common feature among most, if not all, formal regulatory (closure) procedures, that is, the statistics of the velocity fluctuations are close to Gaussian uniformly at all scales. In other words, the description of turbulence by closure theories seem to assume minimal structures in turbulence, or in Kraichnan's words (1959), with maximum stochasticity.

Recently, there have been new attempts of non-perturbative closure theory for the statistics of a passive scalar advected by a white-noise velocity field (Kraichnan 1994). See also discussions of L'vov and Proccacia in this volume.

1.2 Structural Description of Turbulence

It has long been the goal of experimental fluid dynamics community to characterize turbulence in terms of certain basic fluid structures distributed randomly in space and time. There, one seeks the "atoms" that are the fundamental constituents of turbulence. In the seventies and eighties, hopes arise because of the discovery of the so-called "coherent structures" which were thought to control the fundamental processes of turbulence. While the studies have led to detailed knowledge about certain complicated fluid mechanical processes, it is also realized that there are a big variety of the "coherent structures" which are geometrically too complicated to describe by available mathematical tools, and which are too varied to be regarded basic. Nevertheless, there is a general consensus that at the location of very strong vorticity or of very low pressure, the filamentary structures appear. Both laboratory experiments and numerical simulations have confirmed the existence of the so-called vortex filaments (Douady *et al.* 1991, Siggia 1981, She *et al.* 1990, Vincent & Meneguzzi 1991, etc.). The degree of coherence in these regions is relatively greater, compared to other regions of lower vorticity amplitude where structures appear to have much more complex forms. It is fair to say from various observations that turbulence does not contain one kind of "basic" structures, but rather a series of complicated structures.

Note that the attempt to describe turbulence as a superposition of some basic structures (patterns) has an important shortcoming. Since fully developed turbulence is strongly nonlinear at all scales, except at scales smaller than the viscous dissipation cutoff, the existence of any basic structure demands extraordinary condition such that the structure is held against strong disturbance from the mutual interaction with others. We know one example, the Burgers' shock structure which exists because of the infinite compressibility which make the dynamics local and stable. The three-dimensional incompressible Navier-Stokes equations describe however just an oppoite situation of very nonlocal and unstable dynamics, where there is no evidence that any entity holds itself stably. Rather, a quasi-equilibrium holds in a statistical sense with active dynamics of the structures at all scales: stretching, deformation, folding, reconnection, etc. The consequence is that turbulent structures of many kinds co-exist dynamically. Turbulence is unlikely to be a family of one animal, but likely to be a zoo!

2 Scaling of Turbulence: Quantitative Studies

There is little disagreement that turbulence involves a great deal of complexity and it is generally agreed that its properties are only stable and measurable in a statistical context. The problem we are facing is what is the reliable statistical information, before we start to consider how the information is to be used effectively in practical modeling. The study of turbulence in the past has placed overwhelming emphasis on the latter issue which is understandably driven from a practical standpoint. In late eighties and nineties, there appears a new trend in the studies of turbulence with the participation of experimental physicists. The "new" experiments are carefully controlled, the measurements are better calibrated, and more importantly the studies were more physically motivated (e.g. Tong *et al.* 1988, Benzi *et al.* 1994, Noullez *et al.* 1997, Tabeling *et al.* 1996; see also Sreenivasan & Antonia 1996). These studies have led to many accurate measurements, and have provided interesting reliable quantitative outputs. One of the measured quantities of great interest is the scaling exponent.

2.1 Physical Significances of Scaling

The scaling behavior is one of the most intriguing aspects of fully developed turbulence. It refers to the observation that in high Reynolds number flows, moments of the velocity difference across a distance ℓ (the so-called velocity structure function) varies as a power law as ℓ, or moments of the energy dissipation vary with the Reynolds number (R_λ) (to leading order) as a power law. The scaling exponents characterize how fast the moments decrease as $\ell \to 0$, or increase as $R_\lambda \to \infty$. If the exponents are known, then it is possible to predict the moments of the velocity fluctuations at any (smaller) scale based on large scale data, or the moments of the dissipation fluctuations at any (larger) Reynolds number based on moderate Reynolds number data. This is one aspect of the practical interests.

The theoretical interests are more intriguing. Compared to other statistical quantities such as moments themselves or the probability distribution functions (PDF), the scaling exponents only address the relative changes with ℓ or with R_λ, which can be more universal. Kolmogorov (1941) in fact conjectured that the scaling exponents are universal, independent of the statistics of large-scale fluctuations, the mechanism of the viscous damping and the flow environment, when the Reynolds number is sufficiently large. There exist a number of experimental measurements in homogeneous open turbulent flows, i.e. "free" turbulence far from the boundaries (Anselmet *et al.* 1984, Benzi *et al.* 1994), which show evidence of the existence of such universal scaling laws. More interestingly, the experimentally measured values also agree quantitatively with those measured in computer-simulated isotropic Navier-Stokes turbulence with a simplistic boundary conditions, i.e. periodic boundary conditions.

Clearly, a set of dynamical information is contained in the values of the scaling exponents. The study of the scaling laws attracts a great deal of attention in physics in the study of critical phenomena. Here, the turbulent medium is a few degrees more complicated than the equilibrium statistical mechanical systems undergoing phase transitions. Theoretically speaking, turbulence is a nonequilibrium medium with cascade of the energy flux from large to small scales. Observationally, this difference is reflected by a nonlinear dependence of the scaling exponents on the order of the velocity structure functions, the so-called multi-scaling. In particular, the measured scaling exponents deviate from the Kolmogorov 1941 (K41) theory which predicts a linear dependence. Note that the K41 model, until recently, has been the only predictive model of scaling laws with no adjustable parameter. This anomalous scaling problem has attracted much attention, because it is believed that the rich information contained in the entire series of exponents provide an important hint about the self-organization of turbulent structures. Furthermore, only the accurate quantitative results can provide the necessary ground for testing various theoretical descriptions.

2.2 Extended Self-Similarity

One of the important developments in the measurement of scaling exponents is a work of Benzi *et al.* (1993a) about the Extended Self-Similarity (ESS) property of turbulence. It is discovered by Benzi *et al.* that the velocity structure function of any order p (reasonably accurate with a sufficient sample size) depends on the structure function of order q (usually chosen to be 3) and can be much better represented as a power law. In particular, when the structure functions deviate from the power-law behavior as the scale is approaching to the viscous cutoff, the relative power law behavior of p-th moment versus q-th moment holds still remarkably well until at scales very close to the Kolmogorov dissipation cutoff. A number of studies were reported after the initial discovery (see more references in Benzi *et al.* 1996b), and indicated that the ESS property works with varied efficiency in different physical environments, but holds real well in far-field homogeneous flow. These further studies confirm the existence of the ESS phenomenon, and provide important information to a better understanding of the physical origin of ESS. Since the third order structure function is expected to be linearly proportional to the length scale in the inertial range, the relative scaling exponent of the *pth* order structure function with respect to the third order is taken to be a new method of measuring ζ_p which shows improved reliability.

In mathematical terms, the ESS property indicates that as the viscous range is approached, the velocity structure functions of different orders show the same characteristic deviations from a power law behavior in such a way that their relative functional dependence is preserved. More physically speaking, there is one characteristic quantity which controls the deviation of the whole set of the velocity structure functions from the inertial range scaling behavior. The word "extended" indicates possibly the existence of another self-similarity property of turbulence which persists even when this quantity shows its non-power law dependence on the length scale. We will offer more discussions below which support this conclusion.

An even more interesting development is some later work by the same group (Benzi *et al.* 1996a-b) which have reported the experimental evidence of a Generalized ESS property (GESS) satisfied by flows with a variety of physical conditions where the ESS property is not satisfied. Instead of studying the velocity structure functions which are moments, they propose to study the normalized structure functions with respect to a certain order which are generally referred to as the hyper-flatness factors. In the same way, they suggest to evaluate the relative functional dependence of those hyper-flatness (HF) factors. The result is that the HF factors are in a beautiful power-law with each other through the whole range of length scale explored (from the integral scale down to very small scale) and for a variety of different flow conditions such as with or without a shear, near a boundary layer, having relatively small Reynolds number, where the ESS property is known not to work.

The interest of these work is two-fold. First, it leads to a better way to estimate the scaling exponent. The measured scaling exponents do not have the same meaning, strictly speaking, as the original ones proposed by Kolmogorov, in the so-called inertial range. But Benzi *et al.* (1994, 1995) have shown that when the Reynolds number is large enough to show a section of the inertial range in the traditional sense, the exponents estimated by ESS or GESS at smaller Reynolds number are consistent with the estimate of the true inertial range exponents. Therefore, the ESS or GESS provide us with a more accurate way for measuring the inertial range exponents. Second, it points out a more fundamental scaling property of turbulence. This will become clear only if we provide some real physical understanding of the phenomenon. This is the central topic of our discussion below.

3 Models of Anomalous Scalings

During the past thirty years, many theoretical approaches have been suggested to address the anomalous scaling behavior of turbulence. Many scaling models start with a very specific ansatz for the PDF of the coarse-grained energy dissipation at the inertial-range scales. The most famous one is the log-normal model (Kolmogorov 1962). These models violate, in one way or another, an exact inequality (Novikov 1970) for the scalings of high order moments, and therefore, can be considered at best as some approximations but not an overall good description of the inertial-range statistics. A more widely accepted approach is built upon the notion of multifractality of turbulence (Mandelbrot 1974, Parisi & Frisch 1985, Meneveau & Sreenivasan 1987). Statistically, it describes the inertial-range cascade as a (discrete) random multiplicative (RM) process; the probability distribution of the corresponding RM coefficient $\mathcal{W}$, $P(\mathcal{W})$, fully determines the inertial-range scaling exponents ζ_p. Since $P(\mathcal{W})$ can be described in terms of arbitrarily many parameters, the resulting scaling formula may exhibit *a priori* any concave nonlinear dependence on the order p (Parisi & Frisch 1985). The problem arising in this approach is therefore the arbitrariness of the modbel; in other words, the physical, or fluid mechanical meaning of the RM process appears very obscure. Consequently, the parameters in the ansatz $P(\mathcal{W})$ remain purely adjustable parameters.

There have long been approaches which attempted to understand the scaling from a more physical or mathematical basis, e.g., the work of Tennekes (1968), Lundgren (1982), Chorin (1991, 1992), Gilbert (1993), Pullin & Saffman (1993), Saffman and Pullin (1994), among others. Fluid structures which are local solutions of the Navier-Stokes equations are randomly superposed in some way for computing the statistical correlations. There have been many predictions of the energy spectrum, but the scalings of high order correlations are difficult to calculate technically. Moreover, we believe that any local solution cannot encompass the whole complexity of the turbulence statistics, because the strong nonlinearity contradicts the fundamental linear superposition principle. It is likely that long-range correlations of a whole set of local solutions play a dominant role in determining the global state of turbulence.

4 Hierarchical Structure Model of Turbulence

In what follows, we will describe a relatively new approach (She & Leveque, 1994) which has shown features of both the structural approach and the random cascade approach. Based on an assumption of a symmetry, the model predicts the scaling exponents in terms of the properties of the most intermittent structures. The latter corresponds to observable fluid mechanical features of considerable coherence embedded in a disordered turbulent medium. This model, called Hierarchical Structural Model, acknowledges the overwhelming complexity of fully developed turbulence (except the most excited, intermittent structures), but points out a novel simplicity which is the symmetry across length scales and across the amplitude of fluctuations. It was shown (Dubrulle 1994, She & Waymire 1995) that the symmetry is exactly realized by a RM process of a log-Poisson type (and thus also called log-Poisson model).

4.1 Physical Picture

A cartoon picture of turbulence from the viewpoint of the Hierarchical Structure model (She & Leveque 1994) can be described as follows. When turbulence is excited in a three-dimensional domain at the so-called integral scale, the nonlinear interaction spreads the fluctuations to small scales. The dynamical turnover time decreases with scale, so do the root-mean-square (rms) velocity fluctuations. However, with respect to the rms velocity fluctuations, there develop increasingly rare and large amplitude events as the cascade proceeds. Higher are the amplitude of the fluctuations, more coherent are they in spatial configurations, and more phase correlated across length scales. At the statistically steady state, the fluctuations at large and small scales and at large and small amplitudes form a unified hierarchy described by a certain symmetry.

The probability density function (PDF) of velocity fluctuations at the integral scale reflects the motions directly excited by an external mechanism, and thus is not universal. However, the PDFs at smaller scales can be described by a transformation defined by the symmetry which will be a convolution with the integral-scale PDF. It is believed that the transformation (and the symmetry) is intrinsic to the nonlinear dynamics, and can be determined by universal physical principles.

4.2 The Model

In searching to define this transformation, She & Leveque (1994) proposed to study a hierarchy defined through the ratio of the successive moments: $\epsilon_\ell^{(p)} = \langle\epsilon_\ell^{p+1}\rangle/\langle\epsilon_\ell^p\rangle$ $(p = 0, 1, ...)$, where ϵ_ℓ is the coarse-grained energy dissipation at an inertial-range scale ℓ. This hierarchy passes from the mean field described by $\epsilon_\ell^{(0)}$ to the most intermittent structures described by $\epsilon_\ell^{(\infty)} < \infty$ (the upper bound for the field ϵ_ℓ in a finite space-time manifold). While p can be any real number, restricting to the set of integers makes the presentation easy to follow. Since the *pth* order ratio can have a nontrivial scaling: $\epsilon_\ell^{(p)} \sim \ell^{\lambda_p}$, turbulence will generally behave as a multiscaling field. It is interesting to note that $\epsilon_\ell^{(p)}$ represents a sequence of dissipation events with increasing amplitudes when the underlying PDF of ϵ_ℓ exhibits a log-concave tail which is consistent with experimental observations. Therefore, λ_p's describe scaling properties of structures of various amplitude in the physical space.

When $\lambda_0 = \lambda_\infty$, the scaling exponents of all members of the whole hierarchy is identical. The scaling field is said to be a monoscaling field which is statistically equivalent to a fractional Brownian motion. In this case, the scaling exponents τ_p, defined by $\langle\epsilon_\ell^p\rangle \sim \ell^{\tau_p}$, depend linearly on p. The K41 can be recovered as a special case ($\lambda_\infty = \lambda_0 = 0$). Otherwise, it leads to the β-model (Frisch, Nelkin & Sulem 1978). When $\lambda_0 \neq \lambda_\infty$, we have generically a multifractal field. Intuitively, there should be a relation among λ_p's, since the whole hierarchy is the result of a unique dynamics (the Navier-Stokes dynamics) in which low-order and high-order moments are consistently related. She and Leveque (1994) further proposed that there exists a symmetry, that is,

$$\epsilon_\ell^{(p+1)} \sim \epsilon_\ell^{(p)^\beta} \epsilon_\ell^{(\infty)^{(1-\beta)}}; \qquad \lambda_{p+1} = \beta\lambda_p + (1-\beta)\lambda_\infty, \qquad (4.1)$$

where β is a constant independent of p. Under (4.1), the scaling property for isotropic turbulence is uniquely determined by λ_∞, since $\lambda_0 = 0$. Therefore, this theory determines the scaling laws of turbulence in terms of the characteristics of the most intermittent structure. Assuming that they are 1-D filamentary structures which lie in a boundary between two large eddies (of size ℓ_0) with a thickness of the order of the Kolmogorov length scale η, one obtains $\lambda_\infty = -2/3$, and predicts (She & Leveque 1994) the whole set of the scaling exponents ζ_p for the pth order velocity structure function:

$$\langle \delta u_\ell^p \rangle \sim \ell^{\zeta_p}, \qquad \zeta_p = \frac{1}{9}p + 2\left(1 - (\frac{2}{3})^{p/3}\right). \tag{4.2}$$

The formula (4.2) contains no adjustable parameter.

4.3 Comparison

During the last three years, both experimental and numerical studies have been conducted to test the model. Ruiz Chavarria *et al.* (1995a-b) have made the measurements in laboratory flows and have calculated quantities to test specifically the assumption (4.1), the assumption about the symmetry of the hierarchy. They claim that "... the hierarchy of the energy dissipation moment, recently proposed by She & Leveque for fully developed turbulence is in agreement with experimental data ...". Their method of calculation even leads to a direct determination of the parameter β which is in agreement with the proposed one ($\beta = 2/3$) for isotropic turbulence. Furthermore, the measurements of the scaling exponents ζ_p in various flows carried out in several laboratories, e.g., in turbulent wakes (Benzi *et al.* 1994, 1995, 1996a-b), in grid turbulence (Herweijer & van de Water 1994), and in wind tunnel turbulence (Anselmet *et al.* 1984), and in jet turbulence (Noullez *et al.* 1997) are all consistent with (4.2) with remarkable accuracy. Finally, direct numerical simulations of the isotropic Navier-Stokes turbulence also accurately support (4.2) (Cao *et al.* 1996). Some of these comparisons are reported below in Table 1.

There has been skepticism about the meaning of the comparison. In this regard, we would like to make the following remarks. First, the scaling exponents of the longitudinal velocity structure functions in a far-field of fully developed turbulent open flow have been measured in several flow environments, and the results are generally consistent (see Table 1). In other words, these experimental values are robust and stable. Recently, Belin, Tabeling & Willaime (1996) have reported the measurement of the scaling exponents in a closed flow system, which show values somewhat below the above reported ones $\zeta^{(4)}$. The measurements in the Taylor-Couette flow (another closed flow system) by Swinney's group at Texas also seem to show the same trend. The reason for this discrepancy is not yet clear. One possibility is that there is a systematic deviation of the scaling exponents between open and closed system, due to the interaction with the wall-ejected structures and strong rotation (both system develop strong swirls).

Order p	$\zeta_p^{(1)}$	$\zeta_p^{(2)}$	$\zeta_p^{(3)}$	$\zeta_p^{(4)}$	$\zeta_p^{(5)}$	SL Model ζ_p
1	–	0.37	–	–	0.362 ± 0.003	0.364
2	0.71	0.70	0.70 ± 0.01	0.70	0.695 ± 0.003	0.696
4	1.33	1.28	1.28 ± 0.03	1.26	1.279 ± 0.004	1.279
5	1.65	1.54	1.50 ± 0.05	1.50	1.536 ± 0.01	1.538
6	1.8	1.78	1.75 ± 0.1	1.71	1.772 ± 0.015	1.778
7	2.12	2.00	2.0 ± 0.2	1.90	1.989 ± 0.021	2.001
8	2.22	2.23	2.2 ± 0.3	2.08	2.188 ± 0.027	2.211
9	–	–	–	2.19	–	2.407
10	–	2.59	–	2.30	–	2.593

Table 1. Scaling exponents ζ_p of the pth order velocity structure functions measured in a wind tunnel turbulence$^{(1)}$ (Anselmet *et al.*, 1984), in a wake turbulence$^{(2)}$ (Benzi *et al.*, 1994), in a jet turbulence$^{(3)}$ (transverse velocity structure function) (Noullez *et al.* , 1997), in a low temperature helium experiment$^{(4)}$ (Belin *et al.* 1996), and in an isotropic Navier-Stokes turbulence simulation$^{(5)}$ (Cao, Chen & She, 1996). The SL model reads $\zeta_p = p/9 + 2(1 - (2/3)^{p/3})$.

Secondly, the Extended Self-Similarity property in turbulence (Benzi *et al.* 1993; Stolovitzky & Sreenivasan 1993; Benzi *et al.* 1994; Briscolini *et al.* 1994) has greatly enhanced the accuracy of the measurement. Although the mechanism is not yet clear, the fact that it is a useful property in measuring scalings which leads to no detectable distortion of the measured value is widely accepted. So the reported values in Table 1 are quite reliable. The good agreement can hardly be attributed to pure coincidence.

Thirdly, it is fair to regard the comparison as a consistency check which is clearly positive. The fact that there exist other cascade ansatz which produce, with multiple adjustable parameters, a fit of the same quality does not invalidate the present description. The advantage of the present model is its simplicity, the physical connection to flow structures, and its ability to predict non-universal features of scalings as we will discuss later. It is also applicable to a variety of other turbulent systems with cascade dynamics. In short, it is worthy of further study.

4.4 Application to the GOY Shell-Model

While recognizing that a deductive theory of turbulence from the Navier-Stokes turbulence is highly desired, it is also important to examine carefully other systems exhibiting some essential features of the Navier-Stokes turbulent dynamics. These features include, from the present phenomenological understanding, the existence of an inertial range of scales of cascade (driven by the inertial force or the nonlinear convective term). The study of the other systems will allow one to identify the essential ingredients in the NS system such as the conservation laws, etc., which governs the cascade dynamics. More importantly, it will stimulate the development of a general theoretical framework for nonequilibrium systems presenting critical and scale invariance properties. This has been the essential motivation behind the study of a dynamical-system model of turbulence, namely the GOY shell model (see also, Kadanoff *et al.* 1995).

The GOY shell-model is a finite-dimensional dynamical system, which was introduced by Gledzer (1973) with an important extension made by Yamada and Ohkitani (1987, 1989) later introducing phase dynamics with complex variables. The dynamics are governed by the following set of ordinary differential equations:

$$(\frac{d}{dt} + \nu k_n^2)u_n = f_n + (a_n u_{n+1}^* u_{n+2}^* + b_n u_{n+1}^* u_{n-1}^* + c_n u_{n-1}^* u_{n-2}^*). \quad (4.3)$$

Here, $\{u_n\}_{0,1,\ldots N-1}$ is a set of complex variables which model the Fourier space excitations in shells of wavenumbers $k_n = k_0\lambda^n \leq k < k_{n+1}$, f_n is a driving force usually acting on some small wavenumber shells, e.g. $f_n = f_2\delta_{n,2} + f_3\delta_{n,3}$. The term $\nu k_n^2 u_n$ models the viscosity damping with the kinematic viscosity ν.

At very small ν, the dynamics is essentially inviscid at small wavenumber shells where the nonlinear coupling (r.h.s. of (4.3)) make a chain linking the fluctuations at different wavenumber shells. At those shells (small n), the moments of the velocity fluctuations, $\langle |u_n|^p \rangle$, vary with the wavenumber as a power law, $k_n^{-\zeta_p} \sim \ell_n^{\zeta_p}$. A number of studies have shown that the GOY shell model shows a very similar scaling laws as in the 3-D Navier-Stokes equations (see e.g. Kadanoff *et al.* 1995 for more references). Leveque & She (1997) have made a careful study of the scaling exponents in the GOY shell model by a set of long numerical integrations of (4.3). The large sample size of the statistical fluctuation data has enabled a detailed study of the convergence of the moments and the exponents. The scaling exponents ζ_p so measured are compared to the predictions of the Hierarchical Structure model as well as other cascade models (log-Normal, log-Stable, p-model, etc.). The conclusion is evident that the functional dependence of ζ_p are better represented by the Hierarchical Structure model (Leveque & She 1997). In Table 2, we report the comparison of ζ_p.

4.5 Other Predictions and Confirmation

The interesting feature of (4.2) is that the parameters determining the set of exponents ζ_p depends only on the properties of the most intermittent structures. These most intermittent structures at the length scale ℓ are theoretically defined, for e.g. the coarse-grained energy dissipation, by the limit $\lim_{p\to\infty} \epsilon_\ell^{(p)}$. In practice, this limit depends on the sample size of the data which is collected to describe the (spatio-temporal) ensemble of turbulence[1]. In the GOY shell model, we have collected enough samples so that for a certain range of scales, the limit has converged. In many other cases, $\epsilon_\ell^{(\infty)}$ depends on the sample size. However, its scaling exponent λ_∞ ($\epsilon_\ell^{(\infty)} \sim \ell^{\lambda_\infty}$) may depend more weakly on the sample size. Even when this dependence exists, it may be important to discover how it controls the dependence of the measured scaling exponents for high order moments, e.g., ζ_p for large p.

It is difficult to draw a specific line between the reliable ζ_p's and those ζ_p's which have not yet fully converged. Instead of making arbitrary determination, we suggest to use the approximate information regarding the most intermittent structures (λ_p or γ, to be introduced later) to characterize the whole set of measured scaling exponents. The basis for this proposal lies in the fact that the symmetry linking the most intermittent structures and less intermittent ones are more fundamental to turbulence, as we explain now.

The general formula for ζ_p reads

$$\zeta_p = \gamma p + C_0(1 - \beta^p), \tag{4.4}$$

where γ is similar to the smallest Hölder exponent of the velocity field when the scale varies in the inertial range, and C_0 is the co-dimension of the set of spatial points for which this exponent is realized.

---

[1] When the time-ensemble is considered, the limit $\lim_{p\to\infty} \epsilon_\ell^{(p)}$ amounts to calculating

$$\lim_{p\to\infty} \frac{\int_0^T \epsilon_\ell^{p+1}(t)dt}{\int_0^T \epsilon_\ell^p(t)dt}$$

which is well-defined for any finite large T. The sample size dependence will be reflected by the T-dependence.

Order p	ζ_p/ζ_3 (GOY)	$0.125p + 1.49(1 - 0.58^{p/3})$ (SL)
1	0.375 ± 0.005	0.372
2	0.705 ± 0.003	0.703
3	1.000	1.000
4	1.268 ± 0.006	1.268
5	1.512 ± 0.014	1.513
6	1.738 ± 0.026	1.737
7	1.946 ± 0.040	1.946
8	2.141 ± 0.058	2.140
9	2.323 ± 0.078	2.323
10	2.50 ± 0.10	2.50
11	2.66 ± 0.13	2.66
12	2.82 ± 0.15	2.82
13	2.97 ± 0.18	2.97
14	3.12 ± 0.21	3.12
15	3.26 ± 0.25	3.27
16	3.40 ± 0.28	3.41
17	3.54 ± 0.32	3.55
18	3.67 ± 0.36	3.68
19	3.80 ± 0.40	3.82
20	3.94 ± 0.44	3.95

Table 2. Scaling exponents ζ_p of the *pth* order velocity structure functions measured in the GOY shell model (Leveque & She 1997) compared with the Hierarchical Structure model of She and Leveque (SL). The parameters in the Hierarchical Structure model for the GOY shell-model is slightly different from those suggested for the Navier-Stokes turbulence. Those parameters are directly measurable from the data set.

The two parameters γ and C_0 are measurable in any finite sample of data[2], and then characterize the most intermittent structures detected within a finite spatial-temporal domain (where the ensemble is defined). The interesting fact is that there exist quantities which do not depend on γ and C_0. For instance, it is easy to verify that

$$\rho(p,q;p',q') = \frac{\zeta_p - p/p'\zeta_{p'}}{\zeta_q - q/q'\zeta_{q'}} = \frac{q'}{p'}\frac{p'(1-\beta^p) - p(1-\beta^{p'})}{q'(1-\beta^q) - q(1-\beta^{q'})}. \tag{4.5}$$

These quantities $\rho(p,q;p',q')$ describe the relative scaling exponents between the normalized moments (p,p') and (q,q'). The fact that $\rho(p,q;p',q')$ depend only on β which characterizes the symmetry in the hierarchy suggests a method to verify the correctness of the hierarchy without a massive data set. If the symmetry is indeed more fundamental, then we may also observe a strong universality property of $\rho(p,q;p',q')$ compared to ζ_p.

Ruiz Chavarria *et al.* (1995a-b) have carried out an experimental test specifically on the symmetry property of the model, and their results have fully confirmed its correctness. Benzi *et al.* (1996) have reported the Generalized Extended Self-Similarity (GESS) property from experimental results that $\rho(p,q;3,3)$ have a remarkable universal behavior for turbulent flows near a boundary layer, with and without a shear, etc. Their results are fully consistent with the prediction of the Hierarchical Structure Model. We believe that the model gives a plausible physical explanation of the ESS and GESS property.

An important practical interest of this universality result (GESS) is to allow us to differentiate among various turbulent systems. In practical situations, the properties (γ and C_0) of the most intermittent structures vary with the sample size, and also with spatial location and direction when the flows are not homogeneous and isotropic. According to the Hierarchical Structure Model, we can still have some reliable scaling laws among normalized moments reflecting the intrinsic symmetry. These scaling laws will allow us to determine γ or C_0 as a function of spatial location or direction, and hence get relevant physical information about the flow from the measurement of the scaling exponents. Deriving the structure information from the statistical measures is the unique advantage of the Hierarchical Structure Model.

4.6 Fluid Structures and Scalings: More Comments

It needs to be emphasized that the most intermittent structures discussed above are, theoretically speaking, not directly the ones which are visualized in flow experiments and in numerical simulations. Instead, they are defined as the series of structures at the inertial-range scales as follows. First, obtain the coarse-grained dissipation field ϵ_ℓ at scale ℓ. Second, identify $\epsilon_\ell^{(\infty)}$ as

[2] We will discuss the measurability of C_0 later.

the spatial locations which contribute the most to the ratio of the moments $\lim_{p\to\infty}\langle\epsilon_\ell^{p+1}\rangle/\langle\epsilon_\ell^p\rangle$. These spatial points form a set whose volume depends on the scale ℓ. The concept of the co-dimension $d-D$ is an abstraction of the fact the volume changes in ℓ as ℓ^{d-D} where d is the dimension of the space and D is dimension of the set.

As the coarse grained scale decreases $\ell \to \eta$ to be close to or below the dissipation cutoff scale η, we can also identify the set $\epsilon_\eta^{(\infty)}$ as the most intermittent structure at scale η. These structures may be close to the iso-surfaces at a high threshold of the original, non-coarse-grained field ϵ. Furthermore, their geometry may not change significantly as we further coarse-grain the field from scale η to $\ell > \eta$. If these two conditions are satisfied, then the geometry of the visualized structures gives a qualitatively correct estimate of the co-dimension of the most intermittent structures. We can hope that when the turbulent velocity field changes smoothly as ℓ goes through η (e.g. no appreciable oscillations around high peaks of the velocity fluctuations), then these conditions are satisfied.

However, in some arbitrary mathematical problems such as the solution of the Navier-Stokes equation with hyperviscosity, strong oscillations in the physical field may appear at very small scales (around the dissipation cutoff scale η) due to the non-positiveness of the hyperviscous "diffusion". The consequence is that the smooth transition from the inertial-range scale ℓ to η is interrupted. In this case, the relation between the characteristics of the inertial-range most intermittent structures with the visualization of the dissipation range quantity such as the dissipation ϵ becomes elusive.

5 Further Theoretical Development

The Hierarchical Structure model has also been examined from a more theoretical standpoint. The main results are that the symmetry (4.1) can be exactly realized via a simple random cascade process called log-Poisson (Dubrulle 1994, She & Waymire 1995), and it follows also from an invariance property in the transformation of the frame of reference in a new coordinate systems, the amplitude-scale system (Dubrulle and Graner 1996a-b, She and Leveque 1996).

5.1 Log-Poisson Cascade

It is shown by Dubrulle (1994) and independently by She & Waymire (1995) that (4.1) can be exactly realized by a random multiplicative cascade process, called log-Poisson. Let the integral-scale eddies be represented by the coarse-grained energy dissipation ϵ_{ℓ_0} (a random variable). Let the small-scale eddies at any given length scale ℓ be generated by

$$\epsilon_\ell = (\frac{\ell}{\ell_0})^\gamma \beta^n \epsilon_{\ell_0} \tag{5.1}$$

where n is an independent Poisson random variable with a mean λ:

$$P(n) = e^{-\lambda}\frac{\lambda^n}{n}, \qquad n = 0, 1, 2, ... \tag{5.2}$$

It can be deduced from (5.1) that

$$\langle \epsilon_\ell^p \rangle = (\frac{\ell}{\ell_0})^{\gamma p} \sum_n \beta^{np} P(n) \langle \epsilon_{\ell_0}^p \rangle. \tag{5.3}$$

Then, (4.1) follows.

According to (5.1), a large-scale eddy has a number of possibilities when it is transformed to a smaller one. The largest amplitude is achieved at $n = 0$ because $\beta < 1$; other smaller amplitude events are obtained by multiplying an integer number of β factors. The $n = 0$ event is of special interest: it varies with scale as ℓ^γ and the probability of finding it is $e^\lambda \sim \ell^{C_0}$. This is the most intermittent event. When $C_0 > 0$, or the strongly excited events reside in a smaller (fractal) set, $e^\lambda \to 0$ as $\ell \to 0$. This is the intermittency, or anomalous scaling, because the whole space is occupied by less excited events. When the multiplication of β factors acts as a Poisson point process, the symmetry (4.1) is exactly realized.

5.2 Log-Poisson Cascade and Other Cascade Models

Compared to other discrete cascade models proposed earlier (e.g. Meneveau & Sreenivasan 1987), the log-Poisson has the following features: First, the cascade from ℓ_0 to ℓ_1 and then to ℓ_2 is identical to the cascade from ℓ_0 to ℓ_2. This can be shown as follows. Let $\mathcal{W}_{01} = (\ell_1/\ell_0)^\gamma \beta^{n_1}$ and $\mathcal{W}_{12} = (\ell_2/\ell_1)^\gamma \beta^{n_2}$. Then, it can be shown, by working with $\log \mathcal{W}$, that $\mathcal{W}_{02} = \mathcal{W}_{01}\mathcal{W}_{12} = (\ell_2/\ell_0)^\gamma \beta^N$ where N is again a Poisson random variable, and moreover $\langle N \rangle = \langle n_1 \rangle + \langle n_2 \rangle$. This proof is valid for any arbitrary ℓ_1 and ℓ_2, which removes one important arbitrariness in defining the step of cascade ℓ_1/ℓ_0 or ℓ_2/ℓ_1. This arbitrariness is a major shortcoming of the previous discrete cascade models when used for describing such a continuous scaling process as turbulence. In this regard, the log-Poisson process is self-consistent and the reason for its success lies in its log-infinite divisibility property (see She & Waymire 1995).

Secondly, the log-Poisson cascade picture does not have any difficulty which other log-infinitely divisible process such as the log-normal model (Kolmogorov 1962) or the log-stable model (Kida 1991) have, namely, the physically unacceptable behavior of ζ_p for large p. It is true that both the log-normal model and log-stable model, with a certain choice of the parameters (which by the way cannot be estimated based on physical properties of the flow), agree with some experimental values for a moderate range of p. We believe that this is the evidence of a good approximation of the model over a range of p, just as the approximation of a smooth function locally by a quadratic form. However, both models make very strong predictions about the asymptotic behavior of ζ_p at large p, or equivalently, the behavior of the fluctuation events at very large amplitude. This behavior requires that the velocity be unbounded in the limit of vanishing viscosity which creates a mathematical inconsistency with the incompressible Navier-Stokes equation (Frisch 1991, 1995). Strictly speaking, there is no evidence supporting the divergence of the velocity in the limit of vanishing viscosity for the three-dimensional Navier-Stokes equation in a periodic domain under a deterministic forcing at low wavenumbers. And it is virtually impossible for experiments to provide a convincing test of the assertion either.

On the other hand, the measured anomalous, or non-Kolmogorovian scalings at moderately large p, in both the Navier-Stokes flows and laboratory flows, have now quite solid evidence. It seems pointless to base the theory of the anomalous scalings on a model whose strong prediction can "never" be checked, without mentioning its unlikelihood from purely a stochastic process point of view (Mandelbrot 1974). By contrast, the log-Poisson model makes no fixed assertion about the large p behavior. It says that the large p behavior is the property of the most intermittent structures currently in the spatio-temporal domain in question. These behaviors could vary case by case, giving rise to some apparent scattering of the measured scaling exponents. On the other hand, there is a stable symmetry which is built by the log-Poisson cascade between the scaling (if it is there) of the most intermittent structures and of other less excited ones. This symmetry is more intrinsic and visible, and is therefore experimentally detectable already at moderately large p. This description enjoys the simplicity and relies on no speculative basis in "unreachable" asymptotics.

5.3 Application to Navier-Stokes Turbulence

The theory of the log-Poisson cascade has an interesting application, that is, to find the probability density function (PDF) of small scale fluctuations from the PDF of large scale ones. Note first that any small-scale fluctuations ϵ_ℓ is *equal in law* (denoted by $\stackrel{\ell}{=}$) to the large-scale fluctuations ϵ_{ℓ_0} multiplied by an independent random variable $\mathcal{W}$. Therefore,

$$\log \epsilon_\ell \stackrel{\ell}{=} \log \epsilon_{\ell_0} + \log \mathcal{W}. \tag{5.4}$$

Consequently,

$$P(\log \epsilon_\ell) = P(\log \epsilon_{\ell_0}) \otimes P(\log \mathcal{W}), \tag{5.5}$$

where $\otimes$ denotes a convolution.

One particular application is that at large Reynolds numbers and at very small scales ($\ell \to 0$), ϵ_ℓ approaching ϵ is fluctuating much more intermittently than ϵ_{ℓ_0} (the averaged dissipation over a large domain). In this case, $P(\log \epsilon_{\ell_0})$ behaves approximately as a δ-function, compared to $P(\log \mathcal{W})$. In

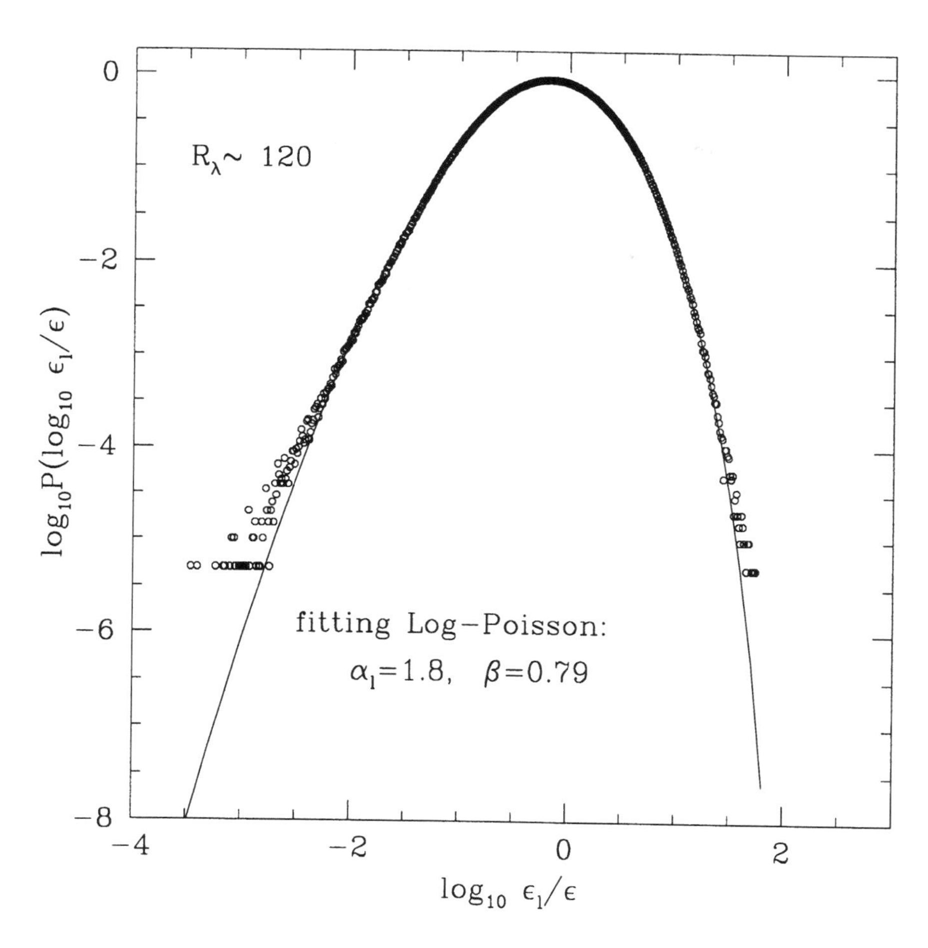

Fig. 1. The probability density function of the energy dissipation measured in a simulation of the Navier-Stokes isotropic turbulence, compared with the PDF obtained by a fit with the log-Poisson model.

this case, $P(\log \epsilon_\ell)$ will mimic closely a log-Poisson form, a smooth version of the discrete (atomic) distribution function $P(\log \mathcal{W})$.

Leveque and She (1996) have tested this prediction. Forced 3-D incompressible Navier-Stokes equations under periodic boundary conditions are integrated numerically using a pseudo-spectral method. With a resolution of 256^3, a sample of stationary isotropic turbulence is generated with a Taylor microscale Reynolds number $R_\lambda \approx 120$. The probability density function $P(\log \epsilon)$ is then measured and compared to a log-Poisson fit. The result is quite satisfactory, as presented in Fig. 1.

Recently, Novikov (1994) introduced a gap argument against (5.1). In fact, there is a big difference between the breakdown coefficient which he used and the experimentalists measured and the random multiplicative coefficients $(\frac{\ell}{\ell_0})^\gamma \beta^n$ used here, because the former cannot be truly random and independent of the large eddies (ϵ_{ℓ_0}). Even a slight dependence will largely affect the theoretical conclusion, especially for large p moments. A more careful analysis taking into account the slight statistical dependence between the breakdown coefficient and the large eddies shows that, the Novikov's equality becomes an inequality which resolves the controversy raised by him. Of course, we can not claim the correctness of the formula (4.4) at this stage, but it seems unlikely that the gap argument presents a serious threat.

5.4 Invariance Principle

Recently, She & Leveque (1996) proposed another derivation of the hierarchical symmetry based on an invariance principle using similar reasoning as in the theory of special relativity. The work was stimulated by an earlier work of Dubrulle and Graner (1996a-b). Instead of addressing the transformation of the PDFs from the large scale to small scale as in the log-Poisson case, we propose to address the transformation of moments. Let ϵ_ℓ denote the coarse-grained energy dissipation where ℓ continuously changes from ℓ_0 to η. At each scale ℓ, the fluctuation of ϵ_ℓ has a maximum amplitude, called $\epsilon_\ell^{(\infty)}$. We make the following fundamental assumptions:

H1 : There is no characteristic length scale other than the integral scale ℓ_0.

H2 : There is no characteristic fluctuation amplitude other than $\epsilon_\ell^{(\infty)}$.

The assumption (H1) is the similarity property in scale, and (H2) is the similarity property in amplitude. Since (H1) and (H2) are so much similar, it is natural to set up a coordinate system in scale and amplitude so that the two similarity properties can be fully explored. Following Dubrulle & Graner (1996a), the new coordinate system is to make the following identification:

$$X(\cdot) = \log E(\cdot); \quad T = \log \ell. \tag{5.6}$$

Here $E(\cdot)$ refers to as the expectation value of a random variable. In this system, the amplitude-scale becomes "space-time". An arbitrary stochastic field will result in a path (or trajectory) in this coordinate system. In particular, a scaling field such as the increment of a fractional Brownian motion of exponent h such that $E(|\Delta x|) = E(|x(t+\ell) - x(t)|) \sim \ell^h$ will be described as a straight line of constant speed h.

The assumption (H1) implies generally that all statistical average quantities behave as a power-law in ℓ/ℓ_0 (at least in leading order). The assumption

(H2) suggests that if we introduce $\Pi_\ell = \epsilon_\ell/\epsilon_\ell^{(\infty)}$, then there exists no characteristic amplitude between 0 and 1 at each ℓ. It then follows that the field Π_ℓ^p is a scaling field, that is, $E(\Pi_\ell^p) = \ell^{V_p}$. Next, we conjecture:

H3 : The field Π_ℓ^p (for any p) defines a so-called inertial frame of reference with an intrinsic speed V_p. In other words, V_p is independent of whether the observer is in a rest or moving frame of reference.

A specific proposal of the meaning of the moving frame of reference is given in She & Leveque (1996). In the moving frame of speed V_p (defined by the scaling field Π_ℓ^p), the space coordinate of another scaling field Π_ℓ^q is proposed to be measured in the following way:

$$X_q^{(p)} = \log\left(E(\Pi_\ell^p \Pi_\ell^q)/E(\Pi_\ell^p)\right). \tag{5.7}$$

In other words, (5.7) suggests that the "space" coordinate measurement in a moving frame corresponds to an average with respect to a weighted probability density function by a factor Π_ℓ^p. As p increases, more weight is put on the high amplitude fluctuations. With this restriction, it can be shown (She & Leveque 1996) that the "space" and "time" coordinates will satisfy a transformation law when the observer moves from one frame of reference to another:

$$X' = X - WT; \quad T' = \left(1 - \frac{W}{V_\infty}\right)T, \tag{5.8}$$

where W is the speed of the moving frame. From (5.8), we can derive a composition law for the relative speeds:

$$V_{p+q}^{(0)} = V_p^{(0)} + V_q^{(p)} - \frac{V_p^{(0)} V_q^{(p)}}{V_\infty}. \tag{5.9}$$

However, (H2) suggests the equivalence of all inertial frames of reference, since they simply correspond to the emphasis on different amplitudes and there is no preferred amplitude. This justifies the second part of (H3) which can be more explicitly stated as $V_q^{(p)} = V_q^{(p')}$ for any p and p'. Substituting $V_q^{(p)} = V_q^{(0)}$ into (5.9) leads to the solution for V_p:

$$V_p = V_\infty(1 - \beta^p), \tag{5.10}$$

where $V_\infty = C_0$ plays the role of the "speed of light" as in the theory of special relativity. Eq. (5.10) is identical to (4.4) in its nonlinear part.

It is interesting to see that the Hierarchical symmetry (4.1) follows from such an elegant argument of the invariance of "speed" $V_q^{(p)} = V_q^{(p')}$ which expresses the fact that all frame of references are equivalent.

5.5 Analogy with the Theory of Special Relativity

Technical resemblance between the discussion in the last section and the theory of special relativity is striking. We can make a few more comments about what we learn from this analogy.

The theory of special relativity challenged the classical picture of Newton which involved an absolute and homogeneous space, within which things changed in an absolute and homogeneous time. In particular, the constancy of the speed of light measured in all moving frames of reference imposes a special connection between the space and time coordinate of things under study. The Lorenz transform merely expressed this "relativity" in concrete mathematical form and gave rise to many predictions testable in physical environment.

The analogy between our present treatment and the theory of special relativity does not imply any change in the understanding of the Newtonian space-time structure for the fluid mechanics. The analogy is seen in the new "space-time" coordinate system for the convenience of describing the statistical structure of turbulence, that is, how the statistical averages (moments) change as the length scale decreases. When the length scales ("time") and the moments ("space") are normalized or scaled properly (setting up the origin), the theory of the last section indicates that turbulent stochastic field (the kinematic of event) is such that the two variables of the length scale and the moment amplitude are related.

It will become more clear if we contrast it with the Kolmogorov 1941 (K41) description within the same framework. In K41, there is no upper characteristic fluctuation amplitude, so $V_p = \zeta_p = hp$ $(h = 1/3)$. It is easy to check that $V_{p+q} = V_p + V_q$, $X_q^{(p)} = (V_{p+q} - V_p)T = X_{p+q} - V_pT$. Therefore, we have a system of Galilean transformation which corresponds to the prescription that Kolmogorov (1941) gave to turbulence field. To put it differently, Kolmogorov described an absolute and homogeneous "space" of amplitude of fluctuations in the sense that big amplitude fluctuations cascade to small scales in identical way as small amplitude fluctuations.

In contrast, we find important to recognize the role of the most singular structures for the evolution of the whole statistical ensemble of fluctuations from large to small scales (the kinematic in the new "space-time" coordinate system). This is closely parallel to recognizing the existence of the speed of light, which consequently make the new "space" and "time" united. The fact that we have a different law of transformation from the Lorenz law is due to the lack of symmetry between the positive speed and negative speed, one specifying divergent changes and other convergent changes as the scale decreases.

6 Universality and Non-Universality

The issue of the universality is one of the most important question for turbulence (see e.g. Nelkin 1994). When the medium becomes increasingly complicated, physical quantities show richer behavior, and theories describing the relationship between these quantities shows inevitably more complexity. Only the discovery of certain universality would lead to the hope for some simple and elegant theoretical description. This is the basic harmony that the complex universe display.

The universal behavior does not invoke detailed mechanisms specific to the dynamics, and should invite a theoretical description of general character. For turbulence, we believe that there are some general principles due to simply the non equilibrium and the strong nonlinearity. Because of the non equilibrium, the energy cascades from large to small scales. Because of the strong nonlinearity, the inertial range is formed. Furthermore, we conjecture that strong nonlinear interaction also leads to a strong phase mixing which is the origin of the symmetry (4.1), or of the new universal physical quantities such as (4.5). At present, the Hierarchical Structure Model is consistent with all existing observations. But it presents only the first step in grasping the universal aspect. More important work will be ahead to elucidate the "first" principle behind this universality.

Next, we argue that the understanding of the universality is the only way to address the non-universal behavior. We believe that ζ_p may not be universal. Experimentally observed values show certain scattering; we think that it is not a simple statistical convergence problem. There is more physical meaning behind. For example, we conjecture that it may reflect the fluctuation of the most intermittent structures captured by the given sample. In any sample set, they are the largest fluctuation events and also the rarest events. They change easily from sample to sample, from environment to environment. Because ζ_p, even for the moderately large p, is sensitive to them, we observe scattering. This explanation, if being correct, indicates that ζ_p is not the most robust measure. The Hierarchical Structure Model suggests to study $\rho(p,q;p',q')$ instead, which describe theoretical the statistical link between the most intermittent structures and less intermittent structures.

The practical interest of studying $\rho(p.q;p',q')$ is to find a better parameterization for ζ_p which is physical. If indeed the property of the most intermittent structures determines ζ_p, then the systematic variation of ζ_p would reveal a change of physical environment, and thus the physical origin of the non-universality is revealed. We believe that a physical theory of the scaling should explain the physical origin of the non-universal behavior of ζ_p (among others). In other words, the theory should contain parameters which can either be direct measurable or can be estimated by plausible theoretical arguments. The Hierarchical Structure Model has offered a plausible candidate.

Acknowledgment This work presented here is a continuation of the collaboration with Dr. E. Leveque and E. Waymire. I have also greatly benefited from discussions with Dr. R. Benzi and S. Chen. I am also grateful to the support by the Office of Naval Research which has enabled a number of scientific interactions relevant to the development of the work.

References

Anselmet, F.,Gagne, Y., Hopfinger, E. J. & Antonia R. (1984): High order velocity structure functions in turbulent shear flows, *J. Fluid Mech.* **140**, 63.

Belin, F., Tabeling, P., Willaime, H. (1996): Exponents of the structure functions in a low temperature helium experiment, *Physica D.*, **93**, 52.

Benzi, R., Ciliberto, S., Ruiz Chavarria, G., & Tripiccione, R. (1993a): Extended self-similarity in the dissipation range of fully developed turbulence. *Europhys. Lett.* **24**, 275.

Benzi, R., Ciliberto, S., Baudet, C., Massaioli, F., Tripicione, R. & Succi, S. (1993b): Extended self-similarity in turbulent flows, *Phys. Rev. E* **48**, 29.

Benzi, R., Ciliberto, S., Baudet, C. & Ruiz Chavarria, G. R. (1994): On the scaling of three dimensional homogeneous and isotropic turbulence, *Phys. D* **80**, 385.

Benzi, R., Biferale, L., Ciliberto, S., Struglia, M. V. & Tripiccione, R. (1995): On the intermittent energy transfer at viscous scales in turbulent flows, *Europhys. Lett.* **32** (9), 709.

Benzi, R., Biferale, L., Ciliberto, S., Struglia, M. V. & Tripiccione, R. (1996a): Scaling property of turbulent flows, *Phys. Rev. E* **53**, 3025.

Benzi, R., Biferale, L., Ciliberto, S., Struglia, M. V. & Tripiccione, R. (1996b): Generalized scaling in fully developed turbulence, *Physica D*, **96**, 162.

Briscolini, M., Santangelo, P., Succi, S. & Benzi, R. (1994): Entended self-similarity in the numerical simulation of three-dimensional homogeneous flows, *Phys. Rev. E* **50**, 1745.

Cao, N., Chen, S. & She Z.-S. (1996): Scalings and relative scalings in the Navier-Stokes turbulence, *Phys. Rev. Lett.* **76**, 3711.

Chorin, A.J. (1991): Equilibrium Statistics of a Vortex Filament with Applications, *Commun. Math. Phys.* **41**, 619.

Chorin, A.J. & Akao, J.H. (1992): Vortex equilibria in turbulence theory and quantum analogies, *Physica D* **52**, 403.

Douady, S., Couder, Y. & Brachet, M. E. (1991): Direct observation of the intermittency of intense vorticity filaments in turbulence, *Phys. Rev. Lett.* **67** 983.

Dubrulle, B. (1994): Intermittency in fully developed turbulence: log-Poisson statistics and scale invariance, *Phys. Rev. Lett.* **73**, 959.

Dubrulle, B. & Graner, F. (1996a): Possible Statistics of invariant Systems, *J. Phys. II France*, **6**, 797.

Dubrulle, B. & Graner, F. (1996b), Scale invariance and scaling exponents in fully developed turbulence, *J. Phys. II France*, **6**, 817.

Frisch, U., Sulem, P. L., Nelkin, M. (1978): A Simple Dynamical Model of Intermittent Fully Developed Turbulence, *J. Fluid Mech.*, **87(4)**, 719.

Frisch, U. (1991): From global scaling, a la Kolmogorov, to local multifractal scaling in fully developed turbulence, *Proc. Roy. Soc. A*, **434**, 89.

Frisch, U. (1995): *Turbulence: The Legacy of A.N. Kolmogorov,* Cambridge University Press, Cambridge, U.K.

Gilbert, A. D. (1993): A cascade interpretation of Lundgren's stretched spiral vortex model for turbulent fine structure, *Phys. Fluids A* **5**, 2831.

Gledzer, E. B. (1973): System of hydrodynamic type admitting two quadratic integrals of motion, *Sov. Phys. Dokl.* **18**, 216.

Herweijer, J. & van de Water, W. (1994): Universal shape of scaling functions in turbulence, *Phys. Rev. Lett.*, **74**, 4651.

Kadanoff, L., Lohse, D., Wang, J. and Benzi, R. (1995): Scaling and dissipation in the GOY shell model, *Phys. Fluids*, **7**, 617.

Kida, S. (1991): Log-stable distribution and intermittency of turbulence. *J. Phys. Soc. Jpn.*, **60(1)**, 5.

Kolmogorov, A. N. (1941): Local structure of turbulence in an incompressible viscous fluid at very large Reynolds numbers, *CR. Acad. Sci. USSR* **30**, 299.

Kolmogorov, A. N. (1962): A refinement of previous hypothesis concerning the local structure of turbulence in a viscous incompressible fluid at high Reynolds numbers, *J. Fluid Mech.* **13**, 82.

Kraichnan, R.H. (1959): The structure of isotropic turbulence at very high Reynolds numbers, *J. Fluid Mech.* **5**, 497.

Kraichnan, R.H. (1994): Anomalous scaling of a randomly advected passive scalar. *Phys. Rev. Lett.* **72**, 1016.

Leveque, E. & She, Z.-S. (1995): Viscous effects on inertial range scalings in a dynamical model of turbulence, *Phys. Rev. Lett.*, **75**, 2690.

Leveque, E. & She, Z.-S. (1996): Log-Poisson Statistics in the Navier-Stokes Turbulence, unpublished.

Leveque, E. & She, Z.-S. (1997): Cascade structures and scaling exponents in dynamical model of turbulence: measurements and comparison, *Phys. Rev. E.*, in press.

Lundgren, T.S. (1982): Strained spiral vortex model for turbulent fine structure, *Phys. Fluids* **25**, 2193.

Mandelbrot, B. B. (1974): Intermittent turbulence in self similar cascades: divergence of high moments and dimensions of the carrier, *J. Fluid Mech.* **62**, 331.

Meneveau, C. & Sreenivasan, K. R. (1987): Simple multifractal cascade model for fully developed turbulence, *Phys. Rev. Lett.* **59**, 1424.

Nelkin, M. (1994): Universality and scaling in fully developed turbulence, *Advances in Physics*, **43**, 143.

Noullez, A., Wallace, G., Lempert, W., Miles, R.B., Frisch, U. (1997): Transverse velocity increments in turbulent flow using the RELIEF technique, *J. Fluid Mech.*, in press.

Novikov, E. A. (1970): Intermittency and scale similarity of the structure of turbulent flow, Novikov, E. A. (1970): *Prikl. Mat. Mech.* **35**, 266.

Novikov, E.A. (1994): Infinitely divisible distributions in turbulence, *Phys. Rev. E*, **50**, S3303.

Parisi, G. & Frisch, U. (1985): On the singularity structure of fully developed turbulence, in *Turbulence and predictability of geophysical of fluid dynamics*, edited by Ghil, M., Benzi, R. & Parisi, G. (North Holland, Amsterdam, 1985), 84.

Pullin, D.I., Saffman, P.G. (1993): On the Lundgren-Townsend model of turbulent fine-scale. *Phys. Fluids* **5** 126.

Ruiz Chavarria, G., Baudet, C. & Ciliberto, S., (1995a): Hierarchy of the energy dissipation moments in fully developed turbulence, *Phys. Rev. Lett.* **74** 1986.

Ruiz Chavarria, G., Baudet, C., Benzi, R. & Ciliberto, S. (1995b): Hierarchy of the velocity structure functions in fully developed turbulence, *J. Phys. II France*, **5**, 485.

Saffman, P.G., Pullin, D.I. (1994): Anisotropy of the Lundgren-Townsend model of fine-scale turbulence. *Phys. Fluids* **6** 802.

She, Z.-S., Jackson, E. & Orszag, S.A. (1990): Intermittent vortex structures in homogeneous isotropic turbulence, *Nature*, **433**, 226.

She, Z.-S., Jackson, E. & Orszag, S.A. (1991): Structure and dynamics of homogeneous turbulence: models and simulations, *Proc. R. Soc. London A*, **434**, 101.

She, Z.-S. & Leveque, E. (1994): Universal scaling laws in fully developed turbulence, *Phys. Rev. Lett.* **72**, 336.

She, Z.-S. & Waymire, E. C. (1995): Quantized energy cascade and log-Poisson statistics in fully developed turbulence, *Phys. Rev. Lett.*, **74**, 262.

She, Z.-S. & Leveque, E. (1996): Invariance principle and hierarchical structures in turbulence, preprint.

Siggia, E. (1981): Numerical study of small scale intermittency in three-dimensional turbulence, *J. Fluid Mech.* **107**, 375.

Sreenivasan, K.R., Antonia, R.A. (1996): The Phenomenology of small scale turbulence, *Ann. Rev. Fluid Mech.*, in press.

Stolovitzky, G. & Sreenivasan, K. R. (1993): Intermittency, the second-order structure function, and the turbulent energy-dissipation rate, *Phys. Rev. E* **48**, 92.

Tabeling, P., Zocchi, G., Belin, F., Maurer, J., Williame, H. (1996): Probability density functions, skewness and flatness in large Reynolds number turbulence, *Phys. Rev. E* **53** 1613.

Tennekes, H. (1968): Simple model for the small-scale structure of turbulence, *Phys. Fluids* **11**, 669.

Tong, P., Goldburg, W.I., Chan, C.K., Sirivat, A. (1988): Turbulent transition by photon correlation spectroscopy, *Phys. Rev. A* **37**, 2125.

Vincent, A., Meneguzzi, M. (1991): The spatial structure and statistical properties of homogeneous turbulence, *J. Fluid Mech.* **225**, 1.

Yamada, M. and Ohkitani, K. (1987): *J. Phys. Soc. Jap.* **56**, 4210; (1989), *Prog. Theo. Phys.* **81**, 329.

Intermittency of passive scalars in delta-correlated flow: Introduction to recent work

U. Frisch and A. Wirth

CNRS, Observatoire de Nice, B.P. 4229
06304 Nice Cedex 4, France

Abstract. Recent work on intermittency for a passive scalar advected by a Gaussian white-in-time velocity field with a power-law energy spectrum has led for the first time to a *systematic* understanding of how intermittency can come about and to testable predictions for anomalous exponents. Some of the key ideas of such work are here presented in a way which does not require any previous knowledge of field-theoretic methods.

1 Introduction

One of the objectives of the physicist is to reveal and to understand natural observed phenomena and to delineate their degree of universality. Such goals are still far from having been reached in fully developed turbulence : the limits of Kolmogorov's 1941 theory are well known (see, e.g.). More specifically, one observes power laws, also called scaling laws, for the structure functions

According to the 1941 Kolmogorov theory, the structure function of order p should be a power law with exponent $p/3$. Actually, experimental data indicate that the structure function of order p, nondimensionalized by a suitable power of the structure function of order two, keeps increasing when the distance between the two points is decreased down to those scales where dissipation begins to be felt.

This phenomenon is called anomalous or intermittent scaling laws. The word "intermittency" refers to the observation that small-scale activity displays increasing clumpiness. The phenomenon of intermittency has led to many interpretative models but has never been systematically related to the underlying Navier–Stokes equation. Intermittency is even more conspicuous when observing the dynamics of a *passive scalar*, for example the temperature when convective effects are absent. The temperature structure functions display also scaling laws with intermittent anomalies, which are even more pronounced than for the velocity. See, for example, Fig. 1, from Ruiz-Chavarria et al. (1996), which compares exponents of structure functions for the velocity and the temperature. The exponents were measured by the Extended Self-Similarity technique (Benzi et al. 1995). Intermittency for passive scalars, in both two and three dimensions, has also been studied in direct numerical

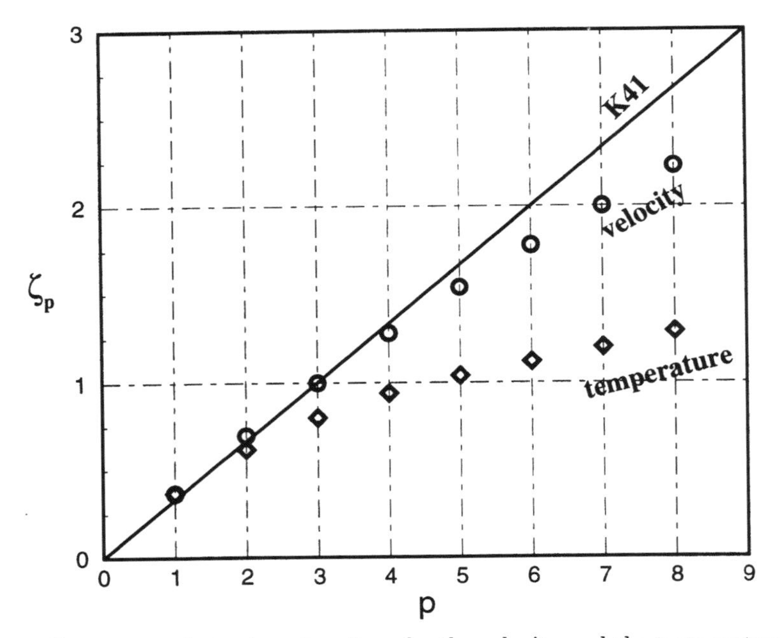

Fig. 1. Exponents of structure functions for the velocity and the temperature, measured by ESS in the wake of a cylinder (Ruiz-Chavarria et al. (1996)).

simulations (see, e.g. Holzer and Siggia 1994; Pumir 1994a; Pumir 1994b). Such simulations often reveal extended regions of almost uniform temperature separated by boundary layers (ramps) with strong gradients.

The equation for a passive scalar in a prescribed velocity field v and with a prescribed inhomogeneous injection term f

$$\partial_t \theta + v \cdot \nabla \theta = \kappa \nabla^2 \theta + f, \tag{1}$$

appears simpler than the Navier–Stokes equation, because it is linear in the unknown θ. Yet, (1) presents, like the Navier–Stokes equation, a closure problem : the equation for the correlation of order p of the passive scalar θ involves correlations of order $p + 1$ with p factors θ and one factor v.

Already in the sixties Kraichnan (1968) observed that the closure problem disappears when the velocity field has a very short correlation time (technically, when it is delta-correlated). Various perturbation methods can then be used to obtain a closed equation for the spectrum of the passive scalar. More recently Kraichnan (1994) proposed to use this delta-correlated model to study intermittency. According to Kraichnan, a Gaussian velocity field, which has no intermittency whatsoever, nevertheless gives rise to anomalous scaling laws for an advected passive scalar.

A number of results about anomalous scaling for the passive scalar have now been obtained by *systematic* methods for Kraichnan's model (Gawędzki and Kupiainen 1996; Chertkov et al. 1995). Our goal, here, is to introduce the nonexpert reader to such work.

In Section 2 we give an elementary introduction to stochastic differential equations, having coefficients which are delta-correlated in time and we show how closed equations for the moments can be obtained. In Section 3 we discuss the spectrum of the passive scalar (which scales normally). In Section 4 we finally consider the fourth-order moments. We give a first insight into the techniques used to show the existence of anomalies. Our presentation will sometimes overlap with that of Gawędzki and Kupiainen (1996), which has however been written for field-theoretic minded readers.

We finally mention some recent work not reviewed hereafter : it has been shown that anomalies exist already at second order when passive magnetic fields are considered (Vergassola 1996). It has been shown that third order anomalies are present when a passive scalar is subject to a uniform velocity gradient (Pumir 1996). A problem similar to (1), but simpler, can be studied using shell models (; Biferale and Wirth 1996).

2 Linear stochastic differential equations with delta-correlated coefficients

Before considering systematically the case of a passive scalar governed by (1), let us rewrite this equation in abstract form (for the moment without the inhomogeneous term) :

$$\partial_t \theta(t) = M\theta(t) + \tilde{M}(t)\theta(t), \tag{2}$$

where M is a linear deterministic operator and $\tilde{M}(t)$ a linear stochastic operator with vanishing mean. In finite dimension, using index notation the problem reads :

$$\partial_t \theta_i(t) = M_{ij}\theta_j(t) + \tilde{M}_{ij}(t)\theta_j(t). \tag{3}$$

We now define the Green's function as the linear operator which maps the initial condition $\theta(t_0)$ into $\theta(t)$:

$$\theta(t) = G(t, t_0)\,\theta(t_0). \tag{4}$$

It is easily verified that :

$$\partial_t G(t,t') = (M + \tilde{M}(t))G(t,t'), \qquad G(t',t') = I. \tag{5}$$

We now suppose that $\tilde{M}(t)$ is a Gaussian operator with vanishing mean. We then derive the so called pre-master equation (Frisch and Bourret (1970)),

$$\partial_t \langle\theta(t)\rangle = M\langle\theta(t)\rangle + \int_{t_0}^{t} \langle\tilde{M}'(t)G(t,t')\tilde{M}(t')\theta(t')\rangle dt', \tag{6}$$

where $\tilde{M}(t)$ and $\tilde{M}'(t)$ are independent and have the same statistical properties.

To prove (6) we start by taking the average of (2):

$$\partial_t \langle \theta(t) \rangle = M \langle \theta(t) \rangle + \langle \tilde{M}(t) \theta(t) \rangle. \tag{7}$$

We then use Gaussian integration by parts (Donsker 1964; Furutsu 1963; Novikov 1964) (see also , Section 4.1), which can be formulated in the following way: If m and m' are independent identically distributed Gaussian random variables with vanishing mean and if $f(m)$ is a deterministic differentiable function of m, we have

$$\langle m f(m) \rangle = \langle m' \partial_\alpha f(m + \alpha m')|_{\alpha=0} \rangle, \tag{8}$$

where ∂_α denotes the derivative with respect to the real variable α.

Note that $\theta(t)$, the solution of (2) with the initial condition $\theta(t_0)$, is a functional of $\tilde{M}(s)$ at all times $t_0 \le s \le t$. It is this functional which plays here the role of $f(.)$ in (8). We thus obtain:

$$\langle \tilde{M}(t) \theta(t) \rangle = \langle \tilde{M}'(t) \partial_\alpha \theta(t; \tilde{M}(.) + \alpha \tilde{M}'(.))|_{\alpha=0} \rangle. \tag{9}$$

The determination of $\partial_\alpha \theta|_{\alpha=0}$ is tantamount to obtaining the coefficient of α in the perturbation of θ when changing $\tilde{M}$ into $\tilde{M} + \alpha \tilde{M}'$. It follows that $\partial_\alpha \theta|_{\alpha=0}$ satisfies

$$\partial_t \partial_\alpha \theta(t) = M \partial_\alpha \theta(t) + \tilde{M}(t) \partial_\alpha \theta(t) + \tilde{M}'(t) \theta(t), \tag{10}$$

This inhomogeneous variant of (2) is solved by using the Green's function:

$$\partial_\alpha \theta(t) = G(t, t_0) \partial_\alpha \theta(t_0) + \int_{t_0}^{t} G(t, t') \tilde{M}'(t') \theta(t') dt'. \tag{11}$$

The term $\partial_\alpha \theta|_{\alpha=0}$ vanishes because there is no perturbation of the initial condition. Using this in (7) and (9), we obtain the announced result (6).

Equation (11) can thus be written as

$$\partial_t \langle \theta_i(t) \rangle = M_{ij} \langle \theta_j(t) \rangle + \int_{t_0}^{t} \langle \tilde{M}'_{i\ell}(t) G_{\ell m}(t, t') \tilde{M}'_{mn}(t') \theta_n(t') \rangle dt'. \tag{12}$$

By using the independence of the operators $\tilde{M}$ and $\tilde{M}'$ we obtain

$$\partial_t \langle \theta_i(t) \rangle = M_{ij} \langle \theta_j(t) \rangle + \int_{t_0}^{t} \langle \tilde{M}_{i\ell}(t) \tilde{M}_{mn}(t') \rangle \langle G_{\ell m}(t, t') \theta_n(t') \rangle dt'. \tag{13}$$

Observe that (13) is not closed because it connects the mean of θ to the mean of a product of a Green's function and θ.

In fact this premaster equation closes in the limit where $\tilde{M}$ becomes delta-correlated in time (called the white-noise limit). More precisely, we perform the following substitution:

$$\tilde{M}(t) \longrightarrow \frac{1}{\epsilon}\tilde{M}(\frac{t}{\epsilon^2}), \qquad \epsilon \to 0, \tag{14}$$

where $\tilde{M}(t)$ is from now on stationary. This has the consequence that the integral of the auto-correlation $\int_0^\infty \langle \tilde{M}(s)\tilde{M}(0)\rangle\, ds$ stays finite while having a correlation time $O(\epsilon^2)$ which tends to zero.

The temporal arguments t and t' thus have a separation $O(\epsilon^2)$. For $\epsilon \to 0$ the Green's function $G_{\ell m}(t,t')$ is thus reduced to unity:

$$G_{\ell m}(t,t') \longrightarrow \delta_{\ell m}. \tag{15}$$

The equation for $\langle\theta\rangle$ is now closed :

$$\partial_t\langle\theta_i(t)\rangle = M_{ij}\langle\theta_j(t)\rangle + \mathcal{D}_{ij}\langle\theta_j(t)\rangle, \tag{16}$$

where

$$\mathcal{D}_{ij} = \lim_{\epsilon\to 0}\frac{1}{\epsilon^2}\int_{t_0}^{t}\left\langle \tilde{M}_{i\ell}(\frac{t}{\epsilon^2})\tilde{M}_{\ell j}(\frac{t'}{\epsilon^2})\right\rangle dt'$$

$$= \int_0^\infty \langle \tilde{M}_{i\ell}(s)\tilde{M}_{\ell j}(0)\rangle ds. \tag{17}$$

In abstract notation (16) reads

$$\partial_t\langle\theta(t)\rangle = M\langle\theta(t)\rangle + \mathcal{D}\langle\theta(t)\rangle, \tag{18}$$

$$\mathcal{D} = \int_0^\infty \langle \tilde{M}(s)\tilde{M}(0)\rangle ds. \tag{19}$$

In the derivation of (18) we have used a semi-intuitive approach. Note that the limit (14) involves mathematical difficulties. Indeed, the white noise is in fact a random distribution which can be defined as the derivative of a Wiener process. The derivation of (18) can be made rigorous by use of the stochastic differential equation formalism of Ito (1951; see also Karatzas and Shreve 1988).

The same method used to obtain a closed equation for the mean of θ can be used to obtain closed equations for the single-time moments of arbitrary orders.

We start with the moments of second order, that is $\langle\theta(t)\otimes\theta(t)\rangle$. In index notation, it reads $\langle\theta_i(t)\theta_j(t)\rangle$. The crucial point is that the quadratic quantities $\theta_i(t)\theta_j(t)$ (unaveraged) verify a linear equation having the same structure as the initial equation. Indeed, from (2) we immediately obtain

$$\begin{aligned}\partial_t\left(\theta_{i_1}(t)\theta_{i_2}(t)\right) &= \left(M_{i_1 j} + \tilde{M}_{i_1 j}(t)\right)\theta_j(t)\theta_{i_2}(t)\\ &+ \left(M_{i_2 j} + \tilde{M}_{i_2 j}(t)\right)\theta_{i_1}(t)\theta_j(t)\end{aligned} \tag{20}$$

or, in abstract notation

$$\partial_t(\theta \otimes \theta) = (\mathcal{M} + \tilde{\mathcal{M}}(t))(\theta \otimes \theta), \tag{21}$$

with $\mathcal{M} = M \otimes I + I \otimes M$, $\tilde{\mathcal{M}}(t) = \tilde{M}(t) \otimes I + I \otimes \tilde{M}(t)$.

It is clear that if $\tilde{M}(t)$ is delta correlated the same is true for $\tilde{\mathcal{M}}(t)$. We thus obtain a closed equation for the second order moments, which reads

$$\partial_t\langle\theta \otimes \theta\rangle = \mathcal{M}\langle\theta \otimes \theta\rangle + \mathcal{D}\langle\theta \otimes \theta\rangle, \tag{22}$$

$$\mathcal{D} = \int_0^\infty \langle\tilde{\mathcal{M}}(s)\tilde{\mathcal{M}}(0)\rangle ds. \tag{23}$$

The generalization to higher order moments is immediate.

We finally need a variant with an inhomogeneous forcing term f,

$$\partial_t\theta(t) = (M + \tilde{M})\theta(t) + \frac{1}{\epsilon} f\left(\frac{t}{\epsilon^2}\right), \tag{24}$$

where f is also Gaussian with vanishing mean and independent of $\tilde{M}$. In the limit $\epsilon \to 0$, the forcing term becomes delta-correlated in time. By performing another Gaussian integration by parts it is easy to show that, for $\epsilon \to 0$, (22) must be replaced by

$$\partial_t\langle\theta \otimes \theta\rangle = (\mathcal{M} + \mathcal{D})\langle\theta \otimes \theta\rangle + F, \tag{25}$$

with

$$F = \int_0^\infty \langle f(s) \otimes f(0)\rangle + \langle f(0) \otimes f(s)\rangle ds. \tag{26}$$

3 Spectrum of the passive scalar

Let us return to the passive scalar problem (1) in the limit of a velocity field and a force which become delta-correlated in time. The equation can be written as

$$\begin{aligned}\partial_t\theta(x,t) + \frac{1}{\epsilon}v(x, \frac{t}{\epsilon^2}) \cdot \nabla\theta(x,t) &= \kappa\nabla^2\theta(x,t) + \frac{1}{\epsilon}f(x, \frac{t}{\epsilon^2}),\\ \nabla \cdot v &= 0.\end{aligned} \tag{27}$$

Here, v and f are homogeneous, isotropic, stationary and independent Gaussian random variables with vanishing mean. In the limit $\epsilon \to 0$ we need the time integrals of their correlation functions

$$\int_0^\infty \langle v_i(x+r,t)v_j(x,t)\rangle ds = \Gamma_{ij}(r), \tag{28}$$

$$\int_0^\infty \langle f(x+r,t)f(x,t)\rangle ds = \frac{1}{2}C(r), \tag{29}$$

which are supposed to be finite. To simplify notation we set

$$\theta(1)\theta(2)\cdots\theta(p) \equiv \theta(x_1,t)\theta(x_2,t)\cdots\theta(x_p,t). \tag{30}$$

From (27), by omitting ϵ for conciseness, we obtain the following equations for the (unaveraged) p-point functions $x_1, x_2, ..., x_p$:

$$\begin{aligned}\partial_t[\theta(1)\theta(2)\cdots\theta(p)] &= \sum_{\alpha=1}^{p}[-v(\alpha)\cdot\nabla_{x(\alpha)} + \kappa\nabla^2_{x(\alpha)}][\theta(1)\theta(2)\cdots\theta(p)] \\ &+ \sum_{\alpha=1}^{p} f(\alpha)\theta(1)\theta(2)\cdots\widehat{\theta(\alpha)}\cdots\theta(p).\end{aligned} \tag{31}$$

(The factor $\widehat{\theta(\alpha)}$ is omitted.) In the limit $\epsilon \to 0$, the method presented in the preceding section gives closed equations for the moments of all orders of the passive scalar. We use the following notation:

$$\theta \equiv \theta(x,t); \quad \theta' \equiv \theta(x',t) \tag{32}$$

$$v \equiv v(x,t); \quad v' \equiv v(x',t). \tag{33}$$

Equation (31) thus reads

$$\partial_t\theta\theta' + (v\nabla + v'\nabla')\theta\theta' = \kappa(\nabla^2 + \nabla'^2)\theta\theta' + f\theta' + f'\theta. \tag{34}$$

It follows that the operators M and $\tilde{M}$ from the previous section are given, respectively, by

$$M = \kappa(\nabla^2 + \nabla'^2) \quad \textit{and} \quad \tilde{M} = -(v\cdot\nabla + v'\nabla'). \tag{35}$$

The equation for the single-time second order moments are thus given by

$$\partial_t\langle\theta\theta'\rangle = \kappa(\nabla^2 + \nabla'^2)\langle\theta\theta'\rangle + \mathcal{D}\langle\theta\theta'\rangle + C\left(\frac{x-x'}{L}\right). \tag{36}$$

The operator $\mathcal{D}$ (called drift term in the theory of stochastic differential equations) is given by

$$\begin{aligned}\mathcal{D} &= \int_0^\infty \langle(v(s)\cdot\nabla + v'(s)\cdot\nabla')(v(0)\cdot\nabla + v'(0)\cdot\nabla')\rangle ds \\ &= \Gamma_{ij}(0)\nabla_i\nabla_j + \Gamma_{ij}(x-x')\nabla_i\nabla'_j \\ &+ \Gamma_{ji}(x'-x)\nabla_j\nabla'_i + \Gamma_{ij}(0)\nabla'_i\nabla'_j.\end{aligned} \tag{37}$$

If we suppose that the scalar field θ is also homogeneous and isotropic, we have

$$\langle\theta(x+r,t)\theta(x,t)\rangle = F_2(r), \quad r = |r|. \tag{38}$$

The drift term then simplifies and can be written, in terms of the velocity structure function, as:

$$\mathcal{D}F_2(r) = S_{ij}(r)\nabla_i\nabla_j F_2(r), \tag{39}$$

where

$$S_{ij}(r) \equiv 2[\Gamma_{ij}(0) - \Gamma_{ij}(r)]$$

$$= \int_0^\infty \langle v_i(x+r,s) - v_i(x,s))v_j(x+r,0) - v_j(x,0))\rangle ds. \tag{40}$$

Following Kraichnan (1994) we suppose that the spatial structure function of the velocity is a power law at *small scales*. Following essentially the formulation of Gawędzki and Kupiainen (1996) we thus write

$$S_{ij} \simeq Dr^\xi \left[(d-1+\xi)\delta_{ij} - \xi\frac{r_i r_j}{r^2}\right], \quad (r \to 0). \tag{41}$$

The expression between brackets satisfies incompressibility and isotropy in d dimensions. If we now suppose that the passive scalar reaches a statistically stationary state, (36) reduces to an ordinary second order differential equation for the function $F_2(r)$:

$$0 = \frac{d-1}{r^{d-1}}\partial_r[Dr^{d-1+\xi} + \kappa r^{d-1}]\partial_r F_2(r) + C\left(\frac{r}{L}\right). \tag{42}$$

This equation can be integrated explicitly. By using the boundary conditions $F_2(\infty) = 0$ and $\partial_r F_2(0) = 0$, we obtain (Gawędzki and Kupiainen 1996)

$$F_2(r) = \frac{1}{d-1}\int_r^\infty \frac{\int_0^\rho C(l/L) l^{d-1} dl}{\kappa\rho^{d-1} + D\rho^{d-1+\xi}} d\rho. \tag{43}$$

Introducing the second order structure function of the passive scalar

$$S_2(r) \equiv \langle(\theta(x+r,t) - \theta(x,t))^2\rangle = 2[F_2(0) - F_2(r)], \tag{44}$$

we note that $S_2(r)$, contrary to $F_2(r)$, has a finite limit when the injection scale $L \to \infty$, limit given by

$$S_2(r) = \frac{4\varepsilon}{(d-1)d}\int_0^r \frac{\rho^d d\rho}{\kappa\rho^{d-1} + D\rho^{d-1+\xi}}. \tag{45}$$

Here, $\varepsilon = (1/2)C(0)$ is the mean dissipation of the variance of the passive scalar.

From (45) we can deduce two asymptotic ranges: (i) the inertial range which corresponds to the limit $\kappa \to 0$ taken first and then r small, for which we have $S_2(r) \propto r^{2-\xi}$ and a passive scalar spectrum of $E_\theta(k) \propto k^{-3+\xi}$; (ii) the inertial-diffusive range, where we keep $\kappa > 0$ finite and then let r become small, for which we obtain

$$S_2(r) = \frac{2\varepsilon}{(d-1)d\kappa}r^2 - \frac{4\varepsilon D}{(d-1)d\kappa^2(2+\xi)}r^{2+\xi} + O(r^{2+2\xi}). \tag{46}$$

The second nonanalytic term, gives a contribution $k^{-3-\xi}$ to the spectrum. For more details, see Frisch and Wirth (1996).

The inertial range with $S_2(r) \propto r^{2-\xi}$ corresponds to a *normal* scaling law, that is a scaling law which can be deduced by simple dimensional inspection of (42).

4 The four point function and the zero modes

Let us proceed to the (single-time) four point function

$$\Psi(1234) \equiv \langle \theta(1)\theta(2)\theta(3)\theta(4) \rangle. \tag{47}$$

One easily shows that it verifies the equation

$$\partial_t \Psi(1234) + \mathcal{M}_4 \Psi(1234) = \sum_{\alpha<\beta} C\left(\frac{x(\alpha) - x(\beta)}{L}\right) \langle \theta(\gamma)\theta(\delta) \rangle, \tag{48}$$

where $\alpha, \beta, \gamma, \delta$ are distinct integers between 1 and 4 and where

$$\mathcal{M}_4 \equiv -\kappa \sum_{\alpha=1}^{4} \nabla^2_{x(\alpha)} + \sum_{\alpha<\beta} S_{ij}(x(\alpha) - x(\beta)) \nabla_{x_i(\alpha)} \nabla_{x_j(\beta)}. \tag{49}$$

This means essentially a combination by pairs of points of the operators appearing in the equations for the second order moment. It can be verified that (48) possesses a stationary solution with a scaling exponent $\zeta_4 = 2\zeta_2 = 2(2-\xi)$. The corresponding fourth order structure function

$$S_4(r) \equiv \langle (\theta(x+r) - \theta(x))^4 \rangle, \tag{50}$$

is

$$S_4(\lambda r) = \lambda^{2\zeta_2} S_4(r). \tag{51}$$

We again have a *normal* scaling law which can be predicted by simple dimensional inspection of (48).

This dimensional analysis can become wrong when the operator $\mathcal{M}_4$, restricted to inertial range scales, presents a non trivial null space with a scaling exponent smaller than $2\zeta_2$. The null space, also called zero mode, will then dominate at small scales. The zero mode verifies

$$\mathcal{M}_4 \Psi_{zm} = 0 \quad \text{(inertial scales)}. \tag{52}$$

There can be various zero modes and only those are acceptable which can be matched correctly at the injection and dissipation scales (Vergassola 1996; Gawędzki and Kupiainen 1996). The determination of such zero modes is a difficult problem for operators with more than two points. It can however be solved perturbatively in two limiting cases: when $\xi \downarrow 0$ (Gawędzki and

Kupiainen 1995) and when $d \to \infty$ (Chertkov et al. 1995; see also Shraiman and Siggia 1995; Shraiman and Siggia 1996). [1]

In the following we will give some basic ideas for the former case, used by Gawędzki and Kupiainen (1995). We first remark that when $\mathcal{M}_4$ acts on functions invariant by simultaneous translation of all arguments, we have, for $\xi = 0$ and $\kappa = 0$,

$$\begin{aligned} \mathcal{M}_4(\kappa = 0, \xi = 0) &= 2D(d-1) \sum_{\alpha<\beta} \nabla_{x,(\alpha)} \nabla_{x,(\beta)} \\ &= -D(d-1) \sum_{\alpha=1}^{4} \nabla^2_{x(\alpha)}. \end{aligned} \tag{53}$$

In this case the operator $\mathcal{M}_4$ is essentially a d-dimensional Laplacian. The acceptable zero modes must be invariant in all the permutations of the four points and in translations and rotations. The constants satisfy these conditions but do not contribute to the structure functions. One remarkable point is that, when $\xi = 0$, there are zero modes which are polynomials of degree four and which have precisely the same scaling exponent as for *normal* scaling. The most general form of a fourth degree polynomial with the required symmetries is:

$$\Psi_{zm}(\xi = 0) = aF_1 + bF_2 + cF_3 \tag{54}$$

$$F_1 = \sum_{\alpha,\beta} ((x(\alpha) - x(\beta))^4 \tag{55}$$

$$F_2 = \sum_{\alpha,\beta,\gamma} (x(\alpha) - x(\beta))^2 (x(\alpha) - x(\gamma))^2 \tag{56}$$

$$F_3 = \sum_{\alpha,\beta,\gamma,\delta} (x(\alpha) - x(\beta))^2 (x(\gamma) - x(\delta))^2. \tag{57}$$

When imposing that Ψ_{zm} is in the null space of the operator $\mathcal{M}_4(\kappa = 0, \xi = 0)$, we obtain $10a + 14b + 3c = 0$. For positive and small values of ξ the zero modes above transform continuously into zero modes with fractional exponent ζ_4 close to 4. If $\zeta_4 < 2\zeta_2 = 4 - 2\xi$, such mode dominates at small scales. To proceed, it is useful to write

$$\Psi_{zm}(\xi = 0) = R^4 (af_1 + bf_2 + cf_3) \tag{58}$$

$$R \equiv \frac{1}{2} \sqrt{\sum_{\alpha,\beta} ((x(\alpha) - x(\beta))^2}, \tag{59}$$

[1] Zero modes for the three-point operator have been determined by nonperturbative numerical means in (Gat et al. 1996), a case of interest when the injection term f is not Gaussian; otherwise the three-point function vanishes by symmetry.

where the functions f_1, f_2 and f_3 depend only on angular variables. For $\xi > 0$ small, we are thus searching for zero modes of the form

$$\Psi_{zm} = R^{4+\lambda\xi+O(\xi^2)}(af_1 + bf_2 + cf_3 + g\xi + O(\xi^2)) \tag{60}$$

$$= R^4(1 + \lambda\xi \ln R + O(\xi^2))(af_1 + bf_2 + cf_3 + g\xi + O(\xi^2)). \tag{61}$$

The calculation of λ is presented in Gawędzki and Kupiainen (1995) and will not be given here. The final result is:

$$\zeta_4 = 2\zeta_2 - \frac{4}{2+d}\xi + O(\xi^2). \tag{62}$$

The exponent is thus smaller than $2\zeta_2$ and gives rise to an *anomalous* scaling law for the structure function of order four. The fact that $\zeta_4 < 2\zeta_2$ implies that the flatness $S_4(r)/S_2(r)$, which is a measure of intermittency, tends to infinity when $r \rightarrow 0$. Note that the expansion in powers of $1/d$ by Chertkov et al. 1995 gives

$$\zeta_4 = 2\zeta_2 - \frac{4}{d}\xi + O(1/d^2), \tag{63}$$

which is perfectly consistent with (62).

The method for the fourth order was simplified and generalized to all even orders, with the result (Bernard et al. 1996)

$$\zeta_{2p} = p\zeta_2 - \frac{2p(p-1)}{2+d}\xi + O(\xi^2). \tag{64}$$

Kraichnan (1994), who used a nonperturbative method, found the following expression

$$\zeta_{2p}^{RHK} = \frac{1}{2}\sqrt{4pd\zeta_2 + (d-\zeta_2)^2} - \frac{1}{2}(d-\zeta_2), \quad \zeta_2 = 2 - \xi. \tag{65}$$

Note that such an anomalous exponent cannot be obtained perturbatively since $\zeta_4^{RHK}(\xi = 0) \neq 4$.

Acknowledgements. We have benefitted from discussions with L. Biferale, G. Falkovich, K. Gawędzki, R. Kraichnan, A. Pumir and M. Vergassola. This work was supported by DRET (grant 93811-57).

References

Benzi R., Ciliberto S., Baudet C., Ruiz Chavarria G. (1995): On the scaling of three dimensional homogeneous and isotropic turbulence. Physica D**80**, 385–398.

Bernard D., Gawędzki K., Kupiainen A., Anomalous scaling in the N-point functions of passive scalar. Phys. Rev. E, **54**, 2564–2572.

Biferale L., Wirth A., (1996): A minimal model for intermittency of passive scalars. these Proceedings.

Chertkov M., Falkovich G., Kolokolov I. Lebedev V. (1995): Normal and anomalous scaling of the fourth-order correlation function of a randomly advected passive scalar. Phys. Rev. E **52**, 4924–4941.

Donsker M.D. (1964): On function space integrals, Analysis in Function Space, 17–30, eds. W.T. Martin & I. Segal. MIT Press.

Frisch U. (1995): *Turbulence: the Legacy of A.N. Kolmogorov.* Cambridge University Press (1995).

Frisch U., Bourret R. (1970): Parastochastics, J. Math. Phys. **11**, 364–390.

Frisch U., Wirth A. (1996): Inertial-diffusive range for a passive scalar advected by a white-in-time velocity field. Europhys. Lett. **35**, 683–687.

Furutsu K., (1963): On the statistical theory of electromagnetic waves in a fluctuating medium, J. Res. Nat. Bur. Standards D **67**, 303–323.

Gat O., L'vov V.S., Podivilov E. and Procaccia I. (1996): Non-perturbative zero modes in the Kraichnan model for turbulent advection, submitted to Phys. Rev. Lett..

Gawędzki K., Kupiainen A. (1995): Anomalous scaling of the passive scalar, Phys. Rev. Lett. **75**, 3834–3837.

Gawędzki K., Kupiainen A. (1996): Universality in turbulence: an exactly soluble model, Lecture Notes in Physics **469**, 71–105, Springer-Verlag.

Holzer M., Siggia E.D. (1994): Turbulent mixing of a passiv scalar. Phys. Fluids **6**, 1820–1837.

Ito K. (1951): On stochastic differential equations. Memoirs Amer. Math. Soc. **4**.

Karatzas I., Shreve S.E. (1988): *Brownian Motion and Stochastic Calculus*, Springer Verlag, (1988).

Kraichnan R.H. (1968): Small-scale structure of a scalar field convected by turbulence. Phys. Fluids **11**, 945–953.

Kraichnan R.H. (1994): Anomalous scaling of a randomly advected passive scalar. Phys. Rev. Lett. **72**, 1016–1019.

Novikov E.A. (1964): Functionals and the method of random forces in turbulence theory. Zh. Exper. Teor. Fiz. **47**, 1919-1926.

Pumir A. (1994a): A numerical study of the mixing of a passiv scalar in three dimensions in the presence of a mean gradient. Phys. Fluids **6**, 2118–2132.

Pumir A. (1994b): Small-scale properties of scalar and velocity differences in three dimensional turbulence. Phys. Fluids **6**, 3974–3984.

Pumir A. (1996): Anomalous scaling behaviour of a passive scalar in the presence of a mean gradient. Europhys. Lett. **34**, 25–29.

Ruiz-Chavarria G., Baudet C.,Ciliberto S. (1996): Scaling laws and dissipation scale of a passive scalar in fully developed turbulence. Physica D, **99**, 369–380.

Shraiman B.I., Siggia E.D. (1995): Anomalous scaling of a passive scalar in turbulent flow. C. R. Acad. Sci. Paris, série II **321**, 279–284.

Shraiman B.I., Siggia E.D. (1996): Symmetry and scaling of turbulent mixing. Phys. Rev. Lett. **77**, 3834–3837.

Vergassola M. (1996): Anomalous scaling for passively advected magnetic fields. Phys. Rev. E **53**, R3021–R3024.

Wirth A. Biferale L. (1996): Anomalous scaling in random shell models for passive scalars. Phys. Rev. E **54**, 4982–4989.

A minimal model for intermittency of passive scalars

L. Biferale[1] & A. Wirth[2]

[1] Universitá di Roma "Tor Vergata", Dip. di Fisica, 00133 Roma, Italy
[2] CNRS, Observatoire de Nice, B.P. 4229, 06304 Nice Cedex 4, France

Abstract. A shell-model version of Kraichnan's (1994) passive scalar problem is introduced which is inspired by the model of Jensen, Paladin and Vulpiani (1992). The here introduced shell-model is even simpler than the one studied previously (Wirth and Biferale 1996). As in the original problem of Kraichnan, the prescribed random velocity field is Gaussian, delta-correlated in time and has a power-law spectrum.

Deterministic differential equations for second and fourth-order moments are obtained and then solved numerically. The second-order structure function of the passive scalar has normal scaling, while the fourth-order structure function has anomalous scaling. For $\xi = 1$ the anomalous scaling exponents ζ_p are determined for structure functions up to $p = 11$ by Monte Carlo simulations of the random shell model, using a stochastic differential equation scheme, validated by comparison with the results obtained form the moment equations of the second and fourth-order structure functions.

1 Introduction

A striking property of a passive scalar quantity advected by a fully developed turbulent flow is its spatio-temporal intermittent behavior (see Sreenivasan 1991 for a recent review). The time evolution of the scalar field θ is described by the partial differential equation :

$$\partial_t \theta + (v \cdot \nabla)\theta = \kappa \Delta \theta + f, \tag{1}$$

where v is the advecting velocity field, κ is the molecular diffusivity and f is an external forcing.

Intermittency in passive scalars advected by highly chaotic flows is thought to be connected to the existence of self-similar processes transferring fluctuations from large to small scales.

Self-similarity is observed in the power-law behavior of structure functions, $\tilde{S}_p(r)$, in the inertial range, that is moments of θ-increments at scales where neither external forcing nor molecular damping are acting :

$$\tilde{S}_p(r) \equiv \langle |\theta(x) - \theta(x+r)|^p \rangle \sim r^{\zeta_p}. \tag{2}$$

The set of scaling exponents, ζ_p, fully characterizes intermittency. In particular, deviations from the dimensional (linear) behavior $\zeta_{2p} = p\zeta_2$ are evidence of a nontrivial scalar transfer among scales, in analogy to the energy intermittent cascade in turbulent flows (see Frisch 1995 for a recent review on this subject).

Two years ago Kraichnan (1994) suggested that a passive scalar has an intermittent behavior even if the advecting velocity field is Gaussian. He proposed to study this problem on a model he had introduced in the late sixties (Kraichnan 1968). For more details concerning this questions we invite the reader to consult the paper by Frisch and Wirth in this volume and the references therein.

One problem of the model proposed by Kraichnan is that analytical calculations are difficult and have let to many controversial discussions. Furthermore, numerical simulations of the model proposed by Kraichnan are hardly doable with the present generation of computers. We are thus here proposing a shell-model version of Kraichnan's model where all numerical calculations can be performed on ordinary workstations, see below, and where the anomalous scaling law for the fourth-oder moment can be obtained analytically (see Benzi et al. 1996).

We have thus investigated the intermittency properties of a shell-model for a passive scalar advected by a prescribed stochastic velocity field. Shell models (see Kadanoff 1995 for a pedagogical introduction and references therein) have been already successful in helping to understand many issues connected to fully developed turbulence.

The problem of defining a shell model for the advection of a passive scalar by a deterministic and chaotic velocity field has been already investigated in Jensen et. al (1992).

We here present a simplification of a shell-model (Wirth and Biferale 1996) for passive scalar transport with a prescribed velocity field, which we called the "minimal model":

$$[\frac{d}{dt} + \kappa\lambda^{2m}]\theta_m(t) = \lambda^m[\theta_{m-1}(t)u_{m-1/2}(t) - \lambda\theta_{m+1}(t)u_{m+1/2}(t)] + f(t)\delta_{1,m} \quad (3)$$

All variables are real and λ is the separation between shells and κ the molecular diffusivity. The advection is modeled by the terms involving the velocity u and the force $f(t)$ acts only on the first shell.

The striking feature of the minimal model is that in spite of its simplicity, it gives rise to a complicated intermittent behavior.

2 The Ito equation

Our goal, as in Kraichnan's work (Kraichnan 1994), is to use a *nonintermittent* velocity field and then to find if the passive scalar is nevertheless intermittent. For this, we assume that the velocity variables $u_m(t)$ and the forcing term $f(t)$ are independent *real* Gaussian and white-in-time, that is, delta-correlated. Furthermore, as in Kraichnan (1968), we make a scaling assumption for the spectrum of the $u_m(t)$'s, namely

$$\langle u_m(t)u_m(t')\rangle = \delta(t-t')\lambda^{-\xi m}, \tag{4}$$

Were ξ ranges from zero to two, remark that $\xi = 2/3$ corresponds to a K41 scaling of the velocity field. For $f(t)$ we assume $\langle f(t)f(t')\rangle = \delta(t-t')$.

As long as the velocity variables $u_m(t)$ have a finite correlation time and, hence, smooth sample paths, there is no particular difficulty in giving a meaning to the set (3) of random ordinary differential equations (ODE's). A well-known difficulty arises with ODE's having white-in-time coefficients: the mathematical meaning of the equation is ambiguous. The physicist's and thus our view-point is to define the solution of such "stochastic differential equations" (SDE's) to be the limit, as the the correlation time tends to zero, of the solution of a random ODE with nonwhite (colored) coefficients. One way to overcome this problem is to use the Ito formalism for SDE which gives a precise mathematical meaning to the white-noise limit. For more details concerning this questions we invite the reader to consult the paper by Frisch and Wirth in this volume and the references therein.

The Ito SDE associated to (3) reads:

$$\begin{aligned} d\theta_m(t) = &-(\kappa\lambda^{2m} + \frac{1}{2}(\lambda^{(2-\xi)(m+1)} + \lambda^{(2-\xi)m}))\theta_m(t)dt \\ &+ [\theta_{m-1}(t)\lambda^{(2-\xi)m/2}dW_m(t) - \theta_{m+1}(t)\lambda^{(2-\xi)(m+1)/2}dW_{m+1}(t)] \\ &+ f\delta_{m,1}dW_f(t). \end{aligned} \tag{5}$$

Here, the $W_m(t)$'s and $W_f(t)$ are independent identically distributed real-valued Brownian motion functions, normalized in such a way that $\langle|W_m(t)|^2\rangle = \langle|W_f(t)|^2\rangle = t$. Please remark that both, equations (3) and (5) conserve the total energy $\sum_m \theta_m^2$ when force and dissipation vanish.

3 Equations for the second and fourth-order moments

In this and the following sections we are interested in the scaling behavior of the p-th-order structure functions:

$$\langle(\theta_m^2)^{p/2}\rangle \propto \lambda^{-\zeta_p m}, \tag{6}$$

where ζ_p is called the scaling exponent of order p. If $\zeta_{2p} = p\zeta_2$ the structure functions are said to have a normal scaling. If $\rho_{2p} = p\zeta_2 - \zeta_{2p} > 0$ the scaling of the structure function of order $2p$ is said to be anomalous.

It is well known that from a linear stochastic differential equation with white-noise coefficients it is possible to obtain exact equations for moments of arbitrary order. Higher order quantities such as $\theta\otimes\theta$, $\theta\otimes\theta\otimes\theta\otimes\theta$, ... satisfy also linear stochastic differential equations with white-noise coefficients, from which closed equations can be obtained for $\langle\theta\otimes\theta\rangle$, $\langle\theta\otimes\theta\otimes\theta\otimes\theta\rangle$, etc. (Kraichnan 1968, see also Frisch and Wirth in this volume). In the shell-model context the moment equations become excessively cumbersome beyond order four. We have obtained the closed equations for the second order structure function:

$$E_m = \langle\theta_m^2\rangle, \tag{7}$$

which reads,

$$\begin{aligned}\frac{d}{dt}E_m &= -(2\kappa\lambda^{2m} + \lambda^{(2-\xi)(m+1)} + \lambda^{(2-\xi)m})E_m \\ &+ \lambda^{(2-\xi)m}E_{m-1} + \lambda^{(2-\xi)(m+1)}E_{m+1} \\ &+ f^2\delta_{m,1}.\end{aligned} \tag{8}$$

The equations for the fourth order structure functions,

$$S_{l,m} = \langle\theta_l^2\theta_m^2\rangle \tag{9}$$

are given by,

$$\begin{aligned}\frac{d}{dt}S_{l,m} &= -(2(\kappa\lambda^{2l} + \kappa\lambda^{2m}) + (1+2\delta_{l-1,m})\lambda^{(2-\xi)(l+1)} + (1+2\delta_{l+1,m})\lambda^{(2-\xi)l} \\ &+ (1+2\delta_{l,m-1})\lambda^{(2-\xi)(m+1)} + (1+2\delta_{l,m+1})\lambda^{(2-\xi)m})S_{l,m} \\ &+ (1+2\delta_{lm})(\lambda^{(2-\xi)l}S_{l-1,m} + \lambda^{(2-\xi)(l+1)}S_{l+1,m} \\ &+ \lambda^{(2-\xi)m}S_{l,m-1} + \lambda^{(2-\xi)(m+1)}S_{l,m+1}) \\ &+ f^2(\delta_{l,1}E_m + \delta_{m,1}E_l).\end{aligned} \tag{10}$$

It is important to note that no closed equations for only the set of $S_{l,l}$'s exist, this feature is present for all moments higher than second-order. Equations (8) and (10) can easily be solved numerically.

4 Scaling

We are now interested in stationary solutions of (8) in the inertial range where the direct influence of dissipation κ can be neglected. This drives us to solving the following equations:

$$-\lambda^{(2-\xi)}E_1 + \lambda^{(2-\xi)}(1+\lambda^{(2-\xi)})E_2 = -1, \tag{11}$$

$$+\lambda^{(2-\xi)m}E_m - \lambda^{(2-\xi)m}(1+\lambda^{(2-\xi)})E_{m+1} + \lambda^{(2-\xi)(m+1)}E_{m+2} = 0 \tag{12}$$

If we furthermore suppose that these equations have solutions which follow a scaling law we do the following Ansatz $E_m = C\alpha^m$. In (12) this leads to a quadratic equation with the solutions $\alpha = 1, \lambda^{\xi-2}$ which corresponds to equipartition and to normal (dimensional) scaling, respectively. The equation (11) gives $\alpha = \lambda^{\xi-2}$. In equation (11) the -1 term on the r.h.s. represents injection of passive scalar at the first shell which is then transported to shells with a higher index. This excludes the equipartition scaling which correspond to a absence of energy transfer. If however now energy is injected (r.h.s.= 0) equipartition is the only scaling law possible. Note that both scaling laws are consistent with eq. (12). The physical interpretation of this is, that the scaling behavior depends on whether passive scalar energy is transported from shells with small index to shells with larger index or not.

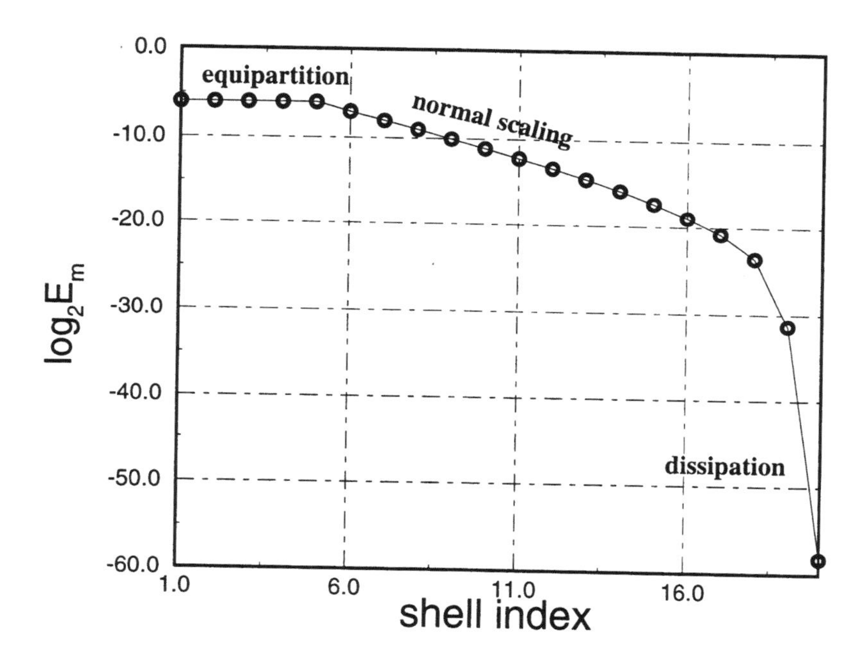

Fig. 1. Energy of a 20-shell simulation where energy is injected at the 5th shell, *lambda* = 2.0 and $\xi = 1.0$

In Fig. 1 both scaling laws can be observed as the injection takes place at the fifth shell which gives rise to a equipartition scaling from the fifth down to the first shell and a normal scaling from the fifth up to the, approximately, $14th$ shell from where on dissipation is no longer negligible. A striking feature of the minimal model is that the scaling laws predicted analytically by just solving a quadratic equation can be validated by very simple numerical calculations (in fact, the moment equation for the second order structure function can be solved by hand).

Determining scaling behavior of $S_{l,l}$ analytically in the inertial range is much more involved as we have to consider equation (10) and determine the scaling laws of $S_{l,m}$ in both indices. The solution to this problem is currently investigated by Benzi and coworkers.

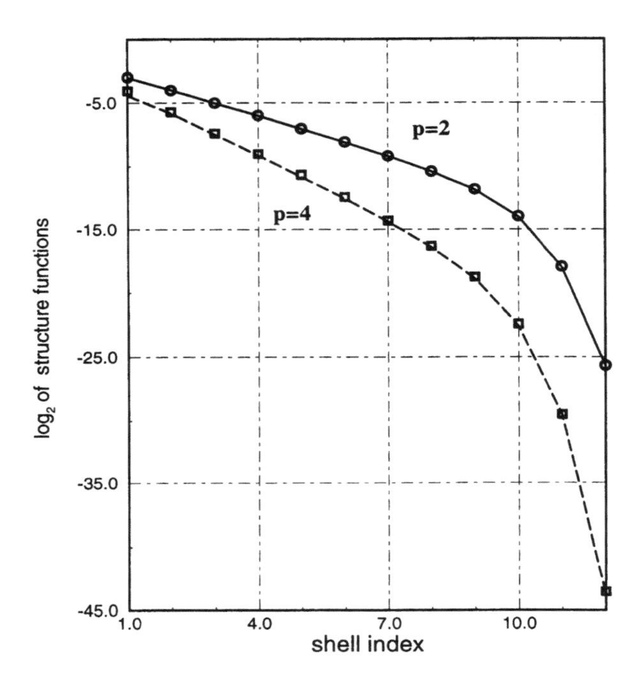

Fig. 2. Comparison between the numerical solution of the moment equations (lines) and the moments obtained by averaging over many realizations (symbols) for the second and fourth order moments (as labeled), for $\lambda = 2.0$, $\xi = 1.0$ and 12 shells.

5 Monte Carlo simulations for structure functions

As we noted, the moment-equation strategy becomes impractical for determining structure functions beyond the fourth order. We therefore resort to Monte Carlo simulations of the stochastic shell model in its Ito version (5). We used the "weak-order-one Euler" scheme. Roughly, this means interpreting the Ito equation (5) as a time-difference equation. This scheme is of order one (in the time step Δt) for averaged quantities such as the structure functions. Higher order schemes are impractical to use as they ask for the numerical approximation of multiple stochastic integrals which would consume a large amount of CPU resources. Averages are calculated as time averages, assuming ergodicity (we checked that changes in the seed of the random generator do not affect the results). Integrating over a large number of realizations is thus equivalent to integrating over many large-eddy turnover times (up to 2^{21}). In Fig. 2 we compared the results from the moment equations ((8) and (10)) and the results from the Monte Carlo simulations for the second and fourth order structure functions. The reader can clearly see that an almost perfect agreement is obtained. Furthermore, the scaling law for the fourth-order structure function is anomalous ($\rho_4 > 0$).

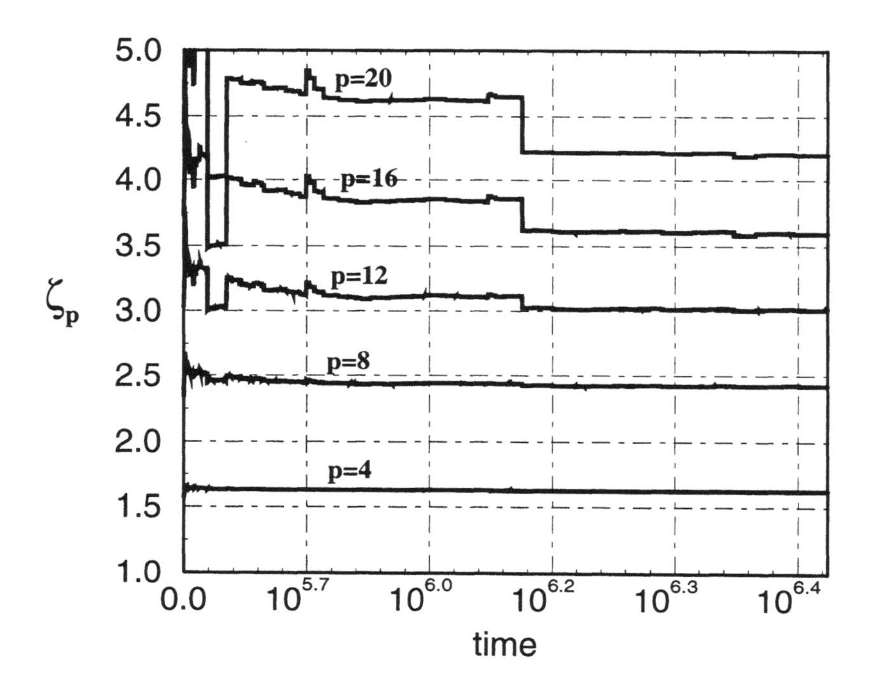

Fig. 3. The averaged values of scaling exponents ζ_p for $p = 4, 8, 12, 16, 20$ as labeled. The scaling is measured between the first and the 6 shell in a 12 shell simulation with $\lambda = 2.0$ and $\xi = 1.0$.

In Fig. 3 the averaged values of scaling exponents ζ_p for $p = 4, 8, 12, 16, 20$ is plotted versus time. While the scaling exponents for the smaller values of p seem to have converged, one clearly sees that very violent events change considerably the scaling exponents for the higher order structure functions even after 10^6 large-eddy turnover times. A striking result of our calculations are that even for our very simple model which allows averaging over a huge number of independent realizations, scaling exponents for structure function of order higher than about 12 could not be obtained.

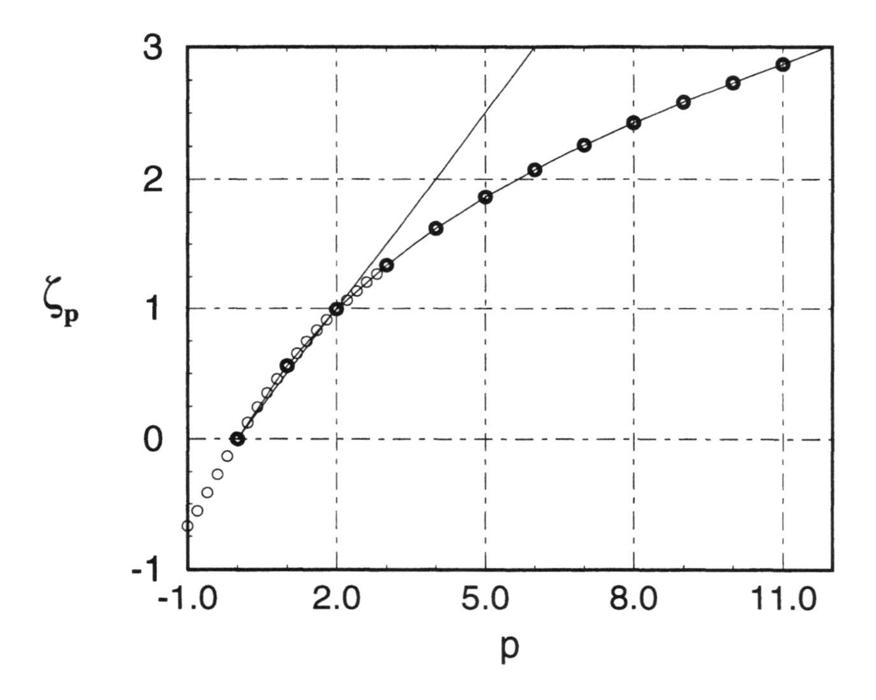

Fig. 4. Anomalous scaling exponents (symbols) vor values of p between -1 and 11 compared with normal scaling (straight line).

In Fig. 4 we give the results of scaling exponents up to order $p = 11$ which could be measured with high accuracy. We finally like to point out that up to the exponents measured ($p \leq 11$) we do not see the graph in Fig. 4 follow a linear law for the larger values of p as it is seen in other models and experiments. We however see a linear behavior in the scaling exponents for $p > 14$, due to insufficient statistics for measuring such high moments.

Acknowledgements We have benefited from extensive discussions with R. Benzi, U. Frisch, A. Noullez and M. Vergassola. This work was supported by fellowships from French Ministère de la Recherche et de la Technologie and DRET (grant 93811-57), and by the GDR Mécanique des Fluides Numérique. L.B. would like to acknowledge partial support by Ministère de l'Enseignement Superieur et de la Recherche (France) and the Observatoire de la Côte d'Azur where this work was completed.

References

Frisch, U. (1995): *Turbulence; The Legacy of A.N. Kolmogorov* Cambridge University Press.
Jensen, M.H., Paladin, G. and Vulpiani, A. (1992): *Phys. Rev. A***45**, 7214.
Kadanoff, L. (1995): Physics Today **48**, No 9, 11 (1995).
Kraichnan, R.H. (1968): Phys. Fluids **11**, 945–953.
Kraichnan, R.H. (1994): Phys. Rev. Lett. **72**, 1016–1019.
Sreenivasan, K.R. (1991): Proc. Roy. Soc. Lond. **A434**, 165.
Wirth, A and Biferale, L. (1996): Phys. Rev. E **54**, 4982–4991.

Generalized scaling in turbulent flows

L. Biferale, R. Benzi, M. Pasqui

Universitá di Roma "Tor Vergata", Dip. di Fisica, 00133 Roma, Italy

Abstract. In this paper we report numerical and experimental results on the scaling properties of the velocity turbulent fields in several flows. The limits of a new form of scaling, named Extended Self Similarity (ESS), are discussed. A generalized version of self scaling which extends down to the smallest resolvable scales is presented. This new scaling is checked in several laboratory and numerical experiment. A possible theoretical interpretation is also proposed. A synthetic turbulent signal showing exactly generalized-ESS scaling is studied.

1 Introduction

In order to characterize the statistical properties of fully developed turbulence one usually studies the scaling properties of moments of velocity differences at the scale r:

$$S_p(r) =< |v(x+r) - v(x)|^p >=< |\delta v(r)|^p > \tag{1}$$

where $< \cdots >$ stands for ensemble average and v is the velocity component parallel to r. At high Reynolds number $Re = U_0 L/\nu$ the $S_p(r)$ satisfies the relation

$$S_p(r) \propto r^{\zeta(p)} \tag{2}$$

for $L > r >> \eta_k$ where L is the integral scale, $\eta_k = (\nu^3/\epsilon)^{1/4}$ is the dissipative (Kolmogorov) scale, ϵ is the mean energy dissipation rate, ν the kinematic viscosity and U_0 the R.M.S. velocity of the flow. The range of length $L > r >> \eta_k$, where the scaling relation (2) is observed, is called the inertial range. The Kolmogorov (K41) theory predicts $\zeta(p) = p/3$, but experimental results of Anselmet et al (1984) and numerical simulations of Vincent and Meneguzzi (1991) show that $\zeta(p)$ deviates substantially from the linear law. This phenomenon is believed to be produced by the intermittent behavior of the energy dissipation which can be taken into account by rewriting eq.(2) in the following way:

$$S_p(r) \propto < \epsilon_r^{p/3} > r^{p/3} \propto r^{\tau(p/3)+p/3} \tag{3}$$

where ϵ_r is the average of the local energy dissipation $\epsilon(x)$ on a volume of size r centered on a point x. A comparison of eq.(1) and eq.(3) leads to the conclusion that the scaling exponents $\tau(p/3)$ of the energy dissipation are related to those of S_p by $\zeta(p) = \tau(p/3) + p/3$.

Since the Kolmogorov (1962) theory many other models, (Frisch et al. 1978), (Parisi and Frisch 1985), (Benzi et al. 1984), (Menevaue and Sreenivasan 1987), (She and Leveque 1994) have been suggested to describe the behavior of the $\zeta(p)$. However, it turns out that the $\zeta(p)$ may be not universal in non homogeneous, anisotropic flow and may depend on the location where measurements are done. Specifically, they may have different values if one measures either far away from boundaries, where turbulence is almost homogeneous and isotropic, or in locations of the flow where a strong mean shear is present. The $\zeta(p)$ depend also on the way in which turbulence is produced, for example 3D homogeneous turbulence, boundary layer turbulence, thermal convection and MHD. Thus there is the fundamental question of understanding in which way all these parameters influence the scaling laws. Furthermore all the above mentioned models assume the existence of two well defined intervals of lengths that are the inertial range and a dissipation range. According to the idea of multiscaling these two ranges may eventually be connected by an intermediate region where the viscosity begins to act, (Frisch and Vergassola 1991). However this idea of a well defined inertial range, where viscosity does not act at all, and the idea of multiscaling turns out to be incompatible with the recently introduced new form of scaling, which has been named Extended Self Similarity (ESS) (Benzi et al 1993a, 1995a).

ESS has been observed in 3D homogeneous and isotropic turbulence both at low and high Re and for a wide range of scales r with respect to scaling (2). In contrast ESS is not observed when a strong mean shear is present, (Stolovitzky and Sreenivasan 1993). All these experimental observations show that also the mechanisms by which energy is actually dissipated in a flow are very poorly understood. Specifically one would like to understand how viscosity acts on different scales. This is clearly an important point in order to safely use large eddy simulations in real applications.

In this paper we summarize all the above mentioned results on scaling. We propose a generalized form of ESS (G-ESS) which has been checked in many different flows. We have also generated a signal which has all the statistical properties of a real turbulent signal. Our interpretation of ESS and of this generalized scaling suggest that there is no sharp viscous cut-off in the intermittent transfer of energy.

The paper is organized as follows: in section 2 we recall the properties of ESS, in section 3 the generalized form of scaling is discussed, in section 4 a possible theoretical interpretation is proposed, in section 5 we present our synthetic signal and some comparison among G-ESS and multiscaling. Finally conclusions are given in section 6.

2 Extended Self Similarity

Extended Self Similarity (ESS) is a property of velocity structure functions of homogeneous and isotropic turbulence, (Benzi et al. 1993a, 1995a). It has been shown using experimental and numerical data, (Briscolini et al. 1994) that the structure functions present an extended scaling range when one plots one structure function against the other, namely:

$$S_n(r) \propto S_m(r)^{\beta(n,m)} \tag{4}$$

where $\beta(n,m) = \zeta(n)/\zeta(m)$. The details of ESS have been reported elsewhere, (Benzi et al. 1995a). In the following we describe only the main features.

The ESS scaling has been checked both on numerical data and in experiments, in a range $30 < R_\lambda < 2000$. A direct consequence of the scaling (4) is that for all p, S_p can be written in the following way:

$$S_p(r) = C_p\ U_o^p \left[\frac{r}{L} f\left(\frac{r}{\eta_k}\right)\right]^{\zeta(p)} \tag{5}$$

with $U_o^3 = S_3(L)$, $L = U_o^3/\epsilon$ being the integral scale and C_p dimensionless constants selected in such a way that $f(x) = 1$ for $x >> 1$. Analysis of experimental and numerical data shows that eq.(4) is satisfied for $5\eta_k < r < L$ (Benzi et al. 1996a).

ESS has been also checked for the temperature and velocity fields in Rayleigh-Benard convection, (Benzi et al. 1994). It turns out that ESS is a very useful tool in order to distinguish between Kolmogorov and Bolgiano scaling, (Benzi et al. 1994). In the case of the Bolgiano scaling it has been found that $\zeta(3) = 2.08$ which is clearly very different from the Kolmogorov value $\zeta(3) = 1$. In spite of this large difference between the values of the exponent $\zeta(3)$, using ESS one discovers (Benzi et al. 1996a) that the ratio $\zeta(p)/\zeta(3)$ in the case of Bolgiano are equal to those of homogeneous and isotropic turbulence. The same property is observed for the $\zeta(p)$ obtained from measurements done on the solar wind, (Grauer 1994).

2.1 Systems where ESS is not observed

In the previous section we have discussed several systems where not only the ESS works but also the exponents $\beta(n,3)$ are universal because they do not depend on the systems and on Re. We want to stress that this kind of universality, observed in different flows, disappears if the system is influenced by the presence of a strong mean shear. In this case ESS does not work, because an extended range of scaling is not present when S_n is drawn as a function of S_3.

Violation of ESS has been observed experimentally in boundary layer turbulence, (Stolovitzky and Sreenivasan 1993), and in the shear behind a cylinder, (Benzi et al 1993b).

2.2 Hierarchy of structure functions

In a recent letter She and Leveque (1994) have proposed an interesting theory to explain the anomalous scaling exponents of velocity structure functions. The theory yields a prediction

$$\zeta(p) = ph_0 + d_0(1 - \beta^{p/3})$$

which is in very good agreement with available experimental data, (Benzi et al 1995a), where h_0 and d_0 describes the singular exponent and the fractal codimension of the most intermittent structure present in the flow, and β is connected to intermittency along the energy cascade. She-Leveque took $h_0 = 1/9$, $d_0 = 2$ because they considered vortex filaments as responsible of the strong intermittent burst; the value $\beta = 2/3$ came from phenomenological dimensional analysis.

The She Leveque model is based upon the fundamental assumption on the hierarchy of the moments, $< \epsilon_r^n >$, of the local energy dissipation. Specifically they consider that:

$$\frac{< \epsilon_r^{n+1} >}{< \epsilon_r^n >} = A_n \left(\frac{< \epsilon_r^n >}{< \epsilon_r^{n-1} >} \right)^{\beta} \left(\epsilon_r^{(\infty)}\right)^{(1-\beta)} \tag{6}$$

where A_n are geometrical constants and $\epsilon_r^{(\infty)} = \lim_{n\to\infty}(\frac{<\epsilon_r^{n+1}>}{<\epsilon_r^n>})$ is associated by She and Leveque with filamentary structures of the flow. On the basis of simple arguments it is assumed that: $\epsilon_r^{(\infty)} \propto r^{-2/3}$. The value of β predicted by She and Leveque is 2/3. Notice that in eq.(6) for n=1, taking into account that $< \epsilon_r >= \epsilon$ is constant in r, one immediately finds that

$$\left(\epsilon_r^{(\infty)}\right)^{(1-\beta)} \propto < \epsilon_r^2 >= \frac{S_6}{S_3^2}. \tag{7}$$

Equation (6), which has been experimentally tested by Ruiz et al (1995a), can be extended to the velocity structure functions, (Ruiz et al. 1995b).

3 A generalized form of ESS

Benzi et al (1996a) have proposed a new form of scaling which should be exactly satisfied at all resolvable scales: from the integral scale down to the viscous subrange. The main idea was to use the phenomenological interpretation of the three free parameters entering in the log-Poisson description h_0, d_0, β in such a way to incorporate all non-universal and viscosity dependence in two of them h_0, d_0 while keeping the other β constant for all flows and at all scales. The main phenomenological motivation was that h_0 and d_0 being connected the statistics of filaments could be strongly affected by possible anisotropies in the flow (at both small and large scales) and by viscosity at small scales.

For this purpose we introduce the dimensionless structure function

$$G_p(r) = \frac{S_p(r)}{S_3(r)^{p/3}} \tag{8}$$

According to Kolmogorov theory (8) should be a constants both in the inertial and in the dissipative range, although the two constant are not necessarily the same.

Let us now study the self scaling properties of $G_p(r)$ or, equivalently:

$$G_p(r) = G_q(r)^{\rho(p,q)} \tag{9}$$

where we have by definition:

$$\rho(p,q) = \frac{\zeta(p) - p/3\ \zeta(3)}{\zeta(q) - q/3\ \zeta(3)} \tag{10}$$

$\rho(p,q)$ is given by the ratio between deviation of $\zeta(p)$ and $\zeta(q)$ from the K41 scaling. It will play an essential role in our understanding of energy cascade. Indeed, it is easy to realize that it is the only quantity that can stay constant along all the cascade process: from the integral to the sub-viscous scales. It is reasonable to imagine that the velocity field becomes laminar in the sub-viscous range, $S_p(r) \propto r^p$, still preserving some intermittent degree parametrized by the ratio between corrections to K41 theory. Relation (9) has been tested by Benzi et al (1996a) in many different experimental set-ups done at different Reynolds numbers and for some direct numerical simulation with and without large scale shear. In all the cases a very good scaling behavior was detected.

4 A theoretical interpretation

The aim of this section is to discuss a possible theoretical interpretation of the experimental and numerical results previously discussed. Our starting point is to revise the concept of scaling in fully developed turbulence.

Let us consider three length scales $r_1 > r_2 > r_3$ and our basic variables to describe the statistical properties of turbulence, namely the velocity difference $\delta v(r_i)$. We shall restrict ourselves to those statistical models of turbulence based on random multiplicative processes.
Thus we shall assume that there exists a statistical equivalence of the form:

$$\delta v(r_i) = a_{ij} \delta v(r_j) \tag{11}$$

where $r_i < r_j$ and a_{ij} is a random number with a prescribed probability distribution P_{ij}.
By definition, we have:

$$a_{13} = a_{12} a_{23} \tag{12}$$

Equation (12) is true no matter what the ratios $\frac{r_1}{r_2}$ and $\frac{r_2}{r_3}$ are. Now we ask ourselves the following question: what is the probability distribution P_{ij} which is functionally invariant under the transformation (12)? This question can be answered by noting that equation (12) is equivalent to:

$$\log a_{13} = \log a_{12} + \log a_{23} \tag{13}$$

(we assume $a_{ij} > 0$). Thus our question is equivalent to asking what are the probability distributions that are stable under convolution. For independently distributed random variables a solution of this problem can be given in a complete form, (see Feller 1966), (Novikov 1994). If the variables are correlated the situation becomes much more difficult to solve, as it is well known from the theory of critical phenomena. For the time being we shall restrict ourselves to independent random variables.

In this case, for instance, the Gaussian and the Poisson distributions are well known examples of probability distributions stable under convolution. These two examples correspond to two turbulence models proposed in literature, namely the log-normal model (Kolmogorov 1962) and the log-Poisson model (She and Leveque 1994), (Dubrulle 1994), (She and Waymire 1995). A more general description can be found in Novikov (1994).

We can have a different point of view on our question which is fully equivalent to the above discussion. A simple solution to our question is given by all probability distribution P_{ij} such that:

$$\langle a_{ij}^p \rangle \equiv \prod_{k=1}^{n} \left(\frac{g_k(r_j)}{g_k(r_i)} \right)^{\gamma_k(p)} \tag{14}$$

for any functions $g_k(r_i)$ and $\gamma_k(p)$ ($\langle \cdots \rangle$ represents average over P_{ij}). Indeed we have:

$$\langle a_{13}^p \rangle = \langle a_{12}^p a_{23}^p \rangle = \prod_{k=1}^{n} \left(\frac{g_k(r_1)}{g_k(r_2)} \right)^{\gamma_k(p)} \prod_{k=1}^{n} \left(\frac{g_k(r_2)}{g_k(r_3)} \right)^{\gamma_k(p)} = \prod_{k=1}^{n} \left(\frac{g_k(r_1)}{g_k(r_3)} \right)^{\gamma_k(p)} \tag{15}$$

We want to remark that equation (14) represents the most general solution to our problem, independent of the scale ratio r_i/r_j.
Let us give a simple example in order to link equation (14) to the case of probability distribution stable under convolution. Let us consider the case of a random log-Poisson multiplicative process, namely:

$$a_{ij} = A_{ij} \beta^x \tag{16}$$

where x is a Poisson process $P(x=N)=\frac{C_{ij}^N e^{-C_{ij}}}{N!}$.
By using (16) we obtain:

$$\langle a_{ij}^p \rangle = A_{ij}^p \exp(C_{ij}(\beta^p - 1)) \tag{17}$$

Equation (17) is precisely of the form (14) if we write

$$A_{ij} = \frac{g_1(r_j)}{g_1(r_i)} \qquad \exp C_{ij} = \frac{g_2(r_i)}{g_2(r_j)} \tag{18}$$

In order to recover the standard form of the She-Leveque model we need to assume that (see also 24):

$$g_1(r_i) \sim r_i{}^h \tag{19}$$

$$g_2(r_i) \sim r_i^2 \tag{20}$$

This example highlights one important point in our discussion, i.e. the general requirement of scale invariant random multiplier (14) does not necessary imply a simple power law scaling as expressed by the equations (19- 20). Moreover, the general expression (14) is compatible only with an infinitively divisible distribution. For instance, the previous random multiplier model for turbulence, such as the random-β model or the p-model, cannot be expressed in the general form (14) independently of the ratio r_i/r_j.

It is worthwhile to review the multifractal language in the light of the previous discussion. In the multifractal language for turbulence, the two basic assumptions are:

I) The velocity difference on scale r shows local scaling law with exponent h, i.e. $\delta v(r) \sim r^h$;

II) the probability distribution to observe the scaling $\delta v(r) \sim r^h$ is given by $r^{3-D(h)}$.

In the multifractal language, therefore, there are two major ansätze: one concerns power law scaling of the velocity difference (assumption I) and the other one concerns a geometrical interpretation (the fractal dimension $D(h)$) of the probability distribution to observe a local scaling with exponent h. How is it possible to generalize the multifractal language in order to take into account equation (14)?

As we shall see, the theory of infinitely divisible distributions is the tool we need to answer the previous question. All published model of turbulence based on infinitively divisible distributions are equivalent to writing $D(h)$ in the form:

$$3 - D(h) = d_0 f\left[\frac{h-h_0}{d_0}\right] \tag{21}$$

where d_0 and h_0 are two free parameters while the function $f(x)$ depends only on the choice of the probability distribution. For instance for log-normal distribution $f(x)=x^2$. Equation (21) allows us to write:

$$\langle \delta v(r)^p \rangle = \int d\mu(h) r^{hp} r^{3-D(h)} = r^{h_0 p + d_0 H(p)} \tag{22}$$

where

$$H(p) = \inf_x (px + f(x)) \tag{23}$$

We can see that equation (22) is equivalent to a random multiplicative process given by:

$$\langle a_{ij}^p \rangle = \left(\frac{r_j}{r_i}\right)^{h_0 p} \left(\frac{r_j}{r_i}\right)^{d_0 H(p)} \tag{24}$$

Equation (24) can be generalized to the form (14) by allowing h_0 and d_0 to depend on r, i.e.

$$\langle a_{ij}^p \rangle = \frac{\left(r_j^{\overline{h}_0(r_j)}\right)^p}{\left(r_i^{\overline{h}_0(r_i)}\right)^p} \left[\frac{r_j^{\overline{d}_0(r_j)}}{r_i^{\overline{d}_0(r_j)}}\right]^{H(p)} \tag{25}$$

where:

$$\overline{h}_0(r) = h_0 s_h(r) \qquad \overline{d}_0(r) = d_0 s_d(r) \tag{26}$$

Equation (25) is equivalent to (14) by using:

$$g_1(r_i) = r_i^{\overline{h}_0(r_i)} \qquad g_2(r_i) = r_i^{\overline{d}_0(r_i)} \tag{27}$$

$$\gamma_1(p) = p \tag{28}$$

$$\gamma_2(p) = H(p) \tag{29}$$

The same results can be obtained by (22), i.e. we have

$$\langle \delta v(r)^p \rangle = r^{\overline{h}_0(r)p + \overline{d}_0(r)H(p)} \tag{30}$$

Note that the saddle point evaluation of (22) is not spoiled by the dependence of h_0 and d_0 on r.

We have seen that (14) can be reformulated in terms of multifractal language for infinitely divisible distribution whose function $D(h)$ can be rewritten as in (21). We can ask the following question: what is the physical meaning of (14) or its multifractal analogue (25- 30)? It is precisely the multifractal language which allows us to answer this question. Indeed, the two basic assumption for the multifractal language can now be replaced in the following way:

I) the velocity difference on scale r behaves as

$$\delta v(r) \sim g_1(r) g_2(r)^x; \tag{31}$$

II) the probability distribution to observe I is $g_2(r)^{f(x)}$.
Then we have

$$\langle \delta v(r)^p \rangle = \int d\mu(x) g_1(r)^p g_2(r)^{px+f(x)} = g_1(r)^p g_2(r)^{H(p)} \tag{32}$$

by employing a saddle point integration. The most clear physical interpretation of (32) is that the probability to observe a given fluctuation of the velocity difference has no more geometrical interpretation linked to the fractal dimension $D(h)$. The probability distributions are controlled by a dynamical variable $g_2(r)$ which at this stage we still need to understand. An insight on the dynamical meaning of $g_2(r)$ can be obtained by the following considerations.

Let us define $\epsilon(r)$ the average of the energy dissipation on a scale r. We can define the eddy turnover time $\tau(r)$ on scale r as:

$$\frac{\delta v^2(r)}{\tau(r)} \sim \epsilon(r) \tag{33}$$

We have seen that all experimental and numerical data suggest that the following relation always holds (see also eq.6):

$$\frac{\epsilon(r)}{\langle\epsilon\rangle} =^s \frac{\delta v^3(r)}{\langle\delta v^3(r)\rangle} \tag{34}$$

where $=^s$ means that all moments on the r.h.s. are equal to l.h.s. By using (33-34) we obtain the definition of length $L(r)$:

$$L(r) \equiv \delta v(r)\tau(r) = \frac{\langle\delta v^3(r)\rangle}{\epsilon} \quad , \tag{35}$$

$L(r)$ cannot be regarded as a real length scale in the physical space. Rather, $L(r)$ should be considered as a dynamical variable entering into the statistical description of turbulence. This is precisely the idea behind ESS which reformulates the scaling properties of turbulence in terms of $L(r)$. Indeed in order to obtain ESS from (32) it is sufficient to state that, within the range of scales where ESS is observed, $g_1(r)^{1/h_0} \sim g_2(r)^{1/d_0} \sim L(r)$. The physical meaning of ESS is strictly linked to (35) and in particular to (34) which is a generalization of Kolmogorov Refined Similarity Hypothesis.

Let us summarize all our previous findings:

(i) we have introduced the idea of scale invariant random multiplier satisfying equation (14);

(ii) we have shown that infinitely divisible distributions are all compatible with(14);

(iii) we have shown that the multifractal language specialized to the case of infinitely divisible distribution gives equation (14) (with $n = 2$ and $\gamma_1(p)$ linear in p) and it is equivalent to scale invariant random multiplier;

(iv) finally we have argued that the correct scaling parameter to describe the statistical properties of small scale turbulent flows is not directly linked to a simple geometrical interpretation, rather it should be considered a dynamical variable.

Our finding (i)-(iv) enables us to have a unified theoretical interpretation of the experimental and numerical results presented at the beginning of this paper. Indeed equation (30) or (32) tells us that the anomalous part of the structure functions:

$$G_p(r) \equiv \frac{\langle \delta v^p(r) \rangle}{\langle \delta v^3(r) \rangle^{p/3}} \tag{36}$$

satisfies the scaling properties:

$$G_p(r) = G_q(r)^{\rho_{p,q}} \tag{37}$$

where $\rho_{p,q} \equiv \frac{\zeta_p - p/3}{\zeta_q - q/3}$. According to our analysis of the experimental and numerical results, the scaling (37) is observed down to the smallest resolved scale.

We have shown that, in the theoretical framework so far exposed, we recover the ESS when $g_1(r)^{1/h_0} \sim g_2^{1/d_0} \sim L(r)$. If $g_1(r)^{1/h_0} \neq g_2(r)^{1/d_0}$ we lose ESS, but its generalized version (37) is still valid.

4.1 Synthetic Turbulence

We can also use (30)and (32) to simulate a synthetic signal according to a random multiplicative process satisfying (25). This can be done by using the algorithm introduced by Benzi et al. (1993c).

Let us consider a wavelet decomposition of the function $\phi(x)$:

$$\phi(x) = \sum_{j,k=0}^{\infty} \alpha_{j,k} \psi_{j,k}(x) \tag{38}$$

where $\psi_{j,k}(x) = 2^{j/2}\psi(2^j x - k)$ and $\psi(x)$ is any wavelet with zero mean. The above decomposition defines the signal as a dyadic superposition of basic fluctuations with different characteristic widths (controlled by the index j) and centered in different spatial points (controlled by the index k). For functions defined on $N = 2^n$ points in in the interval $[0, 1]$ the sums in (38) are restricted from zero to $n - 1$ for the index j and from zero to $2^j - 1$ for k .

Benzi et al. (1993c) have shown that the statistical behavior of signal increments:

$$< |\delta\phi(r)|^p > = < |\phi(x + r) - \phi(x)|^p > \sim r^{\zeta(p)}$$

is controlled by the coefficients $\alpha_{j,k}$. By defining the α coefficients in terms of a multiplicative random process on the dyadic tree it is possible to give an explicit expression for the scaling exponents $\zeta(p)$. For example, it is possible to recover the standard anomalous scaling by defining the α's tree in term of the realizations of a random variable η with a probability distribution $P(\eta)$:

$$\alpha_{0,0}$$

$$\alpha_{1,0} = \eta_{1,0}\,\alpha_{0,0}; \quad \alpha_{1,1} = \eta_{1,1}\,\alpha_{0,0};$$

$$\alpha_{2,0} = \eta_{2,0}\,\alpha_{1,0}; \; \alpha_{2,1} = \eta_{2,1}\,\alpha_{1,0}; \; \alpha_{2,2} = \eta_{2,2}\,\alpha_{1,1}; \; \alpha_{2,3} = \eta_{2,3}\,\alpha_{1,1}, \tag{39}$$

and so on. Let us note that in the previous multiplicative process different scales are characterized by different values of the index j, i.e. $r_j = 2^{-j}$. If the $\eta_{j,k}$ are *i.i.d.* random variable it is straightforward to realize that $\alpha_{j,k}$ are random variables with moments given by:

$$< |\alpha_{j,k}|^p >= r_j^{-\log_2(\overline{\eta^p})} \tag{40}$$

where the "mother eddy' $\alpha_{0,0}$ has been chosen equal to one. In (40) with $\overline{\cdots}$ we intend averaging over the $P(\eta)$ distribution. It is possible to show that that also the signal $\phi(x)$ has the same anomalous scaling of (40). In order to generalize this construction for function showing ESS or generalized-ESS scaling of the form (30) and (37) is now sufficient to take a probability distribution, $P_l(\eta)$, for the random multiplier with the appropriate scale dependency (14). This will be implemented by allowing a dependency of $P(\eta_{jk})$ on the scale $r_j = 2^{-j}$, i.e. the η's random variables will be still independently distributed but not identically distributed with respect to variation of the scaling index j.
According to the previous discussion, ESS corresponds to have only one seed-function defining the multiplicative process, i.e. $g_1(r)^{1/d_0} \neq g_2(r)^{1/h_0}$ in the range of scales where ESS is valid ($r \leq 5\eta_k$). On the other hand, at scales smaller then $5 \sim 6$ Kolmogorov scale, ESS is not more valid because g_2^{1/d_0} begins to deviate substantially from g_1^{1/h_0}: only G-ESS should be observed and we need a multiplicative process defined in terms of two different seed-functions.
Following this recipe we define the signal such that :

$$< \delta v(r)^p >= U_0^p F(r)^{p/3} G(r)^{\zeta(p)-p/3} \tag{41}$$

where

$$F(r) =< \delta v(r)^3 > /U_0^3 \tag{42}$$

The function $G(r)$ is defined in such a way that for $F(r)$ much greater than η_k/L, $G(r) \sim F(r)$ while for very small scales r we have $G(r) \sim \eta_k/L$. In the following we choose the simplest ansätze:

$$G(r) = B + AF(r) \tag{43}$$

with $B = \eta_k/L$ and A is a dimensionless constant.

Let us now take some time in order to clarify the previous definitions. Relation (41) is defined such that experimental results are reproduced with good accuracy and G-ESS scaling (37) is satisfied by definition. By assuming (43) the only unknown function is $F(r) =< \delta v(r)^3 > /U_0^3$. On the other hand, the function $< \delta v(r)^3 >$ is always very well fitted by the Batchelor parameterization:

$$< \delta v(r)^3 >= \frac{U_0^3}{L\eta_k^2} \frac{r^3}{(1 + (r/\eta_k)^2)}. \tag{44}$$

From expression (41) is immediate to extract the expression for the two seed-functions $g_1(r), g_2(r)$ used in the previous sections, namely:

$$g_1(r) = \left(\frac{F(r)}{G(r)}\right)^{1/3} G(r)^{h_0}$$
$$g_2(r) = G(r)^{d_0} \tag{45}$$

Let us note that $g_1(r)$ goes smoothly from the intermittent value, $g_1(r) \sim r^{h_0}$ ($h_0 = 1/9$ for the case of She-Leveque model), assumed in the inertial range to the laminar value, $g_1(r) \simeq r$, characteristic of scales much smaller than Kolmogorov scale.

For the practical point of view we have constructed our signal by using a random process for the multiplier $\eta_{j,k}(r_j)$ with a scale-dependent Log-Poisson distribution. The scale dependency of parameters entering the distribution has been fixed in terms of relations (45) and (18) and such that the $\zeta(p)$ exponents correspond to the She-Leveque expression, namely:

$$\eta_{jk}(r_j) = A_{j,j+1}\beta^{x_{j,j+1}} \tag{46}$$

where $x_{j,j+1}$ is a Poisson variable with mean $C_{j,j+1} = \log(g_2(r_{j+1})/g_2(r_j))$, $A_{j,j+1} = g_1(r_j)/g_1(r_{j+1})$ and $\beta = 2/3$. This choice leads to the standard Log-Poisson scaling in the inertial range:

$$\zeta(p) = h_0 p + (1 - 3h_0)\frac{(1-\beta^{p/3})}{(1-\beta)}$$

and to the following expression for the ratios of deviations to the Kolmogorov law:

$$\rho_{p,q} = \frac{H(p) - p/3}{H(q) - q/3}, \quad H(x) = \frac{1-\beta^{x/3}}{(1-\beta)}.$$

Signal constructed according to this scenario will be referred to as signal-A in the following.

In fig. 1 we show the structure function of order 6 for such a signal plotted versus the separation scale r at moderate Reynolds number. Clearly, for this choice of Reynolds number there is not any inertial range of scale where scaling exponents could be safely measured. On the other hand, our signal shows G-ESS scaling, as can be seen in fig. 2. ESS is violated at scales $r \sim 5\eta$ as it must be as soon as $d_0(r)/h_0(r)$ is different from a constant, (Benzi et al 1995c, Benzi et al. 1996b).

In the following we are interested in making a comparison between the statistics of viscous scales as they emerge from the G-ESS scenario with the Multiscaling interpretation.

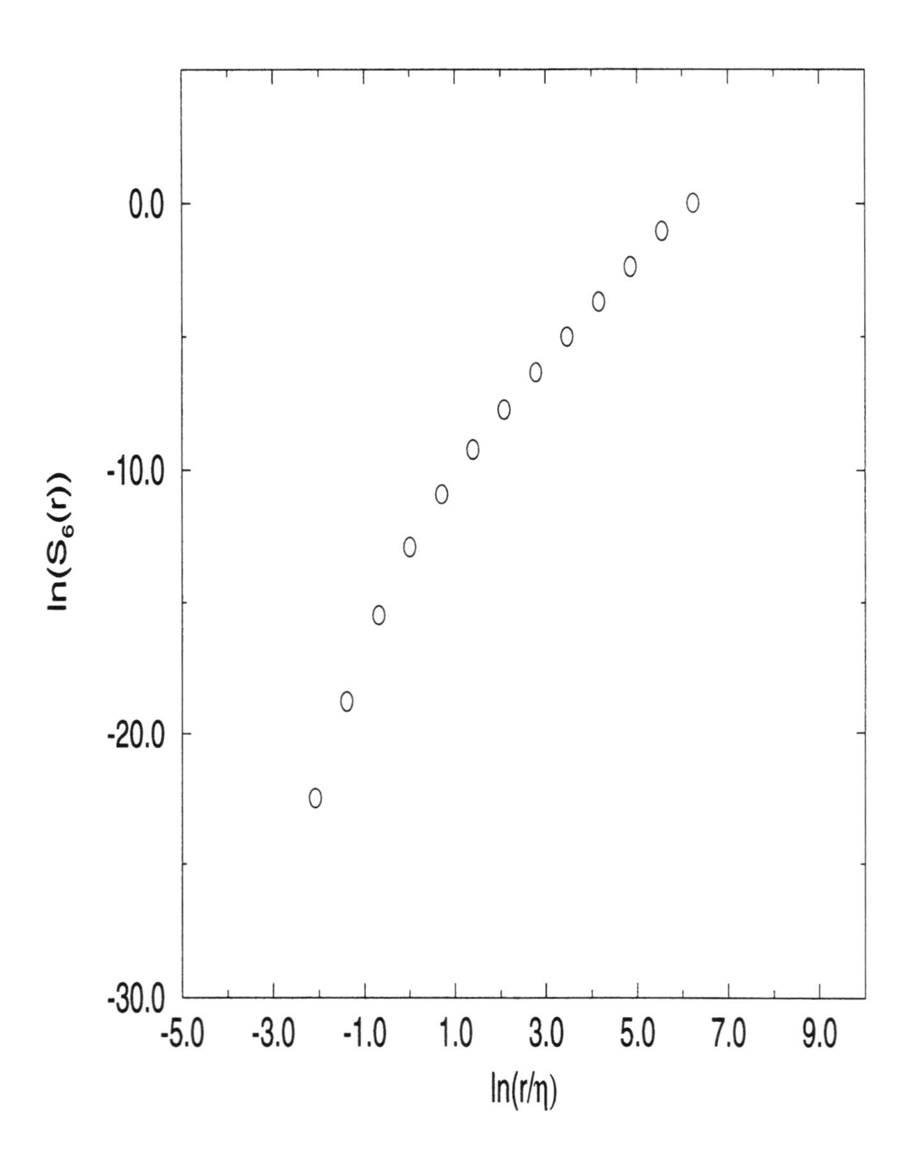

Fig. 1. log-log plot of $S_6(r)$ vs r for our signal at moderate Reynolds number

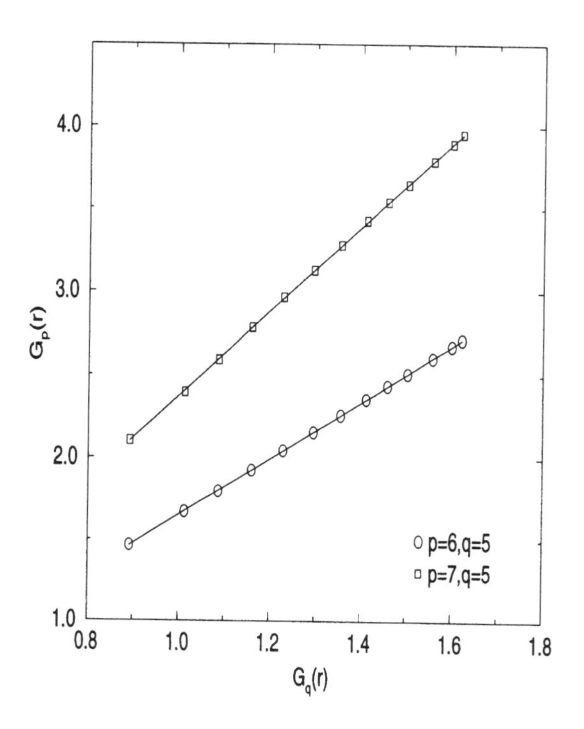

Fig. 2. G-ESS plot for the same signal of fig. 1. $G_6(r)$ vs. $G_5(r)$ (circles) and $G_7(r)$ vs. $G_5(r)$ (squares)

4.2 Multiscaling

It is generally argued that the anomalous scaling can be observed for scales larger than a given viscous cutoff. The physical interpretation of this statement is that non linear, intermittent, transfer of energy is acting only for scales larger than the viscous scale. Below such a scale the structure functions are supposed to show a simple (regular) scaling $\langle \delta v^p(r) \rangle \sim r^p$.

Usually the viscous cutoff is introduced as the scale at which the local Reynolds number is of order one, namely:

$$\frac{\delta v(r) r}{\nu} \sim 1 \tag{47}$$

This condition can be obtained by the requirement that the local energy transfer $\epsilon(r) \sim (\delta v^3(r)/r)$ becomes equal to the energy dissipation $\nu(\delta v^2(r)/r^2)$:

$$\nu \frac{\delta v^2(r)}{r^2} \sim \frac{\delta v^3(r)}{r} \tag{48}$$

which gives equation (47). There is a well defined prediction given by Frisch and Vergassola (1991), based on (47), using the multifractal language. Indeed for any exponent h one can introduce the h-dependent viscous cutoff given by:

$$r_d^{h+1} \sim \nu \tag{49}$$

where $\delta v(r) \sim r^h$.
It follows that $r_d(h)$ is a fluctuating quantity. There are two consequences of this theory. The first one predicts that for the structure functions $\langle \delta v^p(r) \rangle$ there exists a cutoff scale r_p dependent on p and moreover $r_p < r_q$ for $p > q$. The second prediction concerns the moment of the velocity gradients Γ which are:

$$\langle \Gamma^p \rangle \sim \langle r_d(h)^{(h-1)p} r_d(h)^{3-d(h)} \rangle \sim Re^{-z(p)} \tag{50}$$

with $z(p) = \sup_h \left[\left((h-1)p + 3 - D(h) \right) / (1+h) \right]$.
Between the two predictions the first one is qualitatively more peculiar of (49).
In particular, the first prediction states that between the end of the inertial range (i.e. the region where anomalous scaling of $\langle \delta v^p(r) \rangle$ with respect to r is detected) and the dissipation cutoff r_d, the local slope is controlled by $D(h)$. The second prediction is somehow weaker because present experimental data do not distinguish among several models, so far proposed, for the Re-dependence of $\langle \Gamma^p \rangle$.

In order to compare the multiscaling in the dissipation range with our experimental and numerical data, we have produced a synthetic turbulence signal (signal B hereafter) similar to the one already discussed but with d_0 and h_0 independent on r. The effect of dissipation is introduced by using (47). In fig.3 we compare the scaling properties G_4 versus G_6 for signal A and B. As it is clearly shown by the compensated plot given in fig. 3 the signal built by imposing viscous effects "á la multiscaling" presents strong deviation (of the order of 15%) from pure scaling in the viscous sub-range. On the other hand, by definition, signal A shows G-ESS scaling at all scales within numerical precision which is of the order of a few per cent.

The above discussion rules out the effect of multiscaling on the viscous cutoff (49). Previous claims on the validity of multiscaling effects should be considered either wrong or affected by experimental errors. On the other hand our model, used to implement the synthetic signal A, should be considered a very accurate model even for scale close to the regular region where $\delta v(r) \sim r$.

There is, however, the theoretical possibility that multiscaling indeed acts somewhere at very small scales (much smaller then the usual Kolmogorov scale) and that therefore should be taken into account in the G-ESS scenario if one would really like to reach, by using multiplicative models, scales well inside the viscous sub-range.

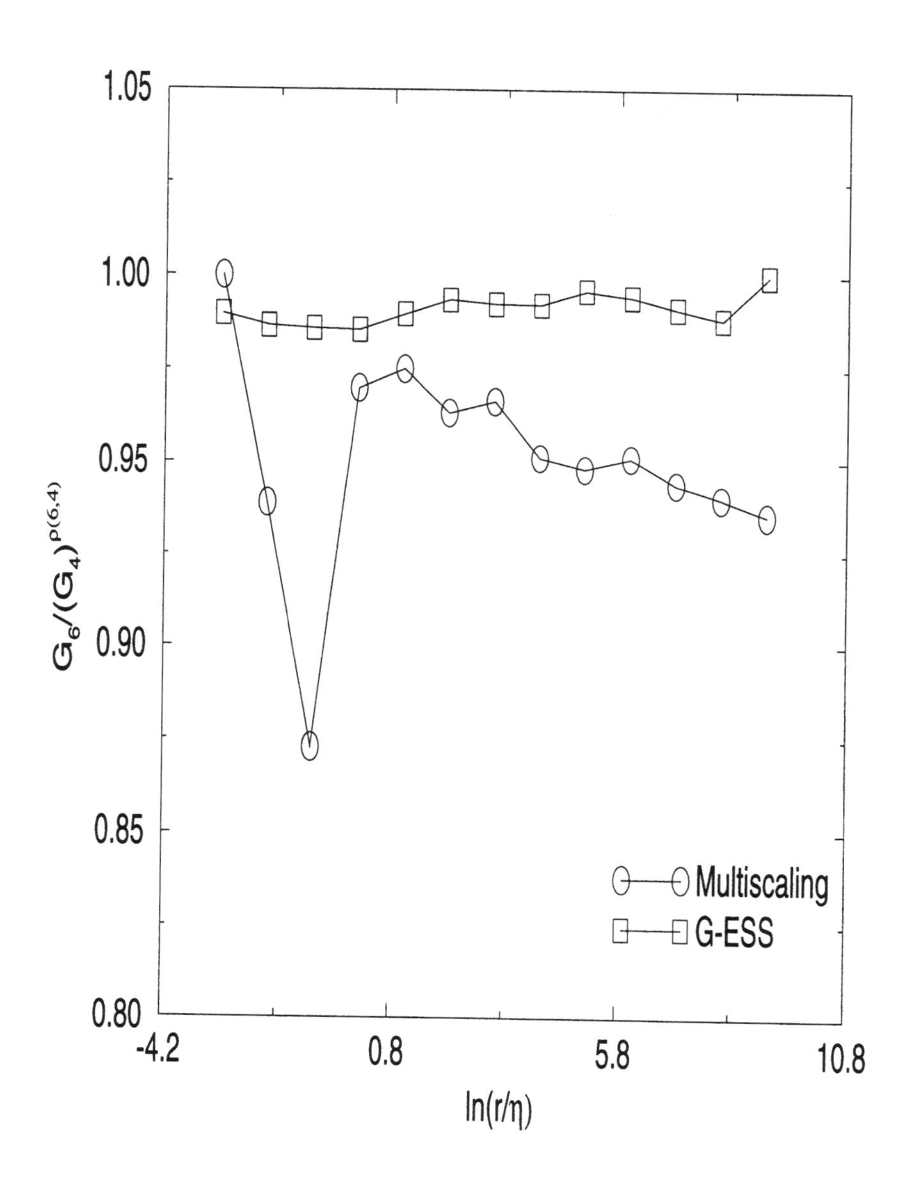

Fig. 3. Compensated plot for $G_6(r)/(G_4(r))^{\rho(6,4)}$ for signal A (squares)and signal B (circles). The exponent $\rho(6,4)$ is fixed by the best fit in the inertial range for both signals

5 Conclusions

In this paper we have reported several new results concerning the scaling behaviour of small scale statistical properties of turbulence. 1) We have reviewed the main results on ESS and in particular we have stressed that in homogeneous and isotropic flows in turbulence, Rayleigh-Benard convection and solar wind magnetohydrodynamics, the ratio $\zeta(p)/\zeta(3)$ seems to have an universal behaviour. This is a rather striking and unexpected result which implies that anomalous violation of dimensional scaling may be explained in an universal way. We do not know any simple phenomenological explanation for our finding.

2) We have mentioned that ESS is not observed when relatively strong shear flows are present. A phenomenological analysis, based on the Kolmogorov equation, shows the relevance of a length scale based on the mean energy dissipation and the shear strength.

3) We have proposed a generalization of ESS. This generalization is supported both by experimental and numerical data and it seems to be not affected by viscous cutoff.

4) We have developed a theory which unifies the previous point. The theory is based on the assumption that the probability distribution is infinitively divisible and predicts the existence of the generalized ESS. The theory can also be used to generate artificial signals which displays all the scaling features observed in real data.

5) We have shown that the original proposal on the multiscaling for the viscous cutoff is incompatible with the turbulence data. The theory formulated in this paper removes this incompatibility and suggests that multiscaling is acting at much smaller scales than previously proposed. The new point on the theory is a change of view in the probability distribution of the original multifractal model which is not directly linked to a geometrical interpretation in terms of fractal dimensions.

References

Anselmet F., Gagne Y., Hopfinger E. and Antonia J., (1984) Fluid Mech. **140**, 63.

Benzi R., Paladin G., G. Parisi G. and Vulpiani A. (1984), J. Phys. A, **17**, 3521.

Benzi R., Ciliberto S., Tripiccione R., Baudet C. and Succi S. (1993a), Phys. Rev. E **48** R29.

Benzi R., Biferale L., Crisanti A., Paladin G., Vergassola M. and Vulpiani A., (1993c) Physica D **65** 352.

Benzi R., Ciliberto S., Baudet C., Ruiz Chavarria G. and Tripiccione R., (1993b) Europhys. Lett. **24**, 275.

Benzi R., Tripiccione R., Massaioli F., Succi S. and Ciliberto S., (1994) Europhys. Lett. **25**, 331.

Benzi R., Ciliberto S., Baudet C and Ruiz Chavarria G. (1995a), Physica D **80**, 385.

Benzi R., Struglia M.V. and Tripiccione R. (1995b)

Benzi R., Biferale L., Ciliberto S., Struglia M.V. and Tripiccione R. (1995c), **32** 709.

Benzi R., Biferale L., Ciliberto S., Struglia M.V. and Tripiccione R., (1996a) Physica D **96** 162.

Benzi R., Biferale L., Ciliberto S., Struglia M.V. and Tripiccione R. (1996b) **RC53** 6.

Briscolini M., Santangelo P., Succi S. and Benzi R., (1994) Phys. Rev. E **50**, 1745.

Dubrulle B., (1994) Phys. Rev. Lett. **73**, 959.

Feller (1966), *An introduction to Probability theory and its applications*, vol.2, Wiley, New York (1966).

Frisch U., Sulem P. and Nelkin M. (1978), J. Fluid Mech. **87**, 719.

Frisch U. and Vergassola M. (1991), Europhys. Lett. **14**, 439.

Grauer R., (1994) Physics Lett. A **195**, 335.

Kolmogorov A. N. (1962), J. Fluid Mech. **13**, 83.

Meneveau C., and Sreenivasan K. R. (1987), Phys. Rev. Lett., **59**, 1424.

Novikov E.A., (1994) Phys. Rev. E **50**, R3303.

Parisi G. and Frisch U., (1985) in: "Turbulence and predictability in geophysical fluid dynamics and climate dynamics", ed. M. Ghil, R. Benzi, G. Parisi(North Holland, Amsterdam), p.84.

Ruiz Chavarria G., Baudet C. and Ciliberto S., (1995a) Phys. Rev. Lett. **74** , 1986.

Ruiz Chavarria G., Baudet C., Benzi R. and Ciliberto S., (1995b) J. Physique **5**, 485.

She Z-S., Leveque E., (1994) Phys. Rev. Lett. **72**, 336.

She Z-S. and Waymire E.C., (1995) Phys. Rev. Lett. **74** 262.

Stolovitzky G. and Sreenivasan K.R., (1993) Phys. Rev. E, **48**, 32.

Vincent A. and Meneguzzi M. (1991), J. Fluid Mech. **225**, 1.

About the Interaction Between Vorticity and Stretching in Coherent Structures

B. Andreotti, S. Douady and Y. Couder

Laboratoire de Physique Statistique
Ecole Normale Superieure
24 Rue Lhomond
75231 Paris Cedex 05
France

Abstract. The main aim of this article is to explore the relationship of vorticity and stretching in flows having a simple configuration forced by their geometry. Using Taylor's four rollers mill experiment the stability of a region of pure strain (a 3D hyperbolic flow having a linear stagnation line) is first investigated. In agreement with previous theoretical predictions this flow is shown to be unstable and to give rise to a periodic pattern of alternate vortices aligned in the direction of stretching. It is demonstrated that the vortices which have been amplified by the stretching react on the strain so that the longitudinal velocity gradient is weakened in their core. This effect is also observed in another experiment where a vortex is formed in a cylindrical tank having a rotating bottom and submitted to an axial pumping. This later experiment demonstrates that the reduction of the stretching can be ascribed to the bidimensionalization induced by the vortex rotation.

1 Introduction

Since G. I. Taylor (Taylor (1938)) it is known that in a 3D turbulent field, vorticity is amplified by stretching. This effect results from the interaction of the two basic quantities formed out of the velocity derivatives : the vorticity and the strain. The aim of this article is to study experimentally the interaction between strain and vorticity in two simple set-ups in which this interaction is forced by the geometrical configuration of the flow.

1.1 Theoretical basis of the interaction of vorticity and strain

We will first recall the theoretical background of this work. The most usual description of a flow is given by equations in terms of velocity v_i and pressure gradient. An alternative way, which better reflects the local structure of the flow, is to use the velocity derivatives. Considering the velocity gradient tensor, one can construct the rate of strain tensor $\sigma_{ij} \equiv (\partial_i v_j + \partial_j v_i)/2$ and the vorticity $\boldsymbol{\omega} \equiv \nabla \times \boldsymbol{v}$. For an incompressible fluid governed by Navier-Stokes equations, the vorticity equation is:

$$\frac{D\omega_i}{Dt} \equiv \frac{\partial \omega_i}{\partial t} + v_j \frac{\partial \omega_i}{\partial x_j} = W_i + \nu \nabla^2 \omega_i \tag{1}$$

where $W_i \equiv \sigma_{ij}\omega_j$ is the vortex stretching vector. This vector, which corresponds to the action of the strain on the vorticity, is the source term of the Equation (1). As for the enstrophy $\omega^2 \equiv \omega_i^2$, it is given by equation :

$$\frac{1}{2}\frac{D\omega^2}{Dt} = \omega_i\omega_j\sigma_{ij} + \nu\omega_i\nabla^2\omega_i \tag{2}$$

A stretching can be defined as: $\gamma \equiv \sigma_{ij}\omega_i\omega_j/\omega^2 = W_i\omega_i/\omega^2$. It is half the temporal rate of change of ω^2.

If the vorticity is considered as a basic quantity, one can express the velocity by using the Biot-Savart inversion of the curl operator (with, however, a gradient flow as an integration constant). In this sense, the velocity must be regarded as a non-local quantity. Similarly, two approaches are possible for the strain dynamics. One can first calculate the velocity from the vorticity field (up to the gradient), and then obtain the strain tensor by derivation. In other words, the strain can be regarded as created by the vorticity distribution. We prefer to consider the strain rate as a second basic quantity which has its own dynamics given by the equation (Ohkitani (1994)):

$$\frac{D\sigma_{ij}}{Dt} = -\sigma_{ik}\sigma_{kj} - \frac{1}{4}(\omega_i\omega_j - \omega^2\delta_{ij}) - \pi_{ij} + \nu\nabla^2\sigma_{ij} \tag{3}$$

where $\pi_{ij} = \partial^2 P/\partial x_i x_j$ is the pressure Hessian tensor. However Equation (3) does not lend itself to a simple physical interpretation. Fortunately, for our present purpose, only the evolution of the vortex stretching vector is necessary, it can be derived from Equation (3) and is given by:

$$\frac{DW_i}{Dt} \equiv \frac{\partial W_i}{\partial t} + v_j\frac{\partial W_i}{\partial x_j} = -\pi_{ij}\omega_j + \nu(\nabla^2 W_i - 2\frac{\sigma_{ij}}{\partial x_k}\frac{\partial\omega_j}{\partial x_k}) \tag{4}$$

Equations (1) and (4) are formally similar, having both an advecting term, a source term and a viscous term. While with our convention, the vortex stretching vector is a local quantity, the stretching induction vector $\Psi_i = \pi_{ij}\omega_j$ i.e. the source term of Equation (4) is non-local (see Ohkitani and Kishiba (1995)) since it can only be obtained by solving the Poisson equation:

$$\nabla^2 P \equiv \pi_{ii} = (\omega^2 - \sigma^2)/2 \tag{5}$$

where $\sigma^2 \equiv 2\sigma_{ij}\sigma_{ij}$ is twice the square of the strain modulus.

This equation is similar to Poisson's law in electrostatics, with the pressure corresponding to the potential resulting from negative and positive charges distributed in proportion to ω^2 and σ^2 respectively. Integrating relation (5) for a whole hydrodynamic structure results into:

$$\int\int\int_V \sigma^2 dt = \int\int\int_V \omega^2 dt \tag{6}$$

which means that the spatial average of ω^2 and σ^2 are equal (in the electrostatic analogy, the medium is neutral). However this does not preclude a local demixing of the two so that the vorticity and the strain of a structure can be spatially separated from each other. For instance, in Burgers-like models of vortices, the vorticity is concentrated in the core of the vortex while the strain is spread in its periphery.

This was more precisely formalized by Raynal (1996), in the case where a fluid is enclosed in a cell containing moving solids. The integral of enstrophy and that of the dissipation may then be different, but are fixed only by the conditions at the solid surfaces. It can be noted that the difference in the integrals is that obtained for a laminar flow having the same boundary conditions. Thus any structure generated by an instability or by turbulence will be superimposed and will then respect strictly the equality of the integrals.

This formalism defines the mechanisms of interaction between the vorticity ω_i and the vortex stretching W_i. If the underlying mechanisms are robust they should exist both in turbulent and non-turbulent flows. In the former they should have a role in the dynamics of turbulence. For this reason we can first recall what is known about the relation of vorticity and strain in this context.

1.2 Interaction of vorticity and strain in turbulent flows

A first way of investigating the importance of vortex stretching in turbulent flows has been to seek the structures corresponding to the regions of large vorticity. This type of work was initiated in numerical simulations by Siggia (1981) and was followed by many other simulations with increasing spatial resolution. In particular, Tanaka and Kida (1993) and Jimenez et al. (1993) investigated in simulated turbulent flows the repartition of the enstrophy ω^2 and of the strain σ^2. The joined density functions show that there exist regions in which the vorticity is large and the strain small. When these regions are singled out, they appear in the form of filaments. Similarly the regions where both ω^2 and σ^2 are large are observed to be in the shape of sheets. Finally, the regions of large σ^2 and small ω^2 are much less probable and appear with no specific spatial structure. As shown by Equation (5), the spatial repartitions of ω^2 and σ^2 are reflected in the structure of the pressure field. This permitted the experimental observation of the low pressure filaments corresponding to the regions of high vorticity and weak strain (Douady et al. 1991). Their results can be summarized in a typical "scenario of a vortex life". The filaments are formed by the roll up of stretched shear layers. They appear very quickly compared to their life time, which is a few turnover times, and finally vanish in a breakdown process. Moreover, they are surprisingly straight at their formation. The difference in the spatial repartition of ω^2 and σ^2 is also responsible for the specific aspect of the fluctuation of the pressure in a point of the flow (Fauve et al. (1993) and Cadot et al. (1995)). The probability distribution function of the pressure exhibits a strong asymmetry with a long exponential tail towards the low pressures and a shorter Gaussian tail towards the high pressures. This is directly related to the existence of well defined high vorticity regions (the filaments) and to the absence of organized high strain regions. Why is there no organized regions of high strain and weak vorticity i.e. no large suppression structures ?

The main progress in the comprehension of the vortex stretching effect in turbulence has been the introduction by Ashurst et al. (1987) and Kerr (1987) of statistics on the angles between the dynamical vectors. These authors were the first to investigate numerically the statistics of alignment between the vorticity ω_i and the eigenvectors of the strain rate tensor σ_{ij}. Their findings were confirmed experimentally by Tsinober et al. (1992) who used a probe with multiple hot wires. They also made statistics on the angle between the vorticity ω_i and the vortex stretching vector W_i. These two studies were completed by Nomura (1995) who measured numerically the angle between the vorticity ω_i and the eigenvectors of the pressure hessian π_{ij}. To summarize the results of these works, there is, in 3D turbulent flows, a strong trend for alignment between the vorticity ω_i, the vortex stretching vector, W_i, one of the eigenvectors of the strain rate tensor σ_{ij} and one of the eigenvectors of the pressure hessian π_{ij}. This can be interpreted as a trend for alignment between the stretching direction and the vorticity. Furthermore, the corresponding σ_{ij} and π_{ij} eigenvalues are statistically small compared to the others. This last result can be interpreted as a tendency for the high vorticity regions to be locally quasi-bidimensional (for a complete discussion, see Andreotti (1997)). This point is confirmed in numerical simulations by Galanti et al. (1996). Since the typical time scales of the evolution of the vorticity and the vortex stretching vector (see Equations (1) and (2)) are of the same order, why are they statistically aligned?

1.3 Interaction of vorticity and strain in analytical solutions of the Navier-Stokes equation

In the present work we seek to explore these relations between vorticity and stretching, trying to answer the questions raised by the previous results. Our basic assumption is that the phenomenology of turbulence exhibits processes which are archetypes of interaction of the strain and of the vorticity which can be observed in well documented flows.

For this purpose we chose Burgers-like models of stretched vortices. These models form a class of exact solutions of Navier-Stokes equations. They consist of vortices submitted to a uniform infinite stretching field of the form:

$$v_x = \gamma x, \quad v_y = -e\gamma y, \quad v_z = -(1-e)\gamma z \ . \tag{7}$$

The stretching γ is thus constant in space and time. Burgers considered the case of the axisymmetrical vortex ($e = 0.5$), Moffat et al. (1994) found the generalization for elliptical vortices and Kerr and Dold (1994) found a solution for the case of a 2D stretching field of the form ($e = 0$). In these models, the vorticity is concentrated by stretching while the viscosity tends to diffuse it. The equilibrium between these two effects determines the size of the core.

We thus sought experiments in which the flow, forced by the geometry, is an approximation of some of the analytical Burgers solutions. In these flows we will investigate the interaction between vorticity and stretching, trying to find if there are some robust mechanisms of interaction between these two quantities. Two questions will be particularly attended:
i) What is the stability of a quasi-pure stretching flow with only residual and incoherent vorticity?
ii) Most authors consider the stretching as an externally fixed quantity. Is this realistic in the complex interplay between vorticity and strain ?

2 Stability of a quasi-pure strain structure

A region of pure strain corresponds to the flow in the vicinity of a stagnation point. It can be for instance the two dimensional flow shown on Figure (1(a)) and defined by:

$$v_x = \gamma x, \quad v_y = -\gamma y, \quad v_z = 0 \ . \tag{8}$$

The stability of such a flow was first investigated theoretically by Aryshev et al. (1981) and by Lagnado et al. (1984) who concluded to its instability. In a different context, there is also a stagnation line in the region separating two eddies of a plane mixing layer. Lin and Corcos (1984) and Neu (1984) showed that in this region there was an amplification of any residual vorticity leading to the formation of the secondary stream wise vortices which are well observed in mixing layers. Finally, Kerr and Dold (1994) revisited the problem and found that taking into account the non-linear terms the instability should lead to steady states characterized by a series of alternate vortices aligned in the direction of the extensional flow. Experimentally however there was no systematic investigation of this instability with the exception of a work by Lagnado and Leal (1990) using Taylor's four rollers mill.

2.1 The approximation of the ideal strain by the four rollers mill

This geometry was introduced by Taylor (1934) in order to study the deformation of viscous drops submitted to a pure strain. Several recent works have demonstrated that with a very viscous fluid the flow in the central region is very close to the ideal given by Equation (7). It is worth noting the reason for which a region of pure strain can be obtained in this configuration. As discussed above, it was pointed out by Raynal (1996) that in a closed cell having moving walls, relation (5) do not necessarily hold for the whole cell. This is precisely the case in this geometry in which, because of the rotation of the cylinders, most of the vorticity is, so to say, concentrated inside of the solid while the dissipation is in the flow. However, the difference in dissipation and enstrophy is fixed by the boundary conditions, so that all the structures which may appear due to the instability of the pure strain flow will preserve the equality between dissipation and enstrophy.

The stability of this flow was investigated by Lagnado and Leal (1990). Over a certain threshold they find the formation of vortices aligned in the stretching direction. But in their experiment this array was limited to two vortices located near the extremities of the cylinders and an intermediate one. This led them to conclude: "the vortices are primarily an end effect and a column of alternate vortices does not appear throughout the entire vertical extent at larger aspect ratio." Our experiments, done in a different experimental set-up, leads to an opposite conclusion.

2.2 Experimental set up

The core of the experiment is formed of four cylinders having parallel axes and placed at the corner of a square. The two cylinders placed on one diagonal rotate at the same velocity, the other two rotating with the opposite angular velocity. We used four cylinders of length $h = 19$ cm and radius $R_o = 3.5$ cm at a distance $R_1 = 6.42$ cm of the axis of the cell (Fig. 1(b)), the whole system being contained in a tank.

The use of Kalleiroscope in preliminary experiments revealed a very complex flow in the whole cell with active structures in the space between the rollers and the container. In this region, if the container is cylindrical, there is formation of Couette-Taylor like structures in the shape of sectors of tori. If the container is rectangular very active recirculations zones are formed in the corners. In both cases the flow in the region of interest (the space located between the cylinders) is strongly disturbed by the existence of these structures.

We thus designed our experiment so that these outer structures would be inhibited. For this purpose the whole system was enclosed inside a rather narrow cylinder of radius $R_2 = 12$ cm. Furthermore four protruding bands parallel to the cylinders were glued onto the inner wall of the cell in order to narrow the minimum gap between the cylinders and the wall (Fig. 1(c)). The formation of structures in this region was thus inhibited, at least at low velocities.

We also observed, as Lagnado and Leal (1990) before us, the formation of intense toroidal circulations induced at the extremities of the cylinders. This derives from the presence of rotating fluid around the cylinder on a static (top or bottom) plate (Eckman pumping). In order to reduce these end effects thin disks of larger radius (4.5 cm) were fixed at the extremities of the cylinders (Fig. 1(b)), so that they rotate as the fluid.

Two fluids were used, a very viscous glycerol solution ($\nu = 3\ cm^2/s$ and $\rho = 1.24\ g/cm^3$) and pure water ($\nu = 0.01 cm^2/s$ and $\rho = 1 g/cm^3$).

We used an ultrasonic Doppler anemometer (DOP 1000 constructed by Signal Processing Co.) to measure various velocity profiles. This apparatus emits ultrasonic pulses and the signal retrodiffused by passive particles gives

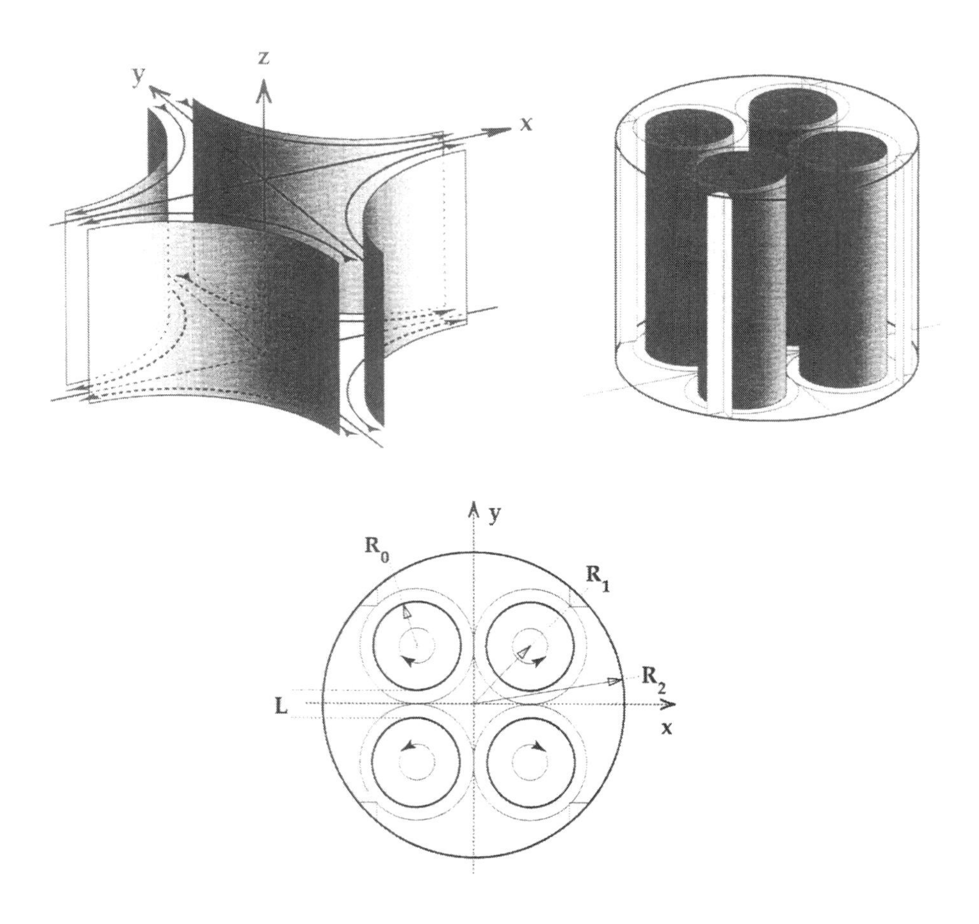

Fig. 1. (a) An ideal stagnation line and the axis of reference used in this work. (b) Scheme of the experimental Taylor's four rollers mill. The end plates of the cylinders are represented. (c) Section of the four rollers mill cell.

the longitudinal component of the velocity in 114 equidistant points located along the ultrasonic beam.

2.3 Results

Observation of the central region of the cell at low Reynolds number shows that a good approximation of a hyperbolic flow is obtained. We measured the velocity gradient in the central zone and found that it was proportional to the angular velocity of the cylinders :

$$\frac{\partial v_x}{\partial x} = \gamma = k\Omega \tag{9}$$

The coefficient of proportionality in our experiment is $k = 0.44$, a value approximately half that found by Taylor (k=0.81). This is certainly due to the fact that our flow is forced through narrow gaps in the outer region of the cell, a hindrance to the efficiency of the stirring of the fluid by the cylinders. For this reason, rather than using the cylinders velocity, we chose to define a Reynolds number from the measured fluid velocity gradient γ:

$$Re = \frac{\gamma L^2}{\nu} \tag{10}$$

where L is the gap between two cylinders ($L = \sqrt{2}R_1 - 2R_o = 2.1$ cm). With this definition we find that below a critical value of $Re_c = 19 \pm 1$ the flow is stable (Fig. 2(a)). Above Re_c the central vertical line destabilizes into a sinusoid. For a slightly larger value the formation of an array of steady alternate vortices is observed (Fig. 2(b)). The distance separating two successive vortices in the vertical direction was of the order of L the gap between the cylinders. As seen on Fig 2(b), lengthwise there are two pairs of very regular vortices and two disturbed structures near the ends. At approximately $Re_c = 45$ the vortices become chaotic (Fig. 2(c)). Well defined vortices are observed in all the range of Reynolds numbers that we explored (up to $Re \sim 10000$).

We measured specifically the velocity profile in the stretching direction $v_x(x)$. Below the instability threshold it is steady and independent of the position in the Oz direction. Using water, high Reynolds numbers are reached and the profile $v_x(x)$ fluctuates strongly in time. However its average is still well defined : it is shown (at $Re \sim 5000$) by the crosses on Figure 3. The velocity gradient induced by the rollers is clearly visible. Since we could visualize the flow during the measurements, we could also manage to trigger the measurement of $v_x(x)$ at the exact time where the axis of one of the turbulent vortices coincided with the axis of the ultrasonic probe. The instantaneous profile thus obtained is shown by the black triangles on Fig. 3 (a and b). In the two cases shown on Fig. 3 the velocity gradient along the axis of the vortex is clearly weaker than the average gradient. This is a general observation and in reverse a systematic recording of the fluctuating profiles led to the conclusion that all the profiles having a weak gradient were those corresponding to a vortex passing along the axis of the probe.

This result demonstrates a new effect valid for concentrated vortices. As predicted by Equation (1) the stretching amplifies the vorticity. But the vortex itself is observed to retroact on the vortex stretching (Equation (4). The resulting reduction of the stretching, reduces in turn the vorticity amplification in Equation (1). In other words there is a negative feedback of the vorticity on the stretching which had amplified this vorticity.

As in all Burgers-like models, in the solutions found by Kerr and Dold (1994), the stretching is an imposed field. As a result, if they are good descriptions of the structures found near the threshold, they fail to describe the situation at high Reynolds number because they do not take into account the retroaction of the vorticity on the stretching and the resulting spatial modulation of the v_x component of the velocity.

This decrease of the stretching in the core of a vortex will also be observed in the second experiment, also aimed at studying the interaction of strain and vorticity, but done in a completely different geometry.

3 Bidimensionalization of a strong laboratory vortex

3.1 The experimental cell

A classical geometry used to create an isolated vortex is the cylindrical tank having a rotating bottom introduced by Escudier 1984 and studied by Morkovin (1962). In this system the rotating disk is a source of vorticity and the toroidal flow (due to the Ekman layer suction) stretches the vorticity in the central region. In this geometry a central vortex is formed which is not stable and undergoes frequent vortex breakdowns. Turner (1966) investigated the vortex generated in a rotating tank in which an axial stretching flow was created by bubbles rising along the cell's axis.

We studied the variant of Escudier's experiment in which the vortex stretching is artificially enhanced by pumping the fluid out of the cell through a hole placed at the centre of the top of the cylinder (Fig. 4). The fluid is re injected in the tank below the rotating disk. It seems *a priori* that in such a configuration the stretching of the vortex is controlled by the flow rate Q imposed by the pump. For large values of this flux the vortex breakdown disappears and a stable structure is obtained. The cylinder has a radius of 7 cm and a height of 12 cm. The disk can be rotated at a frequency up to 20 Hz and the flow rate of the pump was of the order of $Q = 1\ l/mn$. The fluid was pure water.

3.2 Results

We measured the orthoradial velocity profile which is, in the central region, similar to that of a Burgers vortex. At the periphery the flow is close to

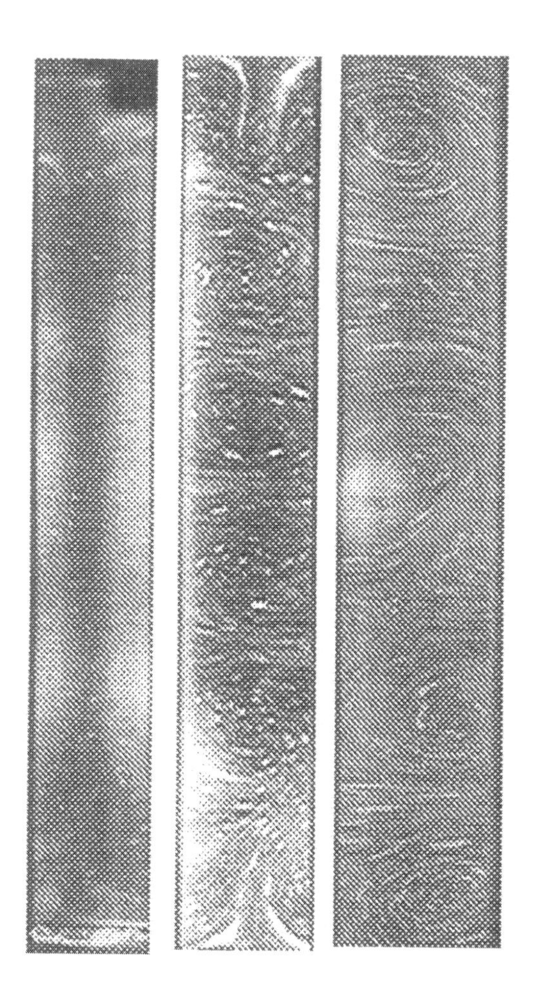

Fig. 2. Photographs of cross sections of the flow in the median plane (Oy, Oz) of the cell (perpendicular to the stretching direction). Three states of the flow are shown (a) below the instability threshold. (b) Above the threshold with a periodic array of alternate vortices. (c) In the turbulent regime.

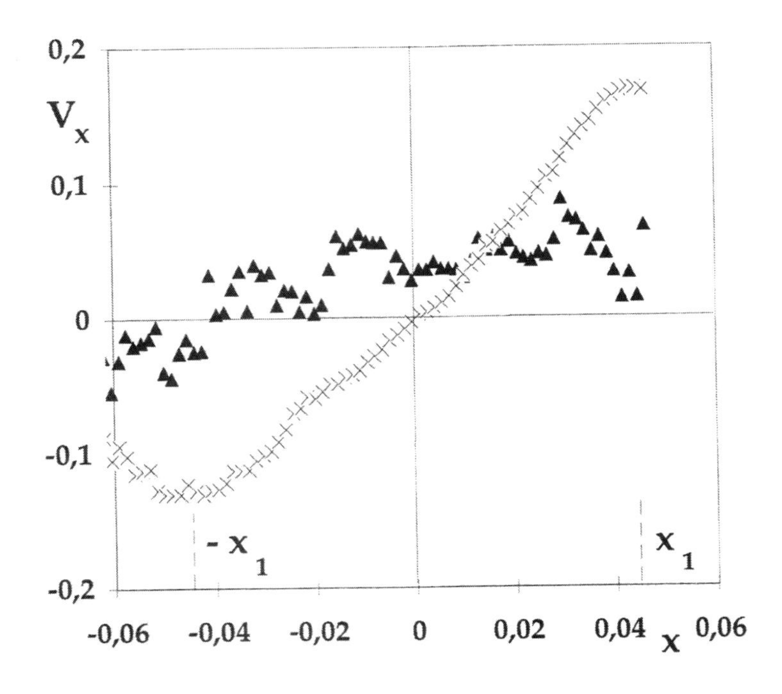

Fig. 3. Two longitudinal velocity profiles $v_x(x)$ in the stretching direction. The crosses (x) show the average profile. The black triangles show an instantaneous profile obtained when the core of a vortex was aligned with the beam of the ultrasonic probe.

a solid body rotation. The structure of the flow in the vertical direction is complex but can be investigated by local injections of ink (Fig. 5). Due to the Ekman pumping, the fluid, in most of the cell, moves down towards the rotating disk. Reaching the disk the fluid is sucked into the boundary layers. In the outer region of the disk it is then ejected towards the periphery from where it will move upwards along the cell's wall. Nearer to the axis it is aspired towards the centre and then directly aspired vertically by the

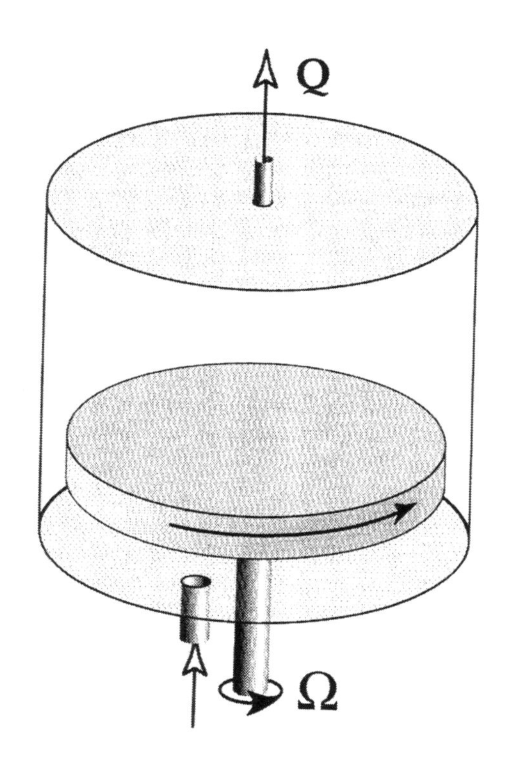

Fig. 4. Scheme of the cylindrical tank with the rotating end wall and the opposite pumping hole. Note that the stretching direction is Oz in contrast with the first experiment where it was Ox

pump through a central column (Fig. 5). This axial column of upward moving fluid is only slightly conical. Its diameter at the top is fixed by that of the pumping hole. In the present series of experiments we first set the disk in rotation and, when a steady regime is reached, we switch on the pump and observe the subsequent transient. The ultrasonic probe is used to measure the longitudinal velocity $v_z(z)$ along the axis of the vortex. Figure 6(a) and 6(b) show successive profiles. The first one corresponds to the situation before the pump was switched on. There is a weak positive velocity $v_z(z)$ due to the

Fig. 5. Photograph of the flow showing the central column in which the rotating fluid is pumped.

spin down effect. The second profile, taken immediately after the pump has been switched on, shows a large velocity gradient localized in the immediate vicinity of the hole. This is the classical flow in the vicinity of a sink. The later evolution shows that the axial velocity spreads downwards so that at time t= 3.9 s there is a large gradient along all the height of the cell. During the later evolution the velocity gradient along the length of the vortex decreases. At t=20 s it has reached its steady state with a finite vertical velocity but a weak gradient along the vortex. Two regions of strong gradients survive in the two boundaries at the top and the bottom of the cell. Simultaneous visual observations show that the transient regime corresponds to the increase of the vorticity in the core and to the simultaneous build-up of the central column.

This observation gives the physical origin of the reduction of the stretching observed in this experiment as well as in the previous one. The stretching of a vortex results into an intensification of its rotation. In turn this rotation tends to make the flow bidimensional in the core and thus to oppose the variations of

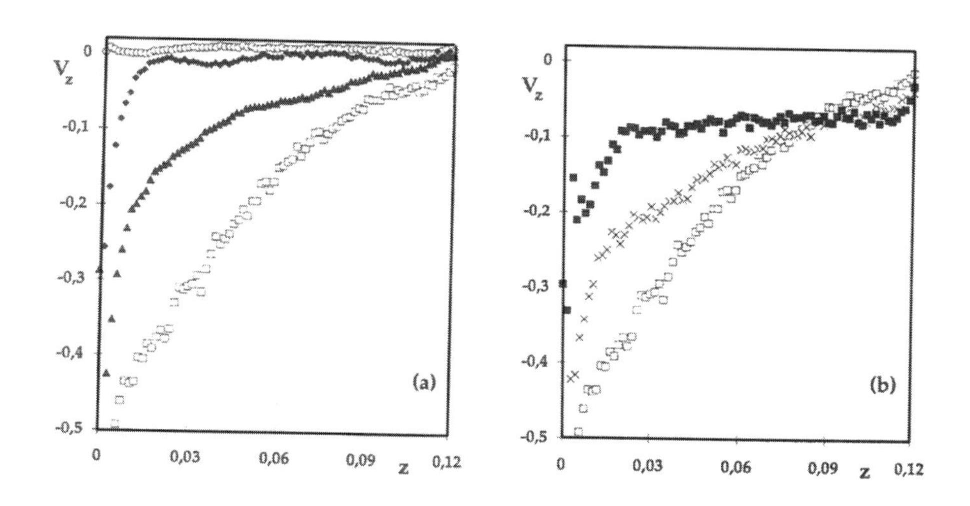

Fig. 6. The temporal evolution of the longitudinal velocity profile in the core of the vortex after the pump has been switched on. (a) Profiles at times: t = 0 s (open circles), 0.2 s (black diamonds), 1.4 s (black triangles), 3.9 s (open squares), (b) Profiles at times: 3.9 s (open squares), 8.7 s (crosses), 20 s (black squares).

velocity in this region. This effect, reducing the velocity gradients, diminishes the stretching.

The specificity of the observed steady flow lies in the coexistence of very large velocities in the orthoradial direction with very large velocities in the axial direction. These velocities both vary very rapidly in the radial direction. But both are in the bulk of the fluid practically constant in the z direction. All the z dependence of the flow is concentrated in the two boundary layers on the top of the tank and on the rotating disk.

4 Conclusion

Two results were presented here, both we believe related to observed characteristics of the turbulent flows. We observed experimentally the intrinsic instability of the regions of pure strain that had been predicted theoretically. It is worth noting that there is a difference in the topology of the regions of pure vorticity or pure strain. The former are vortex tubes with closed flow lines. In contrast the latter are opened hyperbolic regions which suck in perturbations from the outside. The surrounding vorticity is thus captured, stretched and amplified and finally feeds the instability. This occurs even in the case we examined here where most of the vorticity of the flow is frozen in the rotating cylinders. We believe this instability to be responsible for the absence of structures of pure strain and high pressure in turbulence.

We have also demonstrated that there exists a retro-action of the stretching of a vortex on the strain itself. This is a kind of Lenz law which had not been considered before. It can be noted that this effect is linked to the bidimensionalization due to the rotation near the core of axisymmetric vortices. For shear layers in which the vorticity is concentrated in planes no rotation is associated to it and the same feedback on the stretching will not exist.

This reduction of the stretching near the vortices' core has several implications.

In the Burgers-like vortex models the background stretching is assumed to be infinite and constant in space and time. More complex models were introduced by Donaldson and Sullivan (1960) in order to describe vortices of finite size. All these solutions have in common that the stretching is an imposed field in the sense that the vorticity can not retroact on it. Our results show that in experimental situations the stretching being finite the retro-action of the vorticity on the stretching has drastic consequences in the sense that the stretching cannot be imposed. This effect should be taken into account in more realistic models of vortices and will be investigated in further studies.

The bidimensionalization of the vortex due to its stretching provides a physical interpretation to the fact the vortex filaments observed in turbulent flows appear at first as straight structures as also found on theoretical grounds by Constantin and Procaccia (1995) and observed numerically by Galanti et al. (1996). The corresponding reduction of the strain along the axis could then also be responsible for the breakdown of the filaments which occurs ultimately.

We believe these results to be related to many issues in fluid flows, such as the intermittence in turbulence or the discussion about the formation of singularities in finite time.

Acknowledgements. We are grateful to H. Demaie for his help in setting up Taylor's experiment.

References

Andreotti B., (1997): Studying simple models to investigate the significance of some statistical tools used in turbulence. to appear in Physics of Fluids, March 1997.

Aryshev, Y. A., Golovin, V. A. and Ershin, S., A. (1981): Stability of colliding flows. Fluid Dyn. **16(5)** 755-759.

Ashurst W. T., Kerstein A. R., Kerr R. M. and Gibson C. H., (1987): Alignment of vorticity and scalar gradient with strain rate in simulated Navier-Stokes turbulence. Phys. Fluids **30**, 3243-3253.

Cadot, 0., Douady, S. and Couder, Y., (1995): Characterization of the low-pressure filaments in a 3-dimensional turbulent shear flow. Phys. Fluids. **7**, 1-15.

Constantin P. and Procaccia I. (1995): Scaling in fluid turbulence–A geometric theory. Phys. Rev. E **51**, 3207-3222.

Donaldson C. D. and Sullivan R.D., (1960): Behaviors of solutions of Navier Stokes equations for a complete class of three dimensionnal vortices. Proceedings of Heat Transfer and Fluid Mechanics Instabilities, 16-30 Stanford University.

Douady S., Couder Y. and Brachet M. E., (1991): Direct observation of the intermittency of intense vorticity filaments in turbulence. Phys. Rev. Lett. **67**, 983.

Escudier, M. P., (1984): Vortex breakdown : observations and explanations. Exp. Fluids, **2**, 189-196.

Fauve S., Laroche C. and Castaing B., (1993): Pressure fluctuations in swirling turbulent flows. J. Phys. II France **3**, 271-278.

Galanti B., Procaccia I. and Segel D., (1996): Dynamics of vortex lines in Turbulent flows Preprint.

Jimenez J., Wray A. A., Saffman P. G. and Rogallo R. S., (1993): The Structure of intense vorticity in isotropic turbulence. J. Fluid Mech. **255**, 65-90.

Kerr O. and Dold, J. W. (1994): Periodic steady vortices in a stagnation-point flow. Fluid Mech. **276**, 307 -325.

Kerr R. M., (1987): Histograms of helicity and strain in numerical turbulence. Phys. Rev. Lett. **59**. 783.

Lagnado, R. R., Phan Thien, N. and Leal, L. G., (1984): The stability of two–dimensional linear flows. Phys. Fluids , **27**, 1094-1101.

Lagnado, R. R. and Leal, L. G., (1990): Visualization of 3-dimensional flow in a 4-roll mill. Exp. Fluids, **9**, 25-32.

Lin S. J. and Corcos G. M., (1984): The mixing layer: deterministic models of a turbulent flow. Part 3. The effect of plane strain on the dynamics of streamwise vortices. J. Fluid Mech. **141**, 139.

Moffat H. K., Kida S. and Okhitani K., (1994): Stretched vortices–The sinews of turbulence–Large Reynolds number asymptotics J. Fluid Mech. **259**, 241-264.

Mory, M. and Yurchenko N., (1993): Vortex generation by suction in a rotating tank. Eur. J. Mech. B / Fluids **12(6)**, 729-747.

Neu J. C., (1984): The dynamics of stretched vortices. J. Fluid Mech. **143**, 253-276.

Nomura K.K. (1995): On the nature of the pressure Hessian in Homogeneous Turbulence. Bulletin of APS **40(12)**, 1973 .

Ohkitani, K. (1994): Kinematics of vorticity–Vorticity-strain conjugation in incompressible fluid flows. Phys. Rev. E, **50**, 5107-5110.

Ohkitani, K. and Kishiba, S. (1995): Nonlocal nature of vortex stretching in an inviscid fluid. Phys. Fluids, **7**, 411-421.

Raynal F., (1996): Exact relation between spatial mean enstrophy and dissipation in confined incompressible flows. Phys. Fluids, **8**, 2242-2245.

Siggia E. D., (1981): Numerical study of small-scale intermittency in three-dimensional turbulence. J. Fluid Mech., **107**, 375.

Tanaka, M. and S. Kida, (1993): Characterization of vortex tubes and sheets. Phys. Fluids A **5**, 2079-2082.

Taylor, G. I., (1934): The formation of emulsions in definable fields of flow. Proc. Roy. Soc. A **146**, 501-523.

Taylor, G. I. (1938): Production and dissipation of vorticity in a turbulent fluid. Proc. Roy. Soc. A **164**, 15-23.

Tsinober A., Kit E. and Dracos T., (1992): Experimental investigation of the field of velocity gradients in turbulent flows. J. Fluid Mech. **242**, 169-192.

Turner, (1966): The constraints imposed on tornado-like vortices by the top and bottom boundary conditions. J. Fluid Mech. **25**, 377-386.

Pressure and Intermittency in the Inertial Range of Turbulence

O. N. Boratav[1], R. B. Pelz[2]

[1] Mechanical and Aerospace Engineering Department
University of California, Irvine, CA 92697, U.S.A.
[2] Mechanical and Aerospace Engineering Department
Rutgers University, NJ 08855, U.S.A.

Abstract. The intermittency corrections in pressure in the inertial range of turbulence using high resolution decaying turbulence simulations are discussed. The contribution to pressure intermittency of *strain-dominated*, *enstrophy-dominated*, *equal-strain-enstrophy* and the *background-structures* are presented. It is found that the pth order pressure structure function is independent of the order p for a wide range of separation distances r.

1 Introduction

In any generic three-dimensional incompressible fluid flow, strong velocity gradients develop. The equation(s) describing the dynamics (rate of change) of velocity gradients can be obtained by taking the gradient of the Navier-Stokes equations. It is clear from the resulting equation(s) that the coupling between strain (S_{ij}) and rotation/vorticity (R_{ij}) has a crucial role in the velocity gradient dynamics. To understand the interplay between strain and rotation, it is natural to examine the flow characteristics in the following three regions: (i) Where rotation dominates over strain. (ii) Where strain dominates over vorticity. (iii) Where rotation and strain are comparable in magnitude.

In a generic three-dimensional flow field-turbulent or not-, it is expected that regions belonging to all these three classes would coexist. The above-mentioned first class (i) contains points most likely to be located at the central portion (i.e. *eye*) of a vortex (See for example Figure 7b in Boratav and Pelz (1995)). The points belonging to the second category (ii) are likely to be located in regions where oppositely-signed vortices collide canceling their mutual vorticity, forming vorticity null-lines (or planes), along which strain is extremely high, yet, vorticity is negligible. The points in the third category (iii) are likely to be located slightly off the central region of vortices or somewhere on the extremely smashed pancake-like vortices (See Figure 9b in Boratav and Pelz (1995)).

Do these three different categories have equal contribution to intermittency in turbulence ? What is meant by intermittency in this paper is the deviation from 1941 scaling laws (in the inertial range of turbulence). Let us recall Kolmogorov's 1941 scaling laws given by:

$$S_p(r) \equiv < \left(\delta v_{\parallel}(\boldsymbol{x},\boldsymbol{r})\right)^p > = C_p \epsilon^{p/3} r^{p/3} \quad (1)$$

Here, $S_p(r)$ is defined as the pth order longitudinal structure function, $\delta v_{\parallel}(\boldsymbol{x},\boldsymbol{r})$ is the longitudinal velocity increment between two points separated by $\boldsymbol{r}$ (where $r = |\boldsymbol{r}|$), $\boldsymbol{x}$ is the space location of a given point, ϵ is the mean energy dissipation per unit mass (with dimensions length^2/time^3), and $<\quad>$ denotes some relevant averaging. It is believed that any deviation of the scaling exponent of r from $p/3$ is due to intermittency, which is also believed to be related to presence of structures in the flow field.

The question raised here is recently discussed by Boratav and Pelz (1996) in high-resolution decaying turbulent flow simulations. In the next section, we will give an overview of these results. Then, we will focus on the intermittency in the pressure field.

2 Strain, Vorticity, Intermittency

The database used in this work is from three-dimensional Navier-Stokes simulations of decaying turbulence starting from high-symmetry initial conditions suggested by Kida (1985). Pseudospectral method is used with 2/3rd law dealiasing. The resolution is 300^3 points which is equivalent to 1200^3 points in the periodic box when all the symmetries are considered. The maximum wavenumber k_{max} after dealiasing is 400. As the standard resolution check, the product of $k_{max}\eta$ is computed (η is the Kolmogorov lengthscale) and found to be 1.49 at the instant where statistics are collected. The Reynolds number based on Taylor microscale (which is defined as $R_\lambda = \sqrt{(10/3)}E(t)/(\nu\sqrt{\Omega})$ where E and Ω are the volume-averaged energy and enstrophy respectively) is equal to 142 at t=4.0 and decays to 82 at t=6.0. The energy spectrum shows that there is at least a decade of a $k^{-5/3}$ scaling with slight intermittency corrections. The pdf of the velocity field is close to a Gaussian distribution. The pdf's of the velocity derivatives are strongly non-Gaussian and the skewness, flatness and the sixth order moments based on $\partial u/\partial x$ are found to be -0.52, 5.57 and 89.8 at t=6.0 respectively.

Three isotropy tests are perform to conclude that the field is very close to being isotropic at least for the low order moments. (See Boratav and Pelz (1996) for details).

In order to classify the three different categories mentioned in Section 1, Boratav and Pelz (1996) considered a plane whose x and y axes are local enstrophy $R_{ij}R_{ij}$ and strain $S_{ij}S_{ij}$ respectively. The polar coordinates of any point on this plane is given by Δ and Ψ which are named as the *Strain-Enstrophy distance* and *angle* respectively. These are given by:

$$\Delta = \sqrt{(S_{ij}S_{ij})^2 + (R_{ij}R_{ij})^2} \tag{2}$$

$$\Psi = \tan^{-1} \frac{S_{ij}S_{ij}}{R_{ij}R_{ij}} \tag{3}$$

The following four categories are selected for the analysis:
(i) Points with $\Psi < 10^o$. These are named as *'Enstrophy-dominated Structures'* and occupy 4.3% of all the points in the flow field.
(ii) Points with $\Psi > 80^o$. These are named as *'Strain-dominated Structures'* and occupy 16.4% of all the points in the flow field.
(iii) Points with $40^o < \Psi < 50^o$. These are named as *'EQUAL Structures'* and occupy 11.3% of all the points in the flow field.
(iv) Points with $\log \Delta < (\log \Delta)_{rms}$. These are named as the *'Background Structures'* and occupy 51.3% of all points.

Boratav and Pelz (1996) investigated how much each of the above categories contribute to the tails of the pdf's of the velocity structure functions. The velocity structure functions are computed in both longitudinal and lateral direction. Their main conclusions can be summarized as:
(i) *'Enstrophy-dominated Structures'* contribute mostly to the tails of the lateral structure function pdf's.
(ii) *'Strain-dominated Structures'* contribute mostly to the tails of the longitudinal structure function pdf's.
(iii) *'EQUAL Structures'* contribute to the tails of both longitudinal and lateral structure function pdf's.
(iv) The velocity structure function sampled on the *'Background Structures'* have a distribution very close to a Gaussian.
(v) The scaling exponents ζ_p computed in the lateral direction are found to deviate much more from Kolmogorov 1941 scaling than in the longitudinal direction. This implies that the *'Enstrophy-dominated Structures'* have a much more intermittent character, which can *only* be detected in a measurement along the lateral direction.

In Table 1, we present the scaling exponents computed by Boratav and Pelz (1996) (Using Extended self-similarity-ESS- by Benzi et al. 1995). in the longitudinal and lateral direction and compare them to the predictions of a recent, popular intermittency model by She and Lévêque (1994). It is seen that the simulation results in the longitudinal direction are in good agreement with the predictions of the She and Lévêque (1994) model while there are large disagreements between the lateral exponents and the model. There are also problems with consistency of several parameters in the She and Lévêque (1994) model and discussions along these directions will be reported elsewhere (Boratav (1996)).

In the next section, we perform a similar analysis applied to the pressure field of the same simulation.

Order, p		ζ_p	
	Longitudinal (ESS2)	Lateral (ESS2)	She-Lévêque (original)
2	0.6949	0.7194	0.6959
4	1.2795	1.2316	1.2797
6	1.7698	1.5687	1.7778
8	2.1845	1.7870	2.2105
10	2.5462	1.9401	2.5934
12	2.8766	2.0633	2.9383
14	3.1919	2.1757	3.2541

Table 1. Scaling Exponent ζ_p in Structure Functions of Order n=2-14, and comparison to the She-Lévêque model. ESS2 refers to the notation used in Reference Boratav & Pelz (1996).

3 Pressure

It is natural to extend the discussion in the previous section to the pressure field since the Laplacian of it is directly related to enstrophy and strain given by:

$$\nabla^2 P = \frac{1}{2}\omega^2 - S_{ij}S_{ij} \tag{4}$$

In an incompressible field, the volume integral of the difference between the strain and enstrophy add up to zero (in the absence of boundaries with periodic boundary conditions). In the simulations considered here, we observe an *unbalance* in extrema, i.e. the maximum (and points close to maximum) enstrophy larger than maximum (and points close to maximum) strain. Then, the number of points whose strain is greater than enstrophy has to be more than the number of those whose enstrophy is greater than the strain, (to assure neutrality).

Now let us consider the contribution of the three categories defined in Section 1 to the pressure pdf. The distributions are given in Figure 1. We observe that: (a) $\Psi > 80^o$ (*'Strain dominated Structures'*) do *not* show a strong deviation from Gaussianity, (b) $\Psi < 80^o$ (*'Enstrophy dominated Structures'*) have a strongly non-Gaussian, super-exponential distribution (c) Ψ_{40-50} (i.e. $40^o < \Psi < 50^o$ or *'EQUAL Structures'*) also show a strong deviation from Gaussianity, and have an exponential distribution. Their non-Gaussian contribution manifests itself in the negative tails of the pressure pdf and the shape of the distribution around the negative tails is very close to the shape of the distribution of the whole domain.

The conclusions stated above by (a) and (b) are *not* surprising, similar results are already presented by different authors, for example, by Pumir 1994.

On the other hand, the third conclusion (c) is rather unforeseen. The reason why *'EQUAL Structures'* contribute to the negative pressure tails is probably because of the points in the field which have large and comparable (but *not exactly* equal !) enstrophy and strain. Even though their magnitudes are comparable, the difference between enstrophy and strain of such points is *not* negligible. We speculate that such structures are present in any turbulent field and contribute to negative tails of pressure considerably. Experimentally, it would be difficult to visualize such structures since particles used for flow visualization tend to escape (away) from high strain regions.

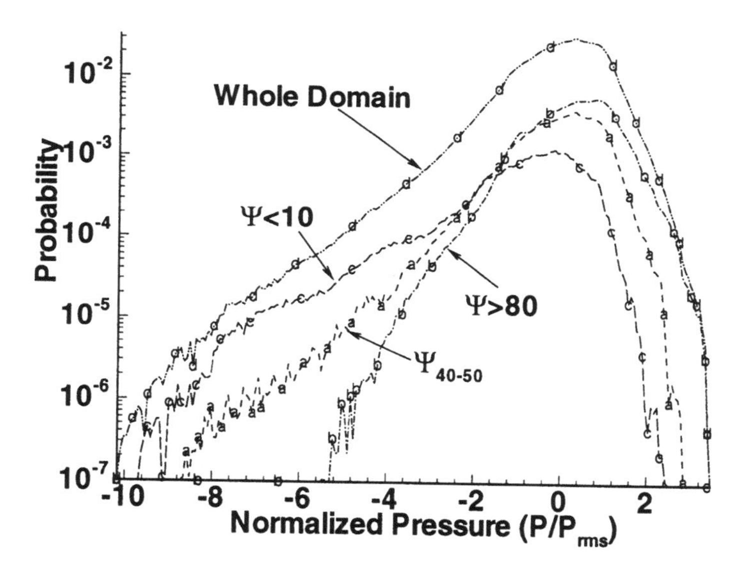

Fig. 1. Pdf of pressure (normalized by its rms) sampled on strain-enstrophy angle Ψ. Sampling sets are: $\Psi < 10°$ (Enstrophy-dominated Structures). $\Psi > 80°$ (Strain-dominated Structures). $40° < \Psi < 50°$ (Equal Structures) and the whole domain. Note that the pdf's here are not normalized so that the areas under each curve gives an indication about the space occupied by each structure category.

3.1 Pressure Structure Function

It is common practice in literature to discuss pressure intermittency effects and structures based on pressure pdf observations. In fact, the relevant quantity consistent with K-41 formalism to be examined should be the pressure structure function, *not* the pressure. It is defined as:

$$D_p(r) \equiv < P(x+r) - P(x) > \qquad (5)$$

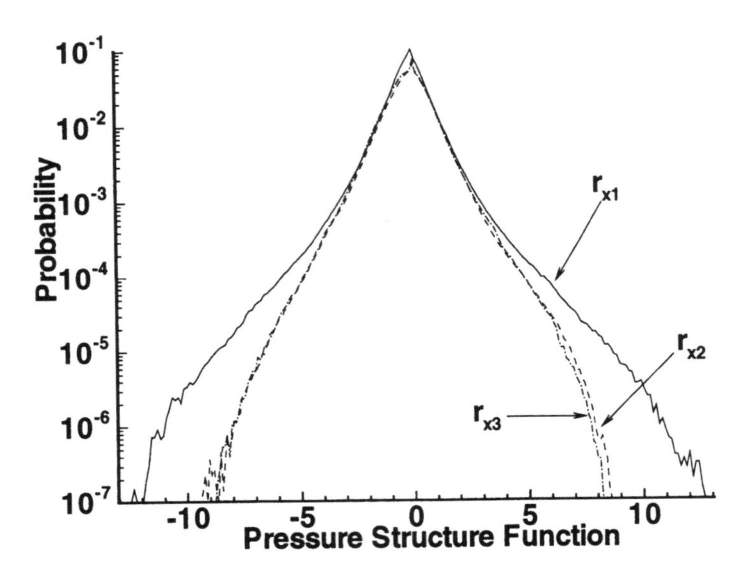

Fig. 2. Pdfs of pressure structure function D_p for three different separations all in the inertial range such that $r_{x3} > r_{x2} > r_{x1}$.

We computed the pressure structure function for different separation distances r which are all located in the inertial range. In Figure 2, we present the distributions corresponding to three separations (all in the inertial range) such that $r_{x3} > r_{x2} > r_{x1}$. We see that all three distributions show deviations from Gaussianity, and are close to an exponential distribution.

We also examined how the different flow categories mentioned before contribute to the tails of the pressure structure function distributions. Because of space limitations, we present the curves corresponding to $r = r_{x1}$ only. In Figure 3, we examine the contribution of strain-dominated and enstrophy-dominated structures to the tails. It is seen from this Figure that both categories contribute almost equally to the tails, with slightly more spread-out wings observed for the enstrophy-dominated structures. We observe that the trend of slight difference to the tail contribution between the two categories of structures diminishes at larger separation (for example r_{x2} or r_{x3}). This is indeed an interesting result since it is considerably different from the trend seen in the pressure pdf's, and also the trends in velocity structure function pdf's given by Boratav and Pelz (1996).

In Figure 4, we examine how much the *'Background structures'* contribute to the tails of pressure structure function corresponding to the separation r_{x1}. The interesting conclusion in this plot is that the *'Background'* field also deviates from Gaussianity (unlike in the velocity structure function case) and has an nonnegligible contribution to the tails.

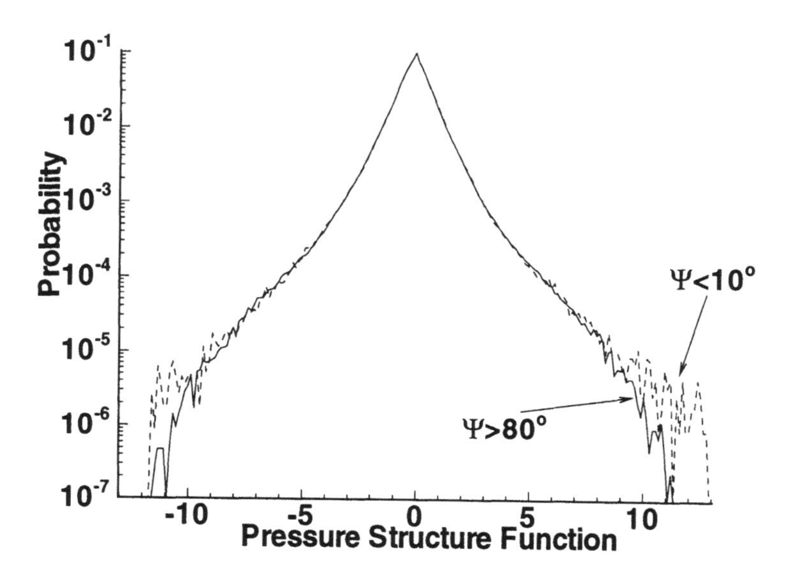

Fig. 3. Pdf of pressure structure function D_p corresponding to the separation r_{x1} sampled on $\Psi > 80^o$ (Strain-dominated Structures) and $\Psi < 10^o$ (Enstrophy-dominated Structures).

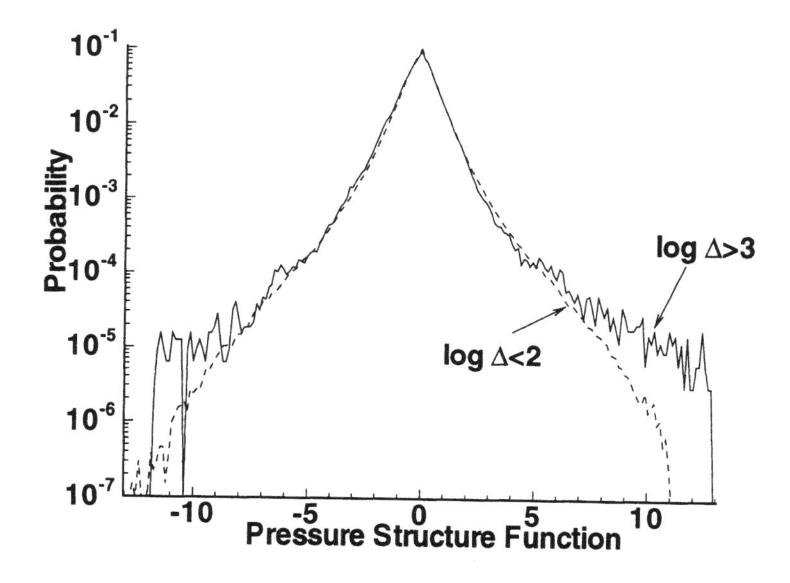

Fig. 4. Pdf of pressure structure function D_p corresponding to the separation r_{x1} sampled on the strain-enstrophy distance Δ: $\log \Delta > 3$ (Points away from origin), $\log \Delta < 2$ (Background). The rms value of $\log \Delta$ is approximately equal to 2.

3.2 Millionshchikov's Hypothesis

It is shown by Batchelor (1951) that the second order structure function of pressure can be expressed as an integral over the fourth order velocity structure function. It is common practice to make use of the Millionshchikov's Hypothesis (Millionshchikov (1941)), to express the fourth order velocity structure functions in terms of the second order structure function. If this assumption is made, the resulting equation for the second order pressure structure function D_{pp} becomes (See, for example Panchev (1971) or Hill and Wilczak (1995)):

$$D_{pp}(r) = \int_0^r y \left(\frac{dD_{LL}}{dy} \right)^2 dy + r^2 \int_r^\infty \frac{1}{y} \left(\frac{dD_{LL}}{dy} \right)^2 dy \tag{6}$$

where D_{LL} is the second order longitudinal velocity structure functions and r is a separation in the inertial range.

Several tests for the verification of Millionshchikov's Hypothesis are proposed by Batchelor (1951) and Uberoi (1953). One test is to examine the ratio which would be 3.0 when the joint-probability distribution of the velocities at two points were normal, given by:

$$\frac{< (u^{'} - u)^4 >}{[< (u^{'} - u)^2 >]^2} \tag{7}$$

which is equivalent to $D_{LLLL}/(D_{LL})^2$ using our previous notation. In Figure 5, we present this ratio as a function of the separation distance r. The inertial range determined using Kolmogorov's four-fifth law (Kolmogorov (1941)) is between the 3rd and 25th cross mark in this figure. It is seen in Figure 5 that there is a considerable deviation from the Millionshchikov's Hypothesis particularly at lower values of the separation distance r.

Finally, we present the comparison of the D_{pp} computed from Equation 6 (which we refer to as Σ) and D_{pp} computed directly (based on the definition of D_p given in Equation 5). The comparison is presented in Figure 6. We see from this figure that there is a large discrepancy between the slopes of each computation for all scales. Due to this discrepancy, we do *not* use results of exponents of pressure structure function which are obtained by using the simplification of the Millionshchikov's Hypothesis in this paper.

3.3 Intermittency Corrections for Pressure Spectra

An alternative expression which relates D_{pp} to integrals of fourth order velocity structure functions is given by Hill and Wilczak (1995). The D_{pp} expression requires the calculation of three fourth order velocity structure functions: The longitudinal D_{LLLL}, the lateral D_{NNNN} and the mixed D_{LLNN}. It is observed in Table (1) that there are corrections to Kolmogorov's 1941 scaling due to intermittency. It is also seen in this table that the lateral deviations

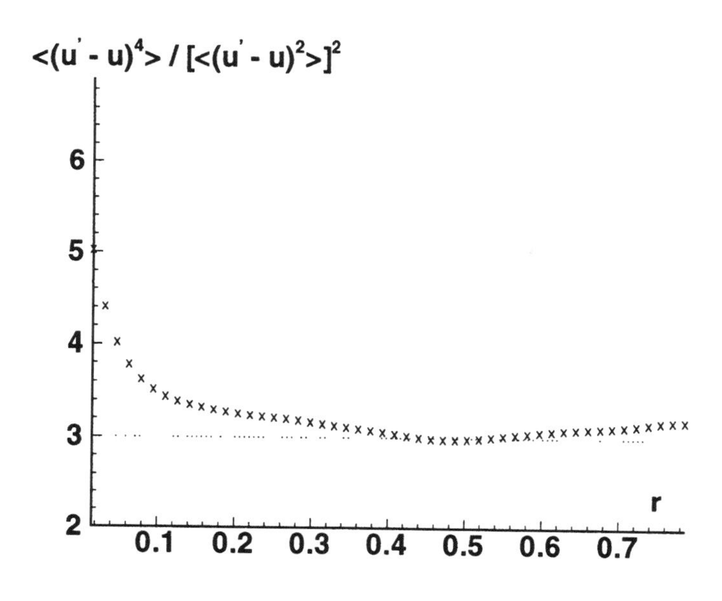

Fig. 5. Ratio of the fourth order velocity structure function to the second order squared as a function of the separation r.

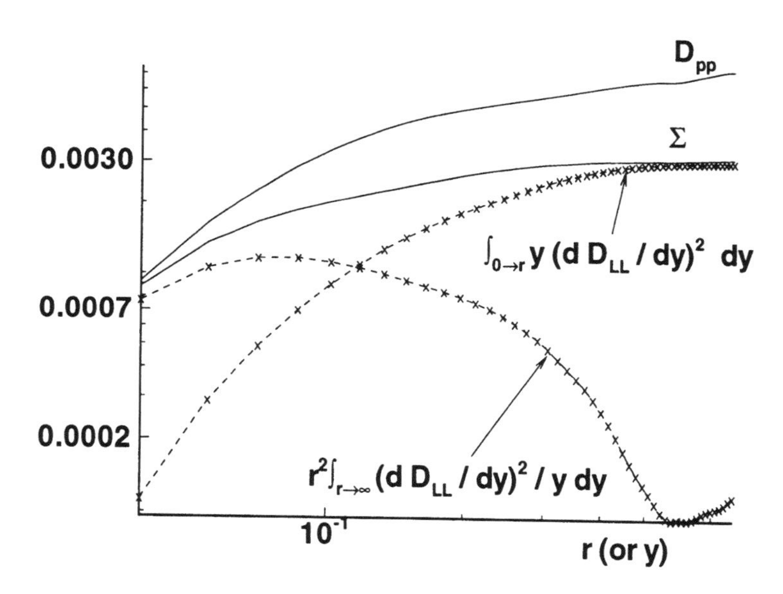

Fig. 6. Comparison of second order pressure structure functions, as a function of separation r. D_{pp}: computed directly, Σ: computed using Equation (6). Two terms in Equation (6) are also shown separately.

are much more than the longitudinal ones. Naturally, questions arise, such as:
Do the intermittency corrections make the pressure spectrum steeper or shallower than the one predicted by Kolmogorov-type formalism which gives a slope of (-7/3) ?.
If one uses Millionshchikov's Hypothesis and the results given by Table (1), one concludes that the pressure spectrum will be predicted to be steeper than the slope (-7/3) since the second order velocity structure function in the exponent is slightly larger than the Kolmogorov prediction (as opposed to all the other powers larger than 3).

On the other hand, if D_{pp} is computed using the fourth order velocity structure functions, it will be shallower than (-7/3). The numerical verification is *not* easy, since the deviations from Kolmogorov's scaling are rather small, particularly at low orders. The pressure spectrum of the simulation computed directly is given in Figure 7. Also plotted on this graph are two lines, one having a slope of -5/3 which is suggested by Van Atta and Wyngaard (1975) another one having a slope of $(-7/3) + \alpha$ where α is the deviation from the Kolmogorov scaling of the fourth order velocity structure function along the direction which gives the largest value. From Table 1, the largest deviation for p=4 is along the lateral direction and α is simply $4/3 - 1.2316 = 0.1017$. One final note before the end of this section is on a

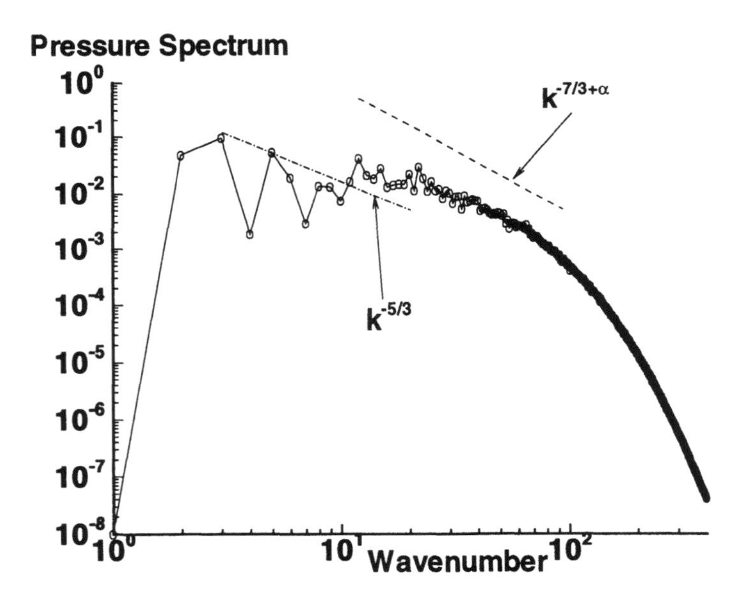

Fig. 7. Pressure spectrum. The two lines drawn have slopes of $-5/3$ and $(-7/3)+\alpha$ where α is the deviation from the Kolmogorov scaling of the fourth order lateral velocity structure function. From Table 1, α is found to be 0.1017.

result given by Hill and Wilczak (1995). Hill and Wilczak (1995) suggested that $D_{pp}/D_{LLLL} \approx 1/3$, which introduces considerable simplification in the D_{pp} expression given in terms of fourth order velocity structure functions. In Figure 8, we present this ratio as a function of separation distance r. We find that there are considerable deviations from the value 1/3 suggested in Reference 8.

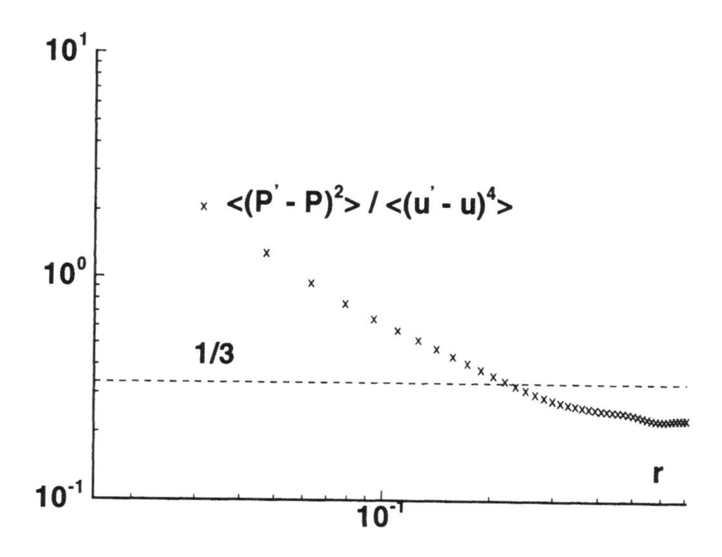

Fig. 8. Ratio of the second order pressure structure function to the fourth order velocity structure function vs separation r.

3.4 Direct Computation of D_{pp}

In Figure 9, we present the ratio of the pth order pressure structure function (p being 2, 3, 4, 6, 8) to the third order pressure structure function. Note that this ratio is constant for a wide range of r implying that pth order (p between 2 and 8) structure function of pressure scales linearly with D_{ppp} independent of the order p of the structure function. Kolmogorov-type scaling gives $< (\Delta P)^p > \sim r^{2p/3}$. In our results, since the scaling exponent p becomes independent of the order, the deviation from the Kolmogorov-type scaling increases as with order p. Hence, the pressure structure function exhibits deviations from a Kolmogorov type scaling.

To our knowledge, there is one result in literature which seems to suggest a similar scaling (i.e. a scaling which becomes independent of the order of

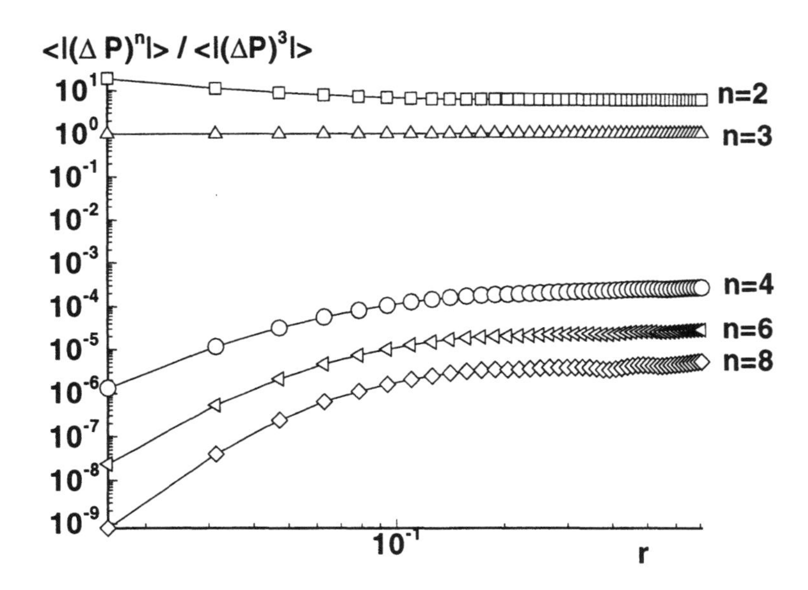

Fig. 9. Ratio of the nth order (n = 2, 3, 4, 6, 8) pressure structure function to the third order vs separation r. Note that the ratio for all orders is constant for a wide range of separation r.

the structure function) which is from a model of the large scale ramp-like behavior of a temperature signal, which gives a structure function linear in r (Van Atta (1977)) indepent of p. In this model, it is suggested that the temperature signal of interest consists of a sequence of ramps of amplitude a and length ℓ separated from one another by quiet periods of lenght s. To check at least whether the pressure distribution in the field follows a similar trend, we selected several enstrophy-dominated isosurfaces. We measured the pressure variation along these isosurfaces and observed that in most of them, the function has more than one dip. A typical variation along one of them is given in Figure 10. A ramp function approximation is one of the possible reasonable candidates.

Acknowledgements. We would like to thank professor Chuck Van Atta for informing us about his work on the temperature structure functions. Computations in this work were performed in Pittsburgh Supercomputing Center.

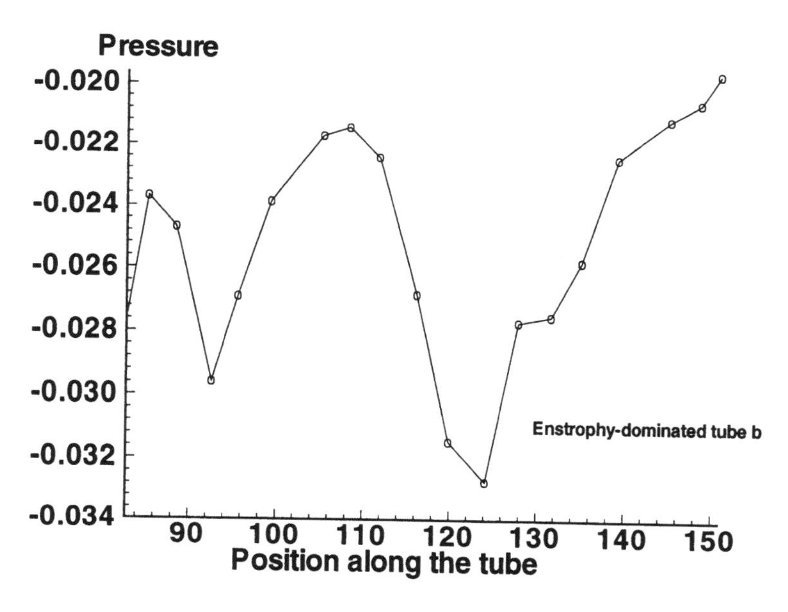

Fig. 10. Variation of pressure along one of the vortex tubes. Here, the vortex tube is defined as an isosurface of the Q-invariant (defined as the difference between enstrophy and mean-squared strain, $0.5\omega^2 - S_{ij}S_{ij}$). The threshold of the isosurface corresponds to twice the rms value of positive Q (enstrophy dominated tube). Notice the multi-dips in pressure along the tube. Similar variation is obtained in 80 different tubes examined in the flow field.

References

Batchelor G. K. (1951): Pressure fluctuations in isotropic turbulence. Proc. Cambridge Philos. Soc. **47**, 359–374.

Benzi R., Ciliberto S., Tripiccione R., Baudet C., Massaioli F., Succi, S. (1993): Extended Self-Similarity in Turbulent Flows. Phys. Rev. E **48(1)**, R29–32.

Boratav O. N., Pelz R. B. (1995): On the Local Topology Evolution of a High-Symmetry Flow. Phys. Fluids, **7(7)**, 1712.

Boratav O. N., Pelz R. B. (1996): Structures and structure functions in the inertial range of turbulence. To appear in Phys. Fluids, May 1997.

Boratav O. N. (1996): On recent intermittency models of turbulence. To appear in Phys. Fluids, May 1997.

Hill R. J., Wilczak, J. M. (1995): Pressure structure functions and spectra for locally isotropic turbulence. J. Fluid Mech. **296**, 247–269.

Jimenez J., Wray A., Saffman P. G., Rogallo R. S.: The Structure of Intense Vorticity in Isotropic Turbulence. J. Fluid Mech., **255**, 65–90.

Jimenez J., Wray A.: On the Dynamics of Small-Scale Vorticity in Isotropic Turbulence. CTR Annual Research Briefs, 1.

Kolmogorov A. N. (1941): The Local Structure of turbulence in an incompressible fluid with very large Reynolds number. C. R. Acad. Sci., **30**, 301.

Kolmogorov A. N. (1941): Dissipation of Energy Under Locally Isotropic Turbulence. C. R. Acad. Sci., **32**, 16.

Kida S.: Three-Dimensional Periodic Flows with High-Symmetry. J. Phys. Soc. Jpn., **54(6)**, 2132.

Millionshchikov M. D. (1941): On the influence of third moments in isotropic turbulence. Doklady Acad. Sci. U.S.S.R. **32(9)**.

Panchev S. (1971): *Random functions and turbulence.* Pergamon Press.

Pumir A. (1994): A numerical study of pressure fluctuations in three-dimensional incompressible, homogeneous, isotropic turbulence. Phys. Fluids **6(6)**, 2071–2083.

She Z-S., Lévêque, E. (1994): Universal Scaling Laws in Fully Developed Turbulence. Phys. Rev. Lett., **72(3)**, 336–339.

Uberoi M. S. (1953): Quadruple velocity correlations and pressure fluctuations in isotropic turbulence. J. Aero. Sci. **3**, 197-204.

Van Atta C. W., Wyngaard, J. C. (1975): On higher-order spectra of turbulence. J. Fluid Mech. **72(4)**, 673–694.

Van Atta C. W. (1977): Effect of coherent structures on structure functions of temperature in the atmospheric boundary layer. Archives of Mechanics. **29(1)**, 161–171.

Time-periodic statistical solutions of the Navier–Stokes equations

Andrei V. Fursikov

Department of Mechanics and Mathematics, Moscow State University
119899 Moscow, Russia

Abstract. Time-periodic space-time statistical solutions of the Navier-Stokes equations with time periodic forcing are constructed. Certain properties of these statistical solutions are established.

1 Introduction

Steady-state statistical solutions (the measure that is invariant with respect to shifts along trajectories of the dynamical system) can be considered as one of the models simulating a climate (see (Dymnikov and Filativ (1994)) . Such statistical solutions exist only if external forces do not depend on time or when external forces are described by a stationary stochastic process. But simulation of climate would be more adequate if we impose on external force (that is the sun radiation in reality) the time-periodicity condition or consider external forces which are described by time-periodic stochastic process. In this case the time-periodic space-time statistical solution is the analog of an invariant measure.

Of course, time-periodic statistical solutions can be applied also in other areas which are not connected with climate theory.

In this paper we give appropriate definitions and construct time periodic statistical solutions of the Navier-Stokes equations. In the bulk of the paper we suppose that the right-hand-side $f(t,x)$ of the Navier-Stokes equations is deterministic (individual) time-periodic vector field, but a case of stochastic time periodic forcing is also considered.

We construct a time-periodic statistical solution with help of Krylov-Bogolubov averaging method (see (Krylov and Bogolubov (1937)). Note that the conception of a space-time statistical solution introduced in (Vishik and Fursikov (1988) is essential for our construction here.

In Section 2 we introduce some preliminary information. Sections 3, 4 are devoted to the construction of time-periodic space-time statistical solutions in the case of 2-D Navier-Stokes equations: the case of deterministic right side f is studied in Section 3 and stochastic f is considered in Section 4. In Section 5 we construct a time-periodic statistical solution for 3-D Navier-Stokes equations with deterministic f.

Besides the existence theorems, certain properties of time-periodic statistical solutions are also established here. We prove that these statistical solutions are concentrated on functions bounded uniformly with respect to $t \in R$. Moreover, the support of the corresponding space statistical solutions depends on t periodically.

2 Formulation of the problem

In this section we give some definitions and remind known assertions, to be used later.

2.1 Navier-Stokes Equations

We consider the Navier-Stokes equations

$$\partial_t y(t,x) + (y,\nabla)y - \nu\Delta y + \nabla p(t,x) = f(t,x) \tag{2.1}$$

$$\operatorname{div} y(t,x) \equiv \sum_{j=1}^{n} \frac{\partial y(t,x)}{\partial x_j} = 0 \tag{2.2}$$

where $y(t,x) = (y_1(t,x),...,y_n(t,x))$ is the velocity vector field, $\nabla p = (\partial p/\partial x_1,...,\partial p/\partial x_n)$ is the gradient of pressure, $f(t,x) = (f_1(t,x),...,f_n(t,x)$ is the vector field of body external forces, $\nu > 0$ is the coefficient of viscosity,

$$\partial_t y = \frac{\partial y}{\partial t}, \qquad (y,\nabla)y = \sum_{j=1}^{n} y_j \frac{\partial y}{\partial x_j} \quad .$$

These equations are supplied by the initial condition

$$y(t,x)|_{t=0} = y_0(x) \quad . \tag{2.3}$$

We consider equations (2.1),(2.2) and initial condition (2.3) in a bounded domain $\Omega \subset R^n$ with C^∞-boundary $\partial\Omega$. On the boundary we impose the adherence condition:

$$y|_{\partial\Omega} = 0 \quad . \tag{2.4}$$

It is convenient for us to write problem (2.1)-(2.4) as an ordinary differential equation with operator-valued coefficients. Such reduction is well-known(see e.g. (Ladyzhenskaya (1969), (Lions (1969),(Temam (1979),(Vishik and Fursikov (1988)). To do it we define the spaces:

$$\mathcal{V} = \{v(x) = (v_1,...,v_n) \in (C_0^\infty(\Omega))^n : \quad \operatorname{div} v = 0\}$$

$$H = \text{closure of } \ \mathcal{V} \ \text{ in } \ (L_2(\Omega))^n , \tag{2.5}$$

and introduce the operator of orthogonal projection

$$\pi : (L_2(\Omega))^n \longrightarrow H \quad .$$

After that applying the operator π to both parts of equations (2.1) we obtain the desired ordinary differential equation:

$$\partial_t y(t,\cdot) + Ay + B(y) = f(t,\cdot)$$

where

$$A = -\pi\nu\Delta \tag{2.6}$$

is the self-adjoint positive operator in the space H which possesses a compact inversef operator, and

$$\tilde{B}(v,w) = \pi[(v,\nabla)w], \qquad B(y) = \tilde{B}(y,y). \tag{2.7}$$

(We assume that $f(t,\cdot) \in L_2(0,T;H)$ where f is the right-hand-side of (2.1)).The details of this reduction can be found in the books mentioned above.

2.2 An abstract equation

So, we consider the abstract analog of the Navier-Stokes system:

$$\partial_t y(t) + Ay + B(y) = f(t) \tag{2.8}$$

where A is a linear self-adjoint positive operator possessing a compact inverse and B is a quadratic map. We define the spaces H^α in the usual way with help of the operator A:

$$H^\alpha = \{y = \sum_{k=1}^{\infty} y_k e_k : ||y||_\alpha^2 = \sum_{k=1}^{\infty} \lambda_k^\alpha y_k^2\}$$

where e_k are the eigen-vectors and $0 < \lambda_1, ..., < \lambda_k < ...$ are the eigen-values of the operator A. In the case where A is the operator (2.6) the equality $H^0 = H$ is true where H is the space (2.5)

Let $\tilde{B}(y,z)$ be the bilinear operator such that $B(y) = \tilde{B}(y,y)$. We suppose that

$$||\tilde{B}(y,v)||_{-\alpha_3} \leq c||y||_{\alpha_1}||v||_{\alpha_2+1} \tag{2.9}$$

where $\alpha_j \geq 0$, $\alpha_1 + \alpha_2 + \alpha_3 > n/2$, and the number n corresponds in the case of the Navier-Stokes equations to the space dimension. For the proof of (2.2) in the case when $\tilde{B}$ is the operator (2.7) see e.g. (Fursikov (1983). We assume also that

$$(\tilde{B}(u,u),u)_{H^0} = 0$$

(Operator $\tilde{B}$ from (2.7) satisfies this property.)

Suppose that for any $T > 0$ the following inclusion takes place:

$$f(t) \in L_2(0,T;H^0)$$

and $f(t)$ is a periodic function with period T_0:

$$f(t+T_0) = f(t) \qquad \forall t \in R \tag{2.10}$$

We set the initial condition

$$y\,|_{t=t_0} = y_0 \tag{2.11}$$

for equation (2.8).

Below and till Section 5 we suppose that (2.8) is the abstract version of the 2-D Navier-Stokes system defined in a bounded domain $\Omega \subset R^2$ and supplied with zero boundary conditions.

In this case, as it is well-known (see e.g. (Ladyzhenskaya (1969), (Temam (1979)), for an arbitrary initial value $y_0 \in H^0$ there exists the unique solution

$$y(t) \in Y(t_0, T) \equiv \{y(t) \in L_2(0,T;H^1) : \partial_t y \in L_2(0,T;H^{-1})\} \tag{2.12}$$

where $T > t_0$ is arbitrary. We denote by $S(t,t_0)$ the resolving operator of the system (2.8):

$$S(t,t_0)y_0 = y(t), \qquad \text{where} \quad y(t) \quad \text{satisfies (2.8),(2.11)} \tag{2.13}$$

LEMMA 2.1. *The solution $y(t)$ of the problem (2.8),(2.11) with a periodic right-hand-side $f(t)$ satisfies the inequalities:*

$$\|y(t)\|_0^2 \le c\left(e^{-\lambda_1(t-t_0)}\|y_0\|_0^2 + \int_0^{T_0} \|f(t)\|_0^2\,dt\right) \tag{2.14}$$

$$T^{-1}\int_0^T \|y(t)\|_1^2\,dt \le c\left(T^{-1}\|y_0\|_0^2 + \int_0^{T_0} \|f(t)\|_0^2\,dt\right) \tag{2.15}$$

where T_0 is the period of the function $f(t)$ (see.(2.10)), λ_1 is the first eigen value of the operator A, $c > 0$ is a constant which does not depend on y_0, f, t T. Besides,

$$T^{-1}\int_0^T \|\partial_t y(t)\|_{-1}^2\,dt \le$$

$$\le c_1\left(1 + c\|y_0\|_0^2 + \int_0^{T_0} \|f(t)\|_0^2\,dt\right)\left(T^{-1}\|y_0\|_0^2 + \int_0^{T_0} \|f\|_0^2\,dt\right)^2 \tag{2.16}$$

where constants c and c_1 do not depend on y_0, f, T.

Lemma 2.1 is a simple corollary of the energy inequality for equation (2.8), of the Gronwall Lemma, and of the function f's periodicity with respect to t.

Lemma 2.1 implies

LEMMA 2.2. *Let $R_1 > 0$ be arbitrary and*

$$y_0 \in B^0_{R_1} \equiv \{u \in H^0, \qquad \|u_0\|_0 \le R_1\} \quad . \tag{2.17}$$

We denote

$$R^2 = 2c \int_0^{T_0} \|f(t)\|_0^2 \, dt \tag{2.18}$$

where c is the constant from (2.14). Then the solution $y(t)$ of problem (2.8),(2.11) satisfies the inclusion

$$y(t) \in B^0_R \qquad for \quad t > \begin{cases} t_0, & when \quad 2R_1^2 < R^2 \\ t_0 + \lambda_1^{-1} \ln(2R_1^2/R^2), & when \quad 2R_1^2 \ge R^2 \end{cases}$$

2.3 Definition of a statistical solution

We define by $\mathcal{B}(X)$ the Borel σ-algebra of sets of Banach space X. Let $\mu_0(\omega_0)$, $\omega_0 \in \mathcal{B}(H^0)$ be an probability measure which determines the probability distribution of initial values (2.11) (at $t = t_0$). Operator (2.13) with t running the segment (t_0, T) , will be denoted as follows: $S_T(t_0) \equiv \{S(t, t_0), \quad t \in (t_0, T)\}$. So, the operator

$$S_T(t_0) : H^0 \longrightarrow Y(t_0, T) \tag{2.19}$$

is continuous.

DEFINITION 2.1. A measure $P(\omega) \equiv P_{(t_0,T)}(\omega)$ is called a space-time statistical solution of equation (2.8) with initial condition μ_0, if

$$P_{(t_0,T)}(\omega) = \mu_0(S_T(t_0)^{-1}\omega), \qquad \forall \omega \in \mathcal{B}(Y(t_0, T)) \tag{2.20}$$

where $S_T(t_0)^{-1}$is the whole preimage of the map (2.19).

We consider a closed set in the space $Y(t_0, T)$ which consists of solutions of equation (2.8):

$$N(t_0, T) = \{y(t) \in Y(t_0, T) : y(t) = S(t, t_0)y_0, \quad \text{where} \quad y_0 \in H^0\} \tag{2.21}$$

It is evident that $S_T(t_0)^{-1}$ is a single-valued operator which is defined on the set $N(t_0, T)$ and for any $y = y(t) \in N(t_0, T)$ $\;S_T(t_0)^{-1}y = y(t_0) = \gamma_{t_0}y$, where $\gamma_{t_0} : Y(t_0, T) \to H^0$ is the operator that restricts a function $y(t) \in Y(t_0, T_0)$ at $t = t_0 : \gamma_{t_0}y = y(t_0)$. Therefore, Definition 2.1 implies the following assertion: PROPOSITION 2.1. *a) A statistical solution $P_{(t_0,T)}$ is concentrated on individual solutions of equation (2.8):*

$$suppP_{(t_0,T)} \subset N(t_0, T)$$

The measures $P_{(t_0,T)}$ and μ_0 are connected by the following relations which are equivalent to (2.20):

$$\gamma_{t_0}^* P(\omega_0) = P(\gamma_0^{-1}\omega_0) = P(y \in N(t_0,T) : \gamma_{t_0} y \in \omega_0) = \mu(\omega_0)$$

$$P(\omega) = P(\omega \cap N_{(t_0,T)}) = \mu_0(\gamma_{t_0}(\omega \cap N_{(t_0,T)})), \quad \forall \omega \in \mathcal{B}(y(t_0,T)) \quad (2.22)$$

In virtue of (2.15),(2.16) the operator (2.19) satisfies the following estimate

$$(T-t_0)^{-1} \|S_T(t_0)y_0\|^2_{Y(t_0,T)} \le c\left(\|y_0\|_0^4 + \|f\|^4_{L_2(0,T_0;H^0)} + 1\right) \quad (2.23)$$

where c does not depend on $\|y_0\|_0, (T-t_0)$ and f. Relations (2.22),(2.23) imply the estimate:

$$(T-t_0)^{-1} \int \|y\|^2_{Y(t_0,T)} P_{(t_0,T)}(dy) \le c(\int \|y_0\|_0^4 \mu_0(dy_0) + \|f\|^4_{L_2(0,T-0,H^0)} + 1) \quad (2.24)$$

where c does not depend on $(T-t_0), \mu_0, f$.

2.4 The case of the semiaxis

We define by $Y(t_0,\infty)$ the space of functions $y(t), t \in (t_0,\infty)$ whose values belong to H^1 such that the seminorms

$$\frac{1}{(T-t_0)} \|y\|_{Y(t_0,T)} \qquad \forall T > t_0$$

are finite.

We define the set $N(t_0,\infty) \subset Y(t_0,\infty)$ with the help of the formula:

$$N(t_0,\infty) = \{y(t) \in Y(t_0,\infty), \quad y(t) = S(t,t_0)y_0 \quad \forall t > t_0, \text{ where } \; y_0 \in H^0\} \quad (2.25)$$

Let $t_0 \le t_1 < T < \infty$ and $\chi_{(t_1,T)}$ be the operator of restriction on to the segment $[t_1,T]$ of a function $y \in Y(t_0,\infty)$. We denote by $\chi^*_{(t_1,T)}$ the operator which is defined on the set of measures $P(\omega), \quad \omega \in Y(t_1,\infty)$, and is generated by the operator $\chi_{(t_1,T)}$. By the definition of the operator $\chi^*_{(t_1,T)}$, the following formula is true:

$$(\chi^*_{(t_1,T)}P)(\omega) = P(\chi^{-1}_{(t_1,T)}\omega) \qquad \forall \omega \in \mathcal{B}(Y(t_1,T)) \quad (2.26)$$

A statistical solution $P_{(t_0,\infty)}(\omega), \quad \omega \subset \mathcal{B}(Y(t_0,\infty))$ of equation (2.8) is defined with the help of the formula analogous to (2.20):

$$P_{(t_0,\infty)}(\omega) = \mu_0(S_\infty(t_0)^{-1}\omega) \qquad \forall \omega \in \mathcal{B}(Y(t_0,\infty)) \tag{2.27}$$

where $S_\infty(t_0) : H^0 \longrightarrow Y(t_0,\infty)$ is the operator generated by operator (2.13), and $S_\infty^{-1}(t_0) : N(t_0,\infty) \longrightarrow H^0$ is the operator inverse of $S_\infty(t_0)$. It is evident that

$$\chi^*_{(t_0,T)} P_{(t_0,\infty)} = P_{(t_0,T)}$$

and therefore in virtue of (2.24) the inequality

$$\frac{1}{(T-t_0)} \int \|y\|^2_{Y(t_0,T)} P_{(t_0,\infty)}(dy) \leq \frac{1}{(T-t_0)} \int \|y\|^2_{Y(t_0,T)} \chi^*_{(t_0,T)} P_{(t_0,\infty)}(dy) \leq$$

$$\leq c\Big(\int \|y_0\|_0^4 \mu_0(dy_0) + \|f\|^4_{L_2(0,T,H^0)} + 1\Big) \tag{2.28}$$

is true where c does not depend on $T - t_0$.

3 Construction of a time-periodic space-time statistical solution

3.1 Operators of time translation

Let us define the operator of time translation:

$$\hat{\tau}\, u(t) = u(t+\tau).$$

It acts continuously on the spaces:

$$\hat{\tau} : Y(t_0,\infty) \longrightarrow Y(t_0 - \tau, \infty)$$

We set for $\tau > 0$:

$$\tilde{\tau} = \chi_{(t_0,\infty)}\tau : Y(t_0,\infty) \longrightarrow Y(t_0,\infty) \tag{3.1}$$

The operator $\tilde{\tau}$ generates the transformation of measures $\tilde{\tau}^*$ in the usual way:

$$\tilde{\tau}^* P(\omega) = P(\tilde{\tau}^{-1}\omega) \qquad \forall \omega \subset \mathcal{B}(Y(t_0,\infty)) \tag{3.2}$$

and both measures P and $\tilde{\tau}^* P$ are defined on the same σ -algebra $\mathcal{B}(Y(t_0,\infty))$.

Let $P_{(0,\infty)} = P$ be a statistical solution, i.e. by definition we have: $P(\omega) = P(\omega \cap N(0,\infty))$. We show that if $\tau = jT_0$, where T_0 is the period of the right side f from (2.8), then $(\widetilde{jT_0})^* P(\omega)$ is also a statistical solution which is defined on $Y(0,\infty)$. Indeed, the following equality holds:

$$(\widetilde{jT_0}) = \chi_{(0,\infty)}(\widehat{jT_0}) = (\widehat{jT_0})\chi_{(jT_0,\infty)} \tag{3.3}$$

and therefore we have for $\omega \in \mathcal{B}(Y(0,\infty))$ that

$$\widetilde{(jT_0)}^{-1}\omega = (\widehat{(jT_0)}\chi_{(jT_0,\infty)})^{-1}\omega = \chi^{-1}_{(jT_0,\infty)}\widehat{(-jT_0)}\omega =$$

$$= \{y(t) \in Y(0,\infty) : \chi_{(jT_0,\infty)}y \in \widehat{(-jT_0)}\omega\} \tag{3.4}$$

Hence, we have:

$$\widetilde{(jT_0)}^{-1}\omega \cap N(0,\infty) = \{y \in N(0,\infty) : \chi_{(jT_0,\infty)}y \in \widehat{(-jT_0)}\omega\} =$$

$$= \widetilde{(jT_0)}^{-1}(\omega \cap N(0,\infty)) \cap N(0,\infty)$$

This equality and (3.2) imply:

$$\widetilde{(jT_0)}^* P(\omega) = P(\widetilde{(jT_0)}^{-1}(\omega \cap N(0,\infty)) = \widetilde{(jT_0)}^* P(\omega \cap N(0,\infty))$$

i.e. the measure $\widetilde{(jT_0)}^* P$ is a statistical solution defined on $Y(0,\infty)$.

3.2 Time-periodic statistical solutions

Let $P = P_{(0,\infty)}$ be a statistical solution defined on $Y(0,\infty)$. As it was indicated above, the measure

$$P_k(\omega) = \frac{1}{T_0 k}\sum_{j=0}^{k} \widetilde{(jT_0)}^* P(\omega) \qquad \forall \omega \cap B(Y(0,\infty)) \tag{3.5}$$

is a statistical solution.

We define formally:

$$\hat{P} = \lim_{k\to\infty} P_k = \lim_{k\to\infty} \frac{1}{T_0 k}\sum_{j=0}^{k} \widetilde{(jT_0)}^* P(\omega) \tag{3.6}$$

THEOREM 3.1. *There exists a sequence of integers $k' \longrightarrow \infty$, such that the weak limit (3.6) exists. The measure $\hat{P}$ defined in (3.6), is concentrated on $Y(0,\infty)$ and is a statistical solution which is periodic with respect to t with period T_0, i.e.*

$$\hat{P} = \tilde{T_0}^* \hat{P} \tag{3.7}$$

This measure $\hat{P}$ satisfy estimates:

$$\int \frac{1}{T}\|y\|^2_{L_2(0,T;H^1)}\hat{P}(dy) \le c\|f\|^2_{L_2(0,T_0;H^0)} \qquad \forall T > 0 \tag{3.8}$$

$$\int \frac{1}{T}\|y\|^2_{Y(0,T)}\hat{P}(dy) \le c(\int \|y_0\|^4\mu_0(dy_0) + \|f\|^4_{L_2(0,T;H^0)}) \tag{3.9}$$

where c does not depend on T.

PROOF. Let $0 < T_1 < kT_0$ and k_1 be the minimum of integers such that $T_1 \leq k_1 T_0$. It is clear that $k_1 T_0 \leq kT_0$. Relation (3.5) and upper bound (2.28) with $t_0 = 0$ yield

$$\frac{1}{T_1}\int ||y||^2_{Y(0,T_1)} P_k(dy) \leq \frac{c}{k_1 T_0}\int ||y||^2_{Y(0,kT_0)} P_k(dy) =$$

$$= \frac{c}{(k_1T_0)(kT_0)}\sum_{j=0}^{k}\int ||y||^2_{Y(0,k_1T_0)} (j\tilde{T}_0)^* P(\omega) =$$

$$= \frac{c}{(k_1T_0)(kT_0)}\sum_{j=0}^{k}\int ||y||^2_{Y(jT_0,(j+k_1)T_0)} P(dy) =$$

$$= \frac{c}{(k_1T_0)(kT_0)}\sum_{l=0}^{k_1-1}\sum_{\substack{j\in[o,k]\\ j\equiv l (mod k_1)}}\int ||y||^2_{Y(jT_0,(j+k_1)T_0)} P(dy) \leq$$

$$\leq \frac{c}{T_0(kT_0)}\int ||y||^2_{Y(0,(k+k_1)T_0)} P(dy) \leq$$

$$\leq c_1\Big(\int ||y_0||^4 \mu_0(dy_0) + ||f||^4_{L_2(0,T_0,H^o)}\Big) \tag{3.10}$$

It follows from (3.10) that

$$\frac{1}{T_1}\int ||y||^2_{Y(0,T_1)} \chi^*_{(0,T_1)} P_k(dy) = \frac{1}{T_1}\int ||y||^2_{Y(0,T_1)} P_k(dy) \leq$$

$$\leq c\Big(\int ||y_0||^4 \mu_0(dy_0) + ||f||^4_{L_2(0,T_0;H^o)}\Big) \tag{3.11}$$

One can derive from (3.11) analogous to (Vishik and Fursikov (1988), Chapter 4 by means of Prokhorov Theorem, that for a certain subsequence $k_1 \subset k$ there exists a weak limit

$$\hat{P}_{(T_1)} = \lim_{k_1\to\infty} \chi^*_{(0,T_1)} P_k \tag{3.12}$$

defined on $L_2(0,T_1;H^0)$. As in (Vishik and Fursikov (1988), Chapter 4 one can establish the following estimate:

$$\frac{1}{T_1}\int ||y||^2_{Y(0,T_1)} \hat{P}_{(T_1)}(dy) \leq \lim_{k_1\to\infty}\frac{1}{T_1}\int ||y||^2_{Y(0,T_1)} \chi^*_{(0,T_1)} P_{k_1}(dy) \leq$$

$$\leq c\Big(\int ||y_0||^4 \mu_0(dy_0) + ||f||^4_{L_2(0,T_0;H^o)}\Big) \tag{3.13}$$

Together with the operator $\chi_{(0,T_1)}$ we consider the analogous operators $\chi_{(0,lT_1)}, l \in N$ and together with (3.12) we define for any $l \in N$ the measure

$$\hat{P}_{(lT_1)} = \lim_{k_l \to \infty} \chi^*_{(0,lT_1)} P_{k_l} \tag{3.14}$$

determined on the space $L_2(0, lT_1; H^0)$ where $\{k_l\} \subset \{k_{l-1}\} \subset ..... \subset \{k_1\}$, i.e. $\{k_l\}$ is a subsequence of the sequence $\{k_{l-1}\}$ for any l. Applying to (3.14) the diagonal process we get that

$$\hat{P}_{(lT_1)} = \lim_{k' \to \infty} \chi^*_{(0,lT_1)} P_{k'} \qquad \forall l \in N \tag{3.15}$$

Moreover, analogous to (3.13) we obtain:

$$\frac{1}{lT_1} \int \|y\|^2_{Y(0,lT_1)} \hat{P}_{(lT_1)}(dy) \leq c(\|f\|^4_{L_2(0,T_0;H^0)} + \int \|y_0\|^4 \mu_0(dy_0)) \tag{3.16}$$

Definition (3.15) implies the equalities:

$$\chi^*_{(0,l_2T_1)} P_{(l_1T_1)} = P_{(l_2T_1)} \qquad \forall l_1 > l_2 \quad . \tag{3.17}$$

On the space $Y(0, \infty)$ we consider the σ-algebra $\hat{\mathcal{B}}(Y(0, lT_1))$ of cylindrical sets possessing the form $\chi^{-1}_{(0,lT_1)}\omega$, where $\omega \in \mathcal{B}(Y(0, lT_1))$. Let $\mathcal{B} = \cup_{l=1}^{\infty} \hat{\mathcal{B}}(Y(0, lT_1))$. In virtue of (3.17) there exists a finite-additive function of sets $\hat{P}$, which is defined on $\mathcal{B}$, such that

$$\chi^*_{(0,lT_1)} \hat{P} = \hat{P}_{(lT_1)} \qquad \forall l \in N \tag{3.18}$$

Relations (3.18) and (3.16) imply that

$$\forall l \in N \quad \frac{1}{lT_1} \int \|y\|^2_{Y(0,lT_1)} \hat{P}(dy) \leq c(\|f\|^4_{L_2(0,T;H^0)} + \int \|y_0\|^4 \mu_0(dy_0)) \tag{3.19}$$

and, hence, the sets function $\hat{P}$ can be extended up to a σ-additive measure defined on the space $Y(0, \infty)$. Inequality (3.9) follows from (3.19) . Since for any k, $suppP_k \subset N(0, \infty)$, then by virtue of (3.15) one can establish as in (Vishik and Fursikov (1988), Chapter 4 that $supp\hat{P}_{(lT_1)} \subset N_{(0,lT_1)}$ and therefore $\hat{P}_{(lT_1)}$ is a statistical solution. Relation (3.18) implies that the measure $\hat{P}$ is concentrated on the space $N(0, \infty)$ and, hence, is a statistical solution.

We prove now relation (3.7). Equalities (3.15),(3.18) imply that

$$\int f(y) \hat{P}(dy) = \lim_{k' \to \infty} \int f(y) (\frac{1}{k'T_0} \sum_{j=0}^{k'} \widetilde{(jT_0)}^* P(dy)) \tag{3.20}$$

and this equality is true for an arbitrary functional $f(y)$ which is continuous and bounded on the space $Y(0, \infty)$ and satisfies the equality:

$$\exists l \in N \qquad f(y_1) = f(y_2) \quad \text{for} \quad \chi_{(0,lT_1)}y_1 = \chi_{(0,lT_1)}y_2 \tag{3.21}$$

The following equalities are true:

$$\int f(y)\tilde{T_0}^* \hat{P}(dy) = \lim_{k'\to\infty} \frac{1}{k'T_0} \sum_{j=0}^{k'} \int f(\tilde{T_0}y)(\widetilde{jT_0})^* P(dy) =$$

$$= \lim_{k'\to\infty} \frac{1}{k'T_0} \sum_{j=0}^{k'} \int f((\widetilde{j+1})T_0)y)P(dy) = \lim_{k'\to\infty} \int f(y) \frac{1}{k'T_0} \sum_{j=0}^{k} (\widetilde{jT_0})^* P(dy) +$$

$$+ \lim_{k'\to\infty} \frac{1}{k'T_0} \sum_{j=0}^{k'} \int [f(((\widetilde{j+1})T_0)y) - f((\widetilde{jT_0})y)]P(dy) =$$

$$= \int f(y)\hat{P}(dy) + \lim_{k'\to\infty} \frac{1}{k'T_0} \int (f(((\widetilde{k'+1})T_0)y) - f(y))P(dy) =$$

$$= \int f(y)\hat{P}(dy) \tag{3.22}$$

because

$$| \frac{1}{k'T_0} \int (f(((\widetilde{k'+1})T_0)y) - f(y))\, P(y)| \leq$$

$$\leq \frac{1}{k'T_0} 2max(|\, f(y)\, |, \, y \in Y(0,\infty)) \to 0 \qquad \text{as} \quad k' \to \infty$$

Relation (3.22) implies (3.7). We repeat estimate (3.10) where the expression $\frac{1}{T_1}||y||^2_{Y(0,T_1)}$ is changed to $\frac{1}{T_1}||y||^2_{L_2(0,T_1;H^1)}$, and taking (2.15) into account. Then we get that for $T_1 \leq k_1T_0 < kT_0$ the following relations are true

$$\frac{1}{T_1} \int ||y||^2_{L_2(0,T_1;H^1)} P_k(dy) \leq \frac{c}{T_0(kT_0)} \int ||y||^2_{L_2(0,(k+k_1)T_0;H^1)} P_k(dy) \leq$$

$$\leq c_1 (\frac{1}{(k+k_1)T_0} \int ||y_0||^4 \mu_0(dy_0) + ||f||^4_{L_2(0,T_0;H^0)}) \tag{3.23}$$

where c_1 does not depend on T_1 and k. If we pass to limit in (3.23) as $k = k' \longrightarrow \infty$ then we get (3.8). □

3.3 The case of the whole axis

As in the case of the semiaxis we define the space $Y(-\infty,\infty)$ as the set of functions $y(t)$, $t \in R$ with values in the space H^1, for which the seminorms

$$\frac{1}{t_2-t_1}\|y\|_{Y(t_1,t_2)} \qquad \forall -\infty < t_1 < t_2 < \infty$$

are finite. We set

$$N(-\infty,\infty) = \{y(t) \in Y(-\infty,\infty): \quad y(t) \quad \text{satisfies (2.8)}\} \ .$$

Using a time periodic space-time statistical solution defined on the space $Y(0,\infty)$, that was constructed in Theorem 3.1, we construct periodic space-time statistical solution defined on the space $Y(-\infty,\infty)$. For integer $k > 0$ we define the σ-algebra $\mathcal{B}_k$ of sets belonging to the space $Y(-\infty,\infty)$ as follows:

$$\mathcal{B}_k = \{\{y \in Y(-\infty,\infty): \chi_{(-kT_0,\infty)}y \in \omega\}, \ \ \omega \in \mathcal{B}(Y(-kT_0,\infty))\} \ . \quad (3.24)$$

On the σ-algebra $\mathcal{B}_k$ we define the measure $P^{(k)}$ by the formula:

$$P^{(k)}(\Omega) = \hat{P}(\chi_{(0,\infty)}\widehat{(-kT_0)}\Omega) \qquad \forall\, \Omega \subset \mathcal{B}_k \quad (3.25)$$

Evidently, (3.24) implies the equality:

$$\mathcal{B}_k \subset \mathcal{B}_l \quad \text{for} \quad k < l \ . \quad (3.26)$$

Let us show that

$$P^{(l)}(\Omega) = P^{(k)}(\Omega) \qquad \forall\, \Omega \subset \mathcal{B}_k \quad \text{for} \quad l > k \quad (3.27)$$

Indeed, the definition yields the equality:

$$\widehat{(-kT_0)}\Omega = \chi^{-1}_{(0,\infty)}\widehat{(-kT_0)}\Omega \qquad \forall\, \Omega \subset \mathcal{B}^k \ . \quad (3.28)$$

Therefore using definition (3.25), formula (3.3) and periodicity property (3.7) for the measure $\hat{P}$ we get the following chain of equalities for $\Omega \subset \mathcal{B}_k$ and $l > k$:

$$P^{(l)}(\Omega) = \hat{P}(\chi_{(0,\infty)}\widehat{(-lT_0)}\Omega) = \hat{P}(\chi_{(0,\infty)}\widehat{(-(l-k)T_0)}\widehat{(-kT_0)}\Omega) =$$

$$= \hat{P}(\chi_{(0,\infty)}\widehat{(-(l-k)T_0)}\chi^{-1}_{(0,\infty)}\widehat{(-kT_0)}\Omega) =$$

$$= \hat{P}(\chi_{(0,\infty)}\chi^{-1}_{((l-k)T_0,\infty)}\widehat{(-(l-k)T_0)}\widehat{(-kT_0)}\Omega) =$$

$$= \hat{P}((l-\widetilde{k})T_0)^{-1}\chi_{(0,\infty)}(-kT_0)\Omega) = \hat{P}(\chi_{(0,\infty)}(\widehat{-kT_0})\Omega) = P^{(k)}(\Omega) \quad . \tag{3.29}$$

Equalities (3.29) prove assertion (3.27).

On the algebra of sets $\mathcal{B} = \cup_{k=0}^{\infty}\mathcal{B}_k$ we define the function $\tilde{P}$ of sets by means of the formula:

$$\tilde{P}(\Omega) = P^{(k)}(\Omega) \qquad \forall \Omega \in \mathcal{B} \ \text{ if } \ \Omega \in \mathcal{B}_k \quad . \tag{3.30}$$

The definition (3.30) is correct in virtue of (3.27). Relations (3.27),(3.25),(3.9) imply the inequality:

$$\int \frac{1}{(t_2-t_1)}||y||^2_{Y(t_1,t_2)}\tilde{P}(dy) \leq c(\int ||y_0||_0^4\mu_0(dy_0) + ||f||^4_{L_2(0,T_0;H^0)}) \tag{3.31}$$

where $-\infty < t_1 < t_2 < \infty$, and c does not depend on t_1, t_2. Inequality (3.31) implies that $\tilde{P}$ can be extended up to the σ-additive measure defined on $\mathcal{B}(Y(-\infty,\infty))$.

We remark the following properties of the measure $\tilde{P}$.

THEOREM 3.1.) *The measure $\tilde{P}$ is periodic with the period T_0, i.e.*

$$\tilde{P}(\omega) = \hat{T_0}^*\tilde{P}(\omega) = \tilde{P}(\hat{T_0}^{-1}\omega) \qquad \forall \omega \in \mathcal{B}(Y(-\infty,\infty)) \quad . \tag{3.32}$$

b) The measure $\tilde{P}$ is concentrated on the set

$$N(-\infty,\infty)\bigcap \Phi_R, \quad \textit{where} \quad \Phi_R \equiv$$
$$\equiv \{y(t) \in Y(-\infty,\infty) : ||y(t)||_0 \leq R \qquad \forall t \in R\} \tag{3.33}$$

and the magnitude R is determined in Lemma 2.2 and depends on the right side f of equation (2.8) only.

PROOF. a) It is sufficient to prove relation (3.32) for a set $\omega \in \mathcal{B}$. So, let $\omega \in \mathcal{B}_k$. Then, the inclusion $\hat{T_0}^{-1}\omega \in \mathcal{B}_{k-1}$ holds and in accordance with (3.30), (3.25) we have

$$\tilde{P}(T_0^{-1}\omega) = P^{(k-1)}(T_0^{-1}\omega) = \hat{P}(\chi_{(0,\infty)}(\widehat{-kT_0})\omega) = P^{(k)}(\omega) = \tilde{P}(\omega)$$

This proves (3.32).

b) First of all we show that $supp\tilde{P} \subset N(-\infty,\infty)$. Let $\omega \in \mathcal{B}_k$. Then for $l > k$ in virtue of (3.30),(3.27),(3.25) and by the inclusion $supp\hat{P} \subset N(0,\infty)$ the following relations are true:

$$\tilde{P}(\omega) = P^{(l)}(\omega) = \hat{P}(\chi_{(0,\infty)}(\widehat{-lT_0})\omega) = \hat{P}((\chi_{(0,\infty)}(\widehat{-lT_0})\omega) \cap N(0,\infty)) =$$
$$= \hat{P}(\chi_{(0,\infty)}(\widehat{-lT_0})(\omega \cap N(-lT_0,\infty)) =$$

$$= \tilde{P}(\omega \cap N(-lT_0,\infty)) \longrightarrow \tilde{P}(\omega \cap N(-\infty,\infty)) \quad \text{as} \quad l \to \infty \tag{3.34}$$

and passing to the limit in (3.34) is possible because the measure $\tilde{P}$ is σ-additive on $\mathcal{B}(Y(-\infty,\infty))$.

We prove that

$$supp\,\tilde{P} \subset \Phi_R \tag{3.35}$$

where Φ_R is defined in (3.33). Let $\mu_0(dy_0)$ be the initial measure of the statistical solution $P = P_{(0,\infty)}$, that was used in the construction of the measure $\tilde{P}$. Suppose that the support of the $\mu_0(dy_0)$ satisfies the inclusion $\text{supp}\mu_0 \subset B^0_{R_1} = \{\|y_0\| \le R_1\}$. We set

$$\Phi_R(t_1,t_2) = \hat{\chi}_{(t_1,t_2)}\Phi_R, \qquad -\infty \le t_1 < t_2 \le \infty \tag{3.36}$$

where Φ_R is the set which was determined in (3.33) . The definition (2.27) with $t_0 = 0$ of the statistical solution P and Lemma 2.2 imply that

$$supp(\widetilde{jT_0})^* P \subset \Phi_R(0,\infty) \quad \text{when} \quad jT_0 > \max(0, \frac{1}{\lambda_1}\ln\frac{2R_1^2}{R^2}) \quad . \tag{3.37}$$

Denote

$$\Theta(y) = \begin{cases} 1, \; y \notin \Phi_R(0,\infty) \\ 0, \; y \in \Phi_R(0,\infty) \end{cases},$$

and let k_0 be the first natural number for which the inequality $k_0T_0 > \max(0, \frac{1}{\lambda_1}\ln\frac{2R_1^2}{R^2})$ holds. Then in virtue of (3.6),(3.37) for any $0 < T_1 < kT_0$ the following relations are true:

$$\mid \int \Theta(y)P_k(dy) \mid = \mid \frac{1}{T_0k}\sum_{j=0}^{k_0} \int \Theta(y)(\widetilde{jT_0})^* P(dy) \mid \le \frac{k_0}{T_0k} \quad .$$

Passing to limit in this inequality as $k \longrightarrow \infty$ we get that

$$\int \Theta(y)\hat{P}(dy) = 0 \quad .$$

This equality implies that

$$supp\hat{P} \subset \Phi_R(0,\infty) \quad . \tag{3.38}$$

We define

$$\tilde{\Phi}_R(-kT_0,\infty) = \{y \in Y(-\infty,\infty) : \chi_{(-kT_0,\infty)}\, y \in \Phi_R(-kT_o,\infty)\} \quad .$$

Then it follows from (3.30),(3.38),(3.25) that for an arbitrary k,

$$\tilde{P}(\tilde{\Phi}_R(-kT_0,\infty)) = P^{(k)}(\tilde{\Phi}_R(-kT_0,\infty)) = \hat{P}(\Phi_R(0,\infty)) = 1 \quad . \tag{3.39}$$

Since

$$\Phi_R = \cap_{k=1}^{\infty} \quad \tilde{\Phi}_R(-kT_0,\infty)$$

then σ-additiveness of the measure $\tilde{P}$ and (3.39) yield (3.35). Relations (3.34),(3.35) imply (3.33). □

3.4 Space statistical solution

Let γ_τ be the operator of evaluation of the function $y(t)$ at $t = \tau : \gamma_\tau y = y(\tau)$. Let $\tilde{P}$ be the space-time statistical solution periodic with respect to t which was constructed above. We define the space statistical solution $\mu(t, \omega)$ by means of the formula:

$$\mu(t,\omega) = \gamma_t^* \tilde{P}(\omega) = \tilde{P}(\gamma_t^{-1}\omega) \qquad \forall \omega \in \mathcal{B}(H^0) \ . \tag{3.40}$$

PROPOSITION 3.1*a) The family of measures $\mu(t,\omega)$ is t-periodic with the period T_0:*

$$\mu(t,\omega) = \mu(t + T_0, \omega) \qquad \forall \omega \in \mathcal{B}(H^0) \tag{3.41}$$

For any $t \in R$ the support

$$K(t) = supp\mu(t, \bullet) \subset H^0$$

of the measure $\mu(t,\cdot)$ is concentrated in the ball $B_R^0 \equiv \{z \in H^0 : ||z||_0 \leq R\}$, where the magnitude R is determined in Lemma 2.2. The family of the sets $K(t)$ is t-periodic with the period T_0:

$$K(t + T_0) = K(t) \qquad \forall t \in R \ . \tag{3.42}$$

PROOF. Relation (3.41) follows directly from definition (3.40) and periodicity property (3.32) of the measure $\tilde{P}$. The assertion

$$supp\mu(t, \bullet) \subset B_R^0$$

follows from (3.33). Relation (3.40) implies (3.42).□

4 Statistical solution for time-periodic stochastic right hand side

We suppose now that the right-hand-side $f(t)$ in (2.8) is stochastic. This means that the distribution of probabilities of this stochastic function, i.e. the probability measure

$$F(\omega), \qquad \omega \in \mathcal{B}(L_{2,loc}(R; H^0)) \tag{4.1}$$

is determined on the space

$$L_{2,loc}(R^1; H^0) =$$
$$= \{u(t) : R \to H^0 : \quad \forall -\infty < t_1 < t_2 < \infty \quad ||u||_{L_2(t_1,t_2,H^0)} < \infty\} \ .$$

We suppose that this measure is periodic with respect to t with the period T_0, i.e.

$$F(\omega)=\hat{T_0}^{*}F(\omega)=F((-\hat{T_0})\omega), \qquad \forall\omega\in\mathcal{B}(L_{2,loc}(R^1;H^0)) \quad . \tag{4.2}$$

Besides, for simplicity of presentation we impose the following boundedness condition on the support of the measure F :

$$supp\hat{\chi}^{*}_{(0,T_0)}F\subset B_{R_0}((L_2(0,T_0;H^0))\equiv$$

$$\equiv\{f(t)\in L_2(0,T_0;H^0):\|f\|_{L_2(0,T_0;H^0)}\le R_0\} \quad . \tag{4.3}$$

We assume also that at initial instant $t=0$ the distribution of probabilities $\mu_0(\omega),\quad \omega\in H^0$, of the stochastic function y_0 is given, and

$$supp\mu_0\subset\mathcal{B}^0_{R_1}=\{\|u\|_0\le R_1\}. \tag{4.4}$$

It is supposed that the stochastic functions f and y_0 are independent. By $S(t,T_0)(u_0,f)$ we denote the operator which acts initial condition y_0 and right-hand-side $f(\tau),\tau\in(t_0,t)$ to the solution $y(t)$ of problem (2.8),(2.11)

$$y(t)=S(t,t_0)(y_0,f) \quad . \tag{4.5}$$

Operator $S(t,t_0)$ acts in the spaces

$$S(t,t_0):H^0\times L_2(t_0,t,H^0)\longrightarrow H^0 \quad . \tag{4.6}$$

Besides, we will consider the operator $S_T(t_0)(u_0,f)=\{S(t,t_0)(u_0,f),\ t\in(t_0,T)\}$, which acts in the spaces

$$S_T(t_0):H^0\times L_2(t_0,T)\longrightarrow Y(t_0,T) \quad . \tag{4.7}$$

The measure $P=P_{(t_0,T)}$ which is defined by the equality

$$P_{(t_0,T)}(\omega)=(\mu_0\times F)(S_T(t_0)^{-1}\omega), \qquad \forall\omega\in\mathcal{B}(Y(t_0,T)) \quad , \tag{4.8}$$

where

$$S_T(t_0)^{-1}\omega=\{(y_0,f)\in H^0\times L_2(t_0,T;H^0):S_T(t_0)(y_0,f)\in\omega\}$$

is called space-time statistical solution of problem (2.8),(2.11) with stochastic y_0 and f. Relation (4.8) means that

$$\int g(y)P_{(t_0,T)}(dy)=\int g(S_T(t_0)(y_0,f))\mu_0(dy_0)F(df) \tag{4.9}$$

for an arbitrary functional g, for which one of the integrals in (4.9) is well-defined. Besides (4.7) we consider the resolving operator of problem (2.8),(2.11) $S_T(t_0),\ T=\infty$:

$$S_\infty(t_0):H^0\times L_{2,loc}(t_0,\infty;H^0)\longrightarrow Y(t_0,\infty) \quad . \tag{4.10}$$

Operator (4.10), and (4.5),(4.7) also satisfy the following estimates which are analogous to (2.14),(2.15) (here, we do not assume that $f(t)$ is periodic with respect to t):

$$||y(t)||_0^2 \le c(e^{-\lambda_1(t-t_0)}||y_0||_0^2 + \int_{t_0}^t e^{-\lambda_1(t-\tau)}||f(\tau)||_0^2 d\tau) \tag{4.11}$$

$$\frac{1}{T-t_0}\int_{t_0}^T ||y(t)||_1^2 dt \le \frac{c}{(T-t_0)}\left(||y_0||_0^2 + \int_{t_0}^T ||f(\tau)||_0^2 d\tau\right) \tag{4.12}$$

where $T_0 > 0$ (in a future T_0 will denote the period of the measure F), λ_1 is the first eigenvalue of the operator A, c is a constant which does not depend on y_0, f, t_0, t, T. Besides, the following inequality is true:

$$\frac{1}{(T-t_0)}\int_{t_0}^T ||\partial_t y(t)||_{-1}^2 dt \le c(1 + ||y_0||^4 +$$

$$+\left(\frac{1}{(T-t_0)}\int ||f||_0^2\, d\tau\right)^2 + \left(\int_{t_0}^T e^{-\lambda_1(t-\tau)}||f(\tau)||_0^2 d\tau\right)^2) \ . \tag{4.13}$$

The measure $P = P_{(0,\infty)}$ which is determined by the formula

$$P_{(0,\infty)}(\omega) = (\mu_0 \times F)(S_\infty(0)^{-1}\omega), \qquad \forall \omega \in \mathcal{B}(Y(0,\infty)) \tag{4.14}$$

is called a statistical solution of problem (2.8),(2.11) defined on the semiaxes $(0,\infty)$.

The analog of formula (4.9) is true for the measure $P_{(0,\infty)}$. Therefore, after integrating of (4.12),(4.13) with respect to the measure $\mu_0 \times F$, using (4.9), definition (2.12) of the space $Y(t_0,T)$, and periodicity condition (4.2) of the measure F , we get the inequality:

$$\int \frac{1}{T-t_0}||y||_{Y(t_0,T)}^2 P_{(0,\infty)}(dy) \le c(1 + \int ||y_0||^4 \mu_0(dy_0) +$$

$$+\int ||f||_{L_2(0,T_0;H^0)}^4 F(df)) \ . \tag{4.15}$$

Later, the statistical solution $P_{(0,\infty)}(\omega)$ will be denoted as $P(\omega)$. Recall that the operators $\chi^*_{(t_1,T)}, \tilde{\tau}, \tilde{\tau}^*$ which act on the measure P are defined by formulae (2.26),(3.1),(3.2).

LEMMA 4.1. *Let P be a statistical solution of problem (2.8),(2.11), T_0 be the period of the measure F, j be a natural number. Then the measure $\widetilde{(jT_0)}^* P$ is the statistical solution also with the same right-hand-side F and initial measure $\mu = \gamma^*_{jT_0}P$.*

The proof of Lemma 4.1 as well as the proof of the assertions formulated below is analogous to the proof of the corresponding assertions from §3. We do not write them here for brevity.

We denote as P_k and $\hat{P}$ the measures that are constructed by the measure $P = P_{(0,\infty)}$ from (4.14) with help of formulae (3.5),(3.6).

THEOREM 4.1. *There exists the sequence of integers $k' \longrightarrow \infty$ for which the weak limit (3.6) exists. The measure $\hat{P}$ defined in (3.6) is concentrated on the space $Y(0,\infty)$ and is the t-periodic statistical solution with the period T_0, i.e. it satisfies (3.7). Besides, the measure $\hat{P}$ satisfies the estimates:*

$$\int \frac{1}{T}\|y\|^2_{L_2(0,T;H^0)}\hat{P}(dy) \leq c\int \|f\|^2_{L_2(0,T_0;H^0)}F(df) \quad ,\forall T > 0 \tag{4.16}$$

$$\int \frac{1}{T}\|y\|^2_{Y(0,T)}\hat{P}(dy) \leq c\Big(\int \|y_0\|^4\mu_0(dy_0) + \int \|f\|^4_{L_2(0,T;H^0)}F(df)\Big) \tag{4.17}$$

where c does not depend on T.

We extend the measure $\hat{P}$ with help of formulae (3.25),(3.30) up to the measure $\tilde{P}$, which is defined on the space $Y(-\infty,\infty)$. The measure $\tilde{P}$ satisfies the inequality:

$$\int \frac{1}{(t_2-t_1)}\|y\|^2_{Y(t_1,t_2)}\tilde{P}(dy) \leq c\Big(\int \|y_0\|^4\mu_0(dy_0) + \int \|f\|^4_{L_2(0,T_0;H^0)}F(df)\Big)$$

This inequality allows to establish σ-additiveness of $\tilde{P}$ $\mathcal{B}(Y(-\infty,\infty))$.

THEOREM 4.2 *a) The measure $\tilde{P}$ is time-periodic with the period T_0, i.e. it satisfies equality (3.32). b) There exists a probability measure $\hat{\mu}(\omega_0)$, $\omega_0 \in \mathcal{B}(H^0)$ such that the following equality for the measure $\hat{P}$ is true:*

$$\hat{P}(\omega) = (\hat{\mu} \times F)(S_\infty(0)^{-1}\omega), \qquad \forall \omega \in \mathcal{B}(Y(0,\infty))$$

and therefore $\hat{P}$ is the statistical solution.

REMARK 4.1. The section b) of Theorem 4.2 explains in what sense $\tilde{P}$ is a statistical solution.

Using the measure $\tilde{P}$ and restriction operator γ_t^* we define by formula (3.40) the space statistical solution $\mu(t,\omega)$, $\omega \in \mathcal{B}(H^0)$.

PROPOSITION 4.1. *a) The family of the neasures $\mu(t,\omega)$ is t-periodic with the period T_0. b) For any $t \in R$ the following inclusion*

$$K(t) \equiv supp\mu(t,\cdot) \subset B^0_\gamma$$

is true when $\gamma < \infty$. Besides, the family of the sets $K(t)$ is t-periodic with period T_0.

5 The case of 3-D Navier-Stokes equations

In this section we construct time-periodic space-time and space statistical solutions of 3-D Navier-Stokes equations with a time-periodic deterministic right-hand-side.

5.1 Galerkin's approximations

The main difficulty of the 3-D case is connected with the fact that unique solvability of the Navier-Stokes equations is not established. Therefore the operator (2.13) is not defined and, hence, the construction of Sections 2,3 fails. That is why at the first stage instead of problem (2.8),(2.11) we consider its Galerkin's approximations.

Let E_k be the subspace of H^0 that is generated by the first k eigenvectors $\{e_1, ..., e_k\}$ of the operator A and $Q_k : H^\alpha \to E_k$ be the operator of orthogonal projection. The Galerkin approximation of problem (2.8),(2.11) is as follows:

$$\partial_t y_k(t) + Ay_k + Q_k B(y_k) = Q_k f(t) \tag{5.1}$$

$$y_k(t) \mid_{t=t_0} = Q_k y_0 \tag{5.2}$$

where $y_k(t) \in E_k$ for any t. Relations (5.1),(5.2) is the Cauchy problem for a system of ordinary differential equations and therefore its solution operator is well-defined. We denote this operator by $S^k(t, t_0)$:

$$S^k(t, t_0)Q_k y_0 = y_k(t), \qquad \text{where} \quad y_k(t) \quad \text{satisfies (5.1),(5.2).}$$

Analogous to (2.12) we denote:

$$\mathcal{L}(t_0, T) \equiv \{y(t) \in L_2(t_0, T; H^1) \cap L_\infty(t_0, T; H^0) : \partial_t y \in L_2(t_0, T; H^{-2})\} \tag{5.3}$$

$$\mathcal{L}_k(t_0, T) \equiv \{y(t) \in L_2(t_0, T; E_k \cap H^1) \cap L_\infty(t_0, T; E_k \cap H^0) : \partial_t y \in L_2(t_0, T; E_k \cap H^{-2})\} \tag{5.4}$$

The norm of the space $\mathcal{L}_k(t_0, T)$ is by definition the same as the norm of the space $\mathcal{L}(t_0, T)$ defined in (5.3). The $\mathcal{L}(t_0, T))$-norm is defined in the usual way.

Note that if we substitute the solution $y_k(t)$ of (5.1), (5.2) into the left hand sides of the inequalities (2.14), (2.15), (2.16) instead of $y(t)$ then these estimates will remain valid. The assertion of Lemma 2.2 remains valid also for the solution $y_k(t)$ of problem (5.1),(5.2). We remark that in this case the constants c and R in relations (2.14),(2.15),(2.16),(2.18) do not depend on k.

We introduce the following analog of the operator $S_T(t_0)$ from Subsection 2.3:

$$S_T^k(t_0) \equiv \{S^k(t, t_0), \quad t \in (t_0, T)\}. \tag{5.5}$$

One can show that the operator

$$S_T^k(t_0) : E_k \cap H^0 \longrightarrow \mathcal{L}_k(t_0, T) \tag{5.6}$$

is continuous as well as operator (2.19) and its norm is bounded uniformly with respect to k.

Let us define by $\mathcal{L}_k(t_0, \infty)$ the space of functions $y_k(t), t \in (t_0, \infty)$ whose values belong to $E_k \cap H^1$, such that the seminorms

$$\frac{1}{(T - t_0)} \|y_k\|_{\mathcal{L}_k(t_0,T)} \quad , \qquad \forall T > t_0$$

are finite. Besides, we define the set $N_k(t_0, \infty) \subset \mathcal{L}_k(t_0, \infty)$ by means of the formula analogous to (2.25):

$$N_k(t_0, \infty) = \{y_k(t) \in \mathcal{L}_k(t_0, \infty), \quad y_k(t) = S^k(t, t_0) Q_k y_0 \quad \forall t > t_0, \\ \text{where} \quad y_0 \in H^0\} \tag{5.7}$$

Let an initial probability measure $\mu_0(\omega_0)$, $\omega_0 \in \mathcal{B}(H^0)$ be given.

A statistical solution $P^k_{(t_0,\infty)}(\omega)$, $\omega \in \mathcal{B}(\mathcal{L}_k(t_0, \infty))$ of equation (5.1) is defined with the help of the formula analogous to (2.27):

$$P^k_{(t_0,\infty)}(\omega) = \mu_0(Q_k^{-1} S^k_\infty(t_0)^{-1} \omega) \qquad \forall \omega \in \mathcal{B}(\mathcal{L}^k(t_0, \infty)) \tag{5.8}$$

where $S^k_\infty(t_0) : E_k \cap H^0 \to \mathcal{L}^k(t_0, \infty)$ is the operator (5.5),(5.6) with $T = \infty$, and $S^k_\infty(t_0)^{-1} : N_k(t_0, \infty) \to E_k \cap H^0$ is the operator inverse of $S^k_\infty(t_0)$.

Analogous to (2.28), the following estimate can be proved:

$$(T - t_0)^{-1} \int \|y\|^2_{\mathcal{L}(t_0,T)} P^k_{(t_0,\infty)}(dy) \leq c(\int \|y_0\|_0^4 \mu_0(dy_0) + \|f\|^4_{L_2(0,T,H^0)} + 1) \tag{5.9}$$

where c does not depend on $T - t_0$ and k.

Using the statistical solution (5.8) and the operator $\tilde{\tau}^*$ defined in the Subsection 3.1 we construct the time-periodic statistical solution of equation (5.1). To do it we apply the analog of formula (3.6):

$$\hat{P}^k = \lim_{m \to \infty} \frac{1}{T_0 m} \sum_{j=0}^{m} \left(\widetilde{jT_0}\right)^* P^k(\omega) \tag{5.10}$$

The following assertion can be proved just by the same way as Theorem 3.1

THEOREM 5.1. *There exists a sequence of integers $m' \to \infty$, such that weak limit (5.10) exists. The measure $\hat{P}^k$ defined in (5.10), is concentrated on $\mathcal{L}^k(0, \infty)$ and is a statistical solution which is periodic with respect to t with period T_0, i.e.*

$$\hat{P}^k = \tilde{T_0}^* \hat{P}^k \quad . \tag{5.11}$$

This measure $\hat{P}^k$ satisfies estimates:

$$\int \frac{1}{T} ||y||^2_{L_2(0,T;H^0)} \hat{P}^k(dy) \leq c||f||^2_{L_2(0,T_0;H^0)} \qquad \forall T > 0 \quad , \tag{5.12}$$

$$\int \frac{1}{T} ||y||^2_{\mathcal{L}(0,T)} \hat{P}^k(dy) \leq c(\int ||y_0||^4 \mu_0(dy_0) + ||f||^4_{L_2(0,T;H^0)}) \quad , \tag{5.13}$$

where c does not depend on T and k.

We define now the space $\mathcal{L}_k(-\infty, \infty)$, as the set of functions $y_k(t)$, $t \in R$ with values in the space $E_k \cap H^1$, for which the seminorms

$$\frac{1}{t_2 - t_1} ||y_k||_{\mathcal{L}_k(t_1,t_2)} \qquad \forall \; -\infty < t_1 < t_2 < \infty$$

are finite. We set

$$N_k(-\infty, \infty) = \{y_k(t) \in \mathcal{L}_k(-\infty, \infty), \quad y_k(t) \quad \text{satisfies} \quad (2.8)\}$$

Using a time periodic space-time statistical solution defined on the space $\mathcal{L}_k(0, \infty)$, that was constructed in Theorem 5.1, we construct a periodic space-time statistical solution defined on the space $\mathcal{L}_k(-\infty, \infty)$.

We do it by means of the construction presented in the Subsection 3.3. The following analog of Theorem 3.2 is true and can be proved just as Theorem 3.2.

THEOREM 5.2.) *The measure $\tilde{P}^k$ is periodic with the period T_0, i.e.*

$$\tilde{P}^k(\omega) = \hat{T_0}^* \tilde{P}^k(\omega) = \tilde{P}^k(\hat{T_0}^{-1} \omega) \quad , \qquad \forall \omega \in \mathcal{B}(\mathcal{L}_k(-\infty, \infty)) \quad . \tag{5.14}$$

b) The measure $\tilde{P}^k$ is concentrated on the set

$$N_k(-\infty, \infty) \bigcap \Phi_R, \quad \textit{where} \quad \Phi_R \equiv$$
$$\equiv \{y(t) \in \mathcal{L}_k(-\infty, \infty) : ||y(t)||_0 \leq R \qquad \forall t \in R\} \tag{5.15}$$

and the magnitude R is determined in Lemma 2.2 and depends only on the right side f of equation (5.1). (It does not depend on k)

Note that the measure $\tilde{P}^k$ satisfies the following analog of inequality (3.31):

$$\int \frac{1}{(t_2 - t_1)} ||y||^2_{\mathcal{L}(t_1,t_2)} \tilde{P}^k(dy) \leq c(\int ||y_0||^4_0 \mu_0(dy_0) + ||f||^4_{L_2(0,T_0;H^0)}) \tag{5.16}$$

where $-\infty < t_1 < t_2 < \infty$, c does not depend on t_1, t_2 and k.

The following assertion is true:

LEMMA 5.1. *Let $\tilde{P}^k(\omega)$, $\omega \in \mathcal{B}(\mathcal{L}_k(-\infty,\infty))$ be the statistical solution constructed in Theorem 5.2. Then for an arbitrary $\beta > 1$*

$$\int \int_{-\infty}^{\infty} \frac{1}{(1+|t|)^{1+\beta}} \|y\|^2_{\mathcal{L}(0\wedge t, 0\vee t)} \tilde{P}^k(dy) \le c_\beta \left(\int \|y_0\|_0^4 \mu_0(dy_0) + \|f\|^4_{L_2(0,T;H^0)} \right) \tag{5.17}$$

where $0 \wedge t = \min(0,t)$, $0 \vee t = \max(0,t)$, *and c_β depends on β only.*

PROOF. We denote the right side of (5.16) by A. Let $t_2 = t > t_1 = 0$. Since $t+1 > t$, estimate (5.16) implies the inequality:

$$\int \frac{1}{(1+t)^{1+\beta}} \|y\|^2_{\mathcal{L}(0,t)} \tilde{P}^k(dy) \le (1+t)^{-\beta} A$$

After integration of this inequality with respect to t we get:

$$\int \int_0^{\infty} \frac{1}{(1+t)^{1+\beta}} \|y\|^2_{\mathcal{L}(0,t)} dt \tilde{P}^k(dy) \le c_\beta A \tag{5.18}$$

Taking $t_2 = 0 > -t = t_1$ we get analogously to (5.18):

$$\int \int_{-\infty}^{0} \frac{1}{(1+|t|)^{1+\beta}} \|y\|^2_{\mathcal{L}(-t,0)} dt \tilde{P}^k(dy) \le c_\beta A \tag{5.19}$$

Adding (5.18) and (5.19) we obtain inequality (5.17). □

5.2 The final results

Since the unique solvability Theorem for the 3-D Navier-Stokes equations is not known, Definition 2.1 of a space-time statistical solution is not correct for the 3-D case. That is why we use another definition of statistical solution which is very close to the definition given in (Vishik and Fursikov (1988).

First of all we define certain functions spaces.

Let $\beta > 1$ Then X^β be the space of functions $R \ni t \to y(t) \in H^0$ possessing the finite norm:

$$\|y\|^2_{X^\beta} = \int_{-\infty}^{\infty} (1+|t|)^{-\beta} \left(\|y(t)\|_1^2 + \|\partial y(t)\|^2_{-1} \right) dt < \infty \tag{5.20}$$

The space Z^β is defined as the set of functions $R \ni t \to y(t) \in H^0$ possessing the finite norm:

$$\|y\|^2_{Z^\beta} = \int_{-\infty}^{\infty} (1+|t|)^{-\beta} \|y(t)\|_0^2 dt < \infty \tag{5.21}$$

DEFINITION 5.1. The time-periodic space-time statistical solution of equation (2.8) is a probability measure $P(\omega)$, $\omega \in \mathcal{B}(Z^\beta)$, $\beta > 1$ such that

1) P is concentrated on the space X^β, $\beta > 0$, i.e. $P(X^\beta) = 1$.
2) There exists a set $W \subset X^\beta$ closed in X^β, such that
   a) $W \in \mathcal{B}(Z^\beta)$,
   b) $P(W) = 1$,
   c) W consists of solutions of equation (2.8).
3) Measure P is periodic with respect to time, i.e.

$$P(\omega) = \hat{T}_0^* P(\omega) = P(\hat{T}_0^{-1}\omega), \quad \forall \omega \in \mathcal{B}(Z^\beta) \quad . \tag{5.22}$$

4) The following inequality is true:

$$\int \left(||y||^2_{X^\beta} + \frac{1}{(t_2 - t_1)} ||y||^2_{\mathcal{L}(t_1,T_2)} \right) P(dy) \le c_\beta \left(1 + ||f||^4_{L_2(0,T_0;H^0)}\right) \tag{5.23}$$

where c_β does not depend on f and $-\infty < t_1 < t_2 < \infty$ and depends on $\beta > 1$ only.

THEOREM 5.3. *Let the right hand side of equation (2.8) $f(t) \in L_{2,loc}(-\infty, \infty; H^0)$ be a time-periodic function with period T_0 and $n = 3$ in (2.9). Then there exists a time-periodic space-time statistical solution of equation (2.8).*

PROOF. We take an arbitrary probability measure $\mu_0(\omega_0)$, $\omega_0 \in \mathcal{B}(H^0)$, satisfying inequality

$$\int ||y_0||_0^4 \mu_0(dy_0) < \infty \quad .$$

After that we consider Galerkin's approximation (5.1),(5.2) of problem (2.8),(2.11) and applying the construction of subsection 5.1 we obtain a time-periodic space-time statistical solution $\tilde{P}^k(\omega)$ of the Galerkin's equation (5.1).

Note that by Fubini Theorem

$$\int_0^\infty \frac{1}{(1+t)^{1+\beta}} \int_0^t ||y(\tau)||_0^2 d\tau dt = \frac{1}{\beta} \int_0^\infty (1+\tau)^{-\beta} ||y(\tau)||_0^2 d\tau,$$

$$\int_{-\infty}^0 \frac{1}{(1+|t|)^{1+\beta}} \int_{-t}^0 ||y(\tau)||_0^2 d\tau dt = \frac{1}{\beta} \int_{-\infty}^0 (1+|\tau|)^{-\beta} ||y(\tau)||_0^2 d\tau.$$

These formulae and (5.17), (5.20) yield:

$$\int ||y||^2_{X^\beta} \tilde{P}^k(dy) \le c_\beta \left(\int ||y_0||_0^4 \mu_0(dy_0) + ||f||^4_{L_2(0,T_0;H^0)} \right) \tag{5.24}$$

Let $1 < \beta < \beta_1$. Then the embedding

$$X^\beta \subset\subset Z^{\beta_1}$$

is compact where norms of spaces X^β, Z^{β_1} are defined as in (5.20),(5.21).This assertion can be proved as Lemma 5.2 from Chapter 7 of (Vishik and Fursikov (1988).

Relations (5.24),(5.25) and Lemma 3.1 from Chapter 2 of (Vishik and Fursikov (1988) yield that there exists a subsequence $\{k'\}$ of $\{k\}$ such that the sequence $\tilde{P}^{k'}(\omega)$ converges weakly in the space Z^β to a probability measure $P(\omega)$:

$$\int f(y)\tilde{P}^{k'}(dy) \longrightarrow \int f(y)P(dy) \qquad \forall f(y) \in C_b(Z_{\beta_1}) \tag{5.26}$$

and $P(\omega)$ satisfies the estimate

$$\int \|y\|^2_{X^\beta} P(dy) \le c_\beta \left(\int \|y_0\|^4_0 \mu_0(dy_0) + \|f\|^4_{L_2(0,T_0;H^0)} \right) \quad . \tag{5.27}$$

(Recall that $C_b(Z_{\beta_1})$ is the space of functions continuous and bounded on Z_{β_1}).

Relations (5.14) and (5.26) imply (5.22). Hence, the measure $P(\omega)$ is time-periodic with the period T_0.

By (5.27) $P(\omega)$ is concentrated on the space X_β. The estimation of the second term in the left side of (5.23) is proved with help of methods of Theorem 5.1 from Chapter 4 of (Vishik and Fursikov (1988).

Finally, the property 2 of Definition 4.1 should be proved by the methods of Theorem 5.2 from Chapter 4 of (Vishik and Fursikov (1988). □

We establish also the following important property of statistical solution $P(\omega)$. Let $\beta > 1$. Define

$$\Phi_R = \{y(t) \in X_\beta : \|y(t)\|_0 \le R \ \ \forall t \in R\} \quad . \tag{5.28}$$

PROPOSITION 5.1.*Let $P(\omega)$ be the statistical solution constructed in Theorem 5.3. Then $P(\omega)$ is concentrated on the set Φ_R and the magnitude R is determined in Lemma 2.2 and depends on the right side of equation (2.8) only.*

PROOF. Let $\varepsilon > 0$ and

$$U_\varepsilon = \left\{ z \in Z_{\beta_1} : \inf_{y \in \Phi_R} \|y - z\|_{Z_{\beta_1}} < \varepsilon \right\}$$

be the ε-neighborhood of the set Φ_R in the space Z_{β_1}. We consider the function $\phi_\varepsilon \in C_b(Z_{\beta_1})$ such that

$$0 \le \phi_\varepsilon \le 1, \quad \phi_\varepsilon(y) \equiv 1 \quad \forall y \in U_\varepsilon; \qquad \phi_\varepsilon(y) \equiv 0 \quad \forall y \in Z_{\beta_1} \setminus U_{2\varepsilon}.$$

In virtue of Theorem 5.2 b) and (5.26) we get

$$1 = \int \phi_\varepsilon(y)\tilde{P}^{k'}(dy) \longrightarrow \int \phi_\varepsilon(y)P(dy) \quad \text{as} \quad k' \to \infty$$

Passing to the limit with help of Lebesgue Theorem we get

$$1 = \int \phi_\varepsilon(y)P(dy) \longrightarrow \int_{\Phi_R} P(dy) \qquad \text{as} \quad \varepsilon \to 0 \quad \square$$

Using the measure $\tilde{P}$ and restriction operator γ_t^* we define by formula (3.40) the space statistical solution $\mu(t,\omega)$, $\omega \in \mathcal{B}(H^0)$. This space statistical solution satisfies Proposition 3.1

References

Dymnikov V.P., Filatov A.N. (1994): Fundamentals of the mathematical theory of climate. VINITI, Moscow (in Russian)

Fursikov A.V. (1983): Properties of solutions of certain extremal problems connected with the Navier-Stokes equations. Math USSR Sbornik, **46**, 3, 323-351

Krylov N.M., Bogolubov N.N. (1937): La théorie générale de la mesure dans son application à l'étude des systémes de la mécanique non linéaires. Ann. of Math., **38**, p.65-113

Ladyzhenskaya O.A. (1969): The mathematical theory of viscous incompressible flow. Gordon and Breach, New-York

LionsJ.L. (1969): Quelques methodes de résolution des problèmes aux limites nonlineares. Dunod Gauthier-Villars, Paris

Temam R. (1979): Navier-Stokes Equations. North-Holland, Amsterdam

Vishik M.I., Fursikov A.V. (1988): Mathematical Problems of Statistical Hydromechanics. Kluwer Ac. Pub., Dodrecht, Boston, London

Local Exact Controllability for the 2-D Navier-Stokes Equations with the Navier Slip Boundary Conditions

Oleg Yu. Imanuvilov

Korea Institute for Advanced Study,
207-43 Chungryangri-dong Dongdaemoon-ku
Seoul, Korea 130-012

Abstract. For distributed controls we get a local exact controllability for the 2-D Navier-Stokes equations in the case where the fluid is incompressible and slips on the boundary in agreement with the Navier slip boundary conditions.

1 Introduction

This paper is devoted to a proof of the local exact controllability of the 2-D Navier-Stokes system with Navier slip boundary conditions, defined in a bounded domain $\Omega \subset R^2$ when the control function is distributed on an arbitrary fixed subdomain $\omega \subset \Omega$. The precise statement of the investigated problems and formulations of the main results are placed in section 1. Here we restrict ourselves only to the description of one typical particular case.

Let $(\hat{v}(x), \nabla\hat{p}(x))$, $x \in \Omega \subset R^2$ be a steady-state solution of the 2-D Navier-Stokes system:

$$-\Delta\hat{v} + (\hat{v}, \nabla)\,\hat{v} + \nabla\hat{p} = f(x), \ \operatorname{div}\hat{v} = 0, \ (\hat{v}, \nu)|_{\partial\Omega} = 0, \ \operatorname{rot}\hat{v} + \sigma(\hat{v}, \tau) = 0. \tag{1.1}$$

We consider the nonstationary Navier-Stokes system

$$\partial_t v(t,x) - \Delta v(t,x) + (v, \nabla)\, v + \nabla p = f(x) + u(t,x), \qquad \operatorname{div} v = 0, \tag{1.2}$$

with initial condition and boundary conditions

$$v\,|_{t=0} = v_0(x), \qquad (v, \nu)|_{\partial\Omega} = 0, \qquad (\operatorname{rot} v + \sigma(v, \tau))|_{\partial\Omega} = 0 \tag{1.3}$$

which is sufficiently close to a given steady-state solution

$$\|v_0 - \hat{v}\|_{(W_2^2(\Omega))^2} \le \varepsilon, \qquad (\text{ parameter } \varepsilon \text{ is sufficiently small}).$$

One has to find a boundary control

$$\operatorname{supp} u \subset \omega, \tag{1.4}$$

such that the solution $v(t,x)$ of boundary value problem (1.2), (1.3), (1.4) at the prescribed instant T coincides with $\hat{v} : v(T,x) \equiv \hat{v}(x)$. Such control is constructed in this work.

To make this result more clear let us assume, that $\hat{v}$ satisfies (1.1) and $\hat{v}$ is an unstable singular point of the dynamical system generated by equation (1.2) in the phase space of solenoidal vector fields with the Navier slip conditions on $\partial\Omega$. Let v_0 be an initial condition from the neighborhood of $\hat{v}$ such that the trajectory of the dynamical system going out v_0 does not converge to v_0 as $t \to \infty$. As we show in this work one can construct boundary control, such that the corresponding trajectory going out v_0 reaches $\hat{v}$ during a finite time. In other words, one can suppress a turbulence rise by means of the boundary control. This result makes more clear the question on connections between turbulence and controllability
(Lions (1990)).

The global approximate controllability of the Navier-Stokes equations with the slip boundary conditions was obtained by J.-M. Coron (Coron (1996a)). The proof was based on ideas which were successfully applied in (Coron J.-M. (1993), Coron J.-M. (1996b)) to solve exact controllability problem for the Euler equation. Unfortunately, expect the case $\overline{\sigma} = 0$ approximate controllability is proved in the sense of weak norm. So here we can not combine this result with the local exact controllability one as it was done in (Coron and Fursikov(1996)). Works of A.V.Fursikov, O.Yu. Imanuvilov (Fursikov and Imanuvilov (1994)-Fursikov and Imanuvilov (1996b)) , A.V. Fursikov (Fursikov (1995)) in this volume. In (Fursikov and Imanuvilov (1995)) the local exact controllability for the Burgers equation was studied. The case of the 2-D and 3-D Navier-Stokes system with control on the whole boundary and $\hat{v} = 0$ was investigated in (Fursikov and Imanuvilov (1994)) and (Fursikov (1995)) respectively. Papers (Fursikov and Imanuvilov (1996c)),(Fursikov and Imanuvilov (1997)) are concerned on local exact controllability of the Boussinesq system. The controllability of 2-D Navier-Stokes equations with slip boundary conditions for the case $\sigma(x) \equiv 0$ was studied in (Fursikov and Imanuvilov (1996c)).

2 Statement of the Problem and Formulation of the Main Results

2.1. In a bounded simply connected domain $\Omega \subset R^2$ with boundary $\partial\Omega \in C^\infty$ we consider the Navier-Stokes system

$$\partial_t v(t,x) - \Delta v(t,x) + (v,\nabla)\, v + \nabla p(t,x) = f(t,x) + u(t,x), \quad \operatorname{supp} u \subset \omega, \tag{2.1}$$

$$\operatorname{div} v = \partial_{x_1} v_1 + \partial_{x_2} v_2 = 0 \quad , \tag{2.2}$$

where $(t,x) \in Q \equiv (0,T) \times \Omega$, $v(t,x) = (v_1(t,x), v_2(t,x))$ is the velocity of fluid, $\nabla p(t,x)$ is a pressure gradient, $\partial_t = \frac{\partial}{\partial t}$, $\partial_{x_j} = \frac{\partial}{\partial x_j}$, $(v,\nabla)v = \sum_{j=1}^2 v_j \partial_{x_j} v$, Δ is the Laplace operator, $f = (f_1, f_2)$ is a density of external forces, $\omega \subset \Omega$ is an arbitrary fixed subdomain and $u(t,x)$ is a control function. We assume that

$$v(t,x)|_{t=0} = v_0(x) \tag{2.3}$$

where $v_0(x) = (v_{01}, v_{02})$ is a given initial condition.

We set on $\Sigma = (0,T) \times \partial\Omega$ the Navier slip boundary conditions

$$(\operatorname{rot} v + \sigma(v,\tau))\,|_\Sigma = 0, \quad (v,\nu)\,|_\Sigma = 0 \tag{2.4}$$

where $\nu = (\nu_1, \nu_2)$ is the vector field of outward unit normals to $\partial\Omega$, $\tau = (\tau_1, \tau_2)$ is the unit tangent vector field on $\partial\Omega$, $\sigma(x) \in C^\infty(\partial\Omega)$ defined by

$$\sigma(x) = \frac{2(1-\overline{\sigma})k(x) - \overline{\sigma}}{1-\overline{\sigma}},$$

where k is the curvature of $\partial\Omega$ defined through the relation $\frac{\partial n}{\partial \tau} = k\tau$. We recall that $(v,\nu) = v_1\nu_1 + v_2\nu_2$, $\operatorname{rot} v = \partial_{x_1} v_2 - \partial_{x_2} v_1$.

To set the problem and formulate the main results we have to introduce the functional spaces. Recall, that $W_p^k(\Omega)$, $k \geq 0$, $1 \leq p < \infty$ is the Sobolev space of functions with finite norm

$$||u||_{W_p^k(\Omega)} = \Big(\sum_{|\alpha| \leq k} \int_\Omega \left|\partial^{|\alpha|} u(x)\,/\partial x_1^{\alpha_1} \ldots \partial x_n^{\alpha_n}\right|^p dx\Big)^{1/p},$$

where $\alpha = (\alpha_1, \alpha_2)$, $|\alpha| = \alpha_1 + \alpha_2$.

We set

$$V^k(\Omega) = \{v(x) = (v_1, v_2) \in (W_2^k(\Omega))^2 : \operatorname{div} v = 0\}, \tag{2.5}$$

$$W^{1,2(k)}(Q) = \{v \in L_2(0,T; W_2^{k+2}(\Omega)) : \partial_t v \in L_2(0,T; W_2^k(\Omega))\}, \tag{2.6}$$

$$V^{1,2(k)}(Q) = \{v(t,x) \in (W^{1,2(k)}(Q)) : \operatorname{div} v = 0\}. \tag{2.7}$$

Since ∇p can be determined easily from (2.1) by f, v, below keeping in mind solutions of system (2.1) we write v instead of $(v, \nabla p)$.

Now we set the exact controllability problem. Let a solution $\hat{v} \in V^{1,2(1)}(Q)$ of equation (2.1), (2.2) as well as initial condition $v_0 \in V^2(\Omega)$ be given. We suppose that $\hat{v}$, v_0 satisfy the inequality

$$||\hat{v}(0,\cdot) - v_0||^2_{V^2(\Omega)} < \varepsilon, \tag{2.8}$$

where $\varepsilon > 0$ is sufficiently small. Assume also that the initial datum v_0 satisfies the compatibility conditions

$$(\operatorname{rot} v_0 + \sigma(v_0, \tau))|_{\partial\Omega} = 0, \quad (v_0, \nu)|_{\partial\Omega} = 0. \tag{2.9}$$

The local exact controllability problem is to find a control $u \in (L^2(Q))^2$, such that the solution $v \in V^{1,2(1)}(Q)$ of (2.1)-(2.4) satisfies at $t = T$ the equation

$$v(t,x)|_{t=T} = \hat{v}(T,x). \tag{2.10}$$

For $\omega \subset \Omega$ we set $Q^\omega = (0,T) \times \omega$.

THEOREM 2.1 *Let $\partial\Omega$ be connected, $\hat{v}(t,x) \in V^{1,2(1)}(Q)$ be a given solution of (2.1), (2.2), (2.4) and $v_0(x) \in V^2(\Omega)$ satisfy (2.8) (2.9) with sufficiently small $\varepsilon > 0$. Then there exists a local distributed control $u(t,x) \in (L_2(Q))^2$, $\operatorname{supp} u \subset Q^\omega$, such that the solution $v(t,x) \in V^{1,2(1)}(Q)$ of problem (2.1)- (2.4) exists and satisfies (2.10).*

3 Reduction to a Linear Controllability Problem

3.1. To get rid of pressure we transform the Navier-Stokes system to the equation for the stream function ψ which is connected with velocity field $v(t,x) = (v_1, v_2)$ by equations

$$\partial_{x_1}\psi = -v_2, \qquad \partial_{x_2}\psi = v_1. \tag{3.1}$$

Application the operator ∂_{x_2} to the first of equations (2.1) and ∂_{x_1} to the second one, adding this two new equations yields the equation for the stream function:

$$\partial_t(-\Delta\psi(t,x)) + \Delta^2\psi + \partial_{x_2}((\partial_{x_1}\psi)\Delta\psi) - \partial_{x_1}((\partial_{x_2}\psi)\Delta\psi) = u + g. \tag{3.2}$$

We substituted $u(t,x)+g(t,x)$, instead of $\operatorname{rot} f$ in the right-hand-side of (3.2) taking into account that $g = \operatorname{rot} f$ and u is a control. We just use this form of right-hand-side below. The first boundary condition from (2.4) by virtue of (3.1) can be rewritten as follows:

$$\left(\Delta\psi + \sigma\frac{\partial\psi}{\partial\nu}\right)\bigg|_\Sigma = 0, \qquad \Sigma = (0,T) \times \partial\Omega. \tag{3.3}$$

The second one is transformed to the equation

$$\partial_\tau\psi|_\Sigma = 0 \tag{3.4}$$

where $\tau = (\tau_1, \tau_2) = (-\nu_2, \nu_1)$ is the vector tangential to the $\partial\Omega$. By this equality

$$\psi|_{\partial\Omega} = \text{const},$$

and since $\partial\Omega$ is a connected set, [1] function ψ can be determined by (3.1) up to a constant. We can assume that

$$\psi|_\Sigma = 0 \tag{3.5}$$

[1] Only here, deducing condition (3.5) we use connectedness of $\partial\Omega$. Therefore below controllability problem for stream function is studied without assumption of $\partial\Omega$ connectedness.

without the lose of generality. By virtue of (3.1), (2.4) instead of initial condition (2.3) we have

$$\psi(t,x)|_{t=0} = \psi_0(x) \tag{3.6}$$

where ψ_0 can be determined by the equalities

$$\partial_{x_1}\psi_0 = -v_{02}, \qquad \partial_{x_2}\psi_0 = v_{01}.$$

According to (2.9), (3.5) the following compatibility conditions should be fulfilled:

$$\psi_0|_{\partial\Omega} = 0, \qquad \left(\Delta\psi_0 + \sigma\frac{\partial\psi_0}{\partial\nu}\right)\Big|_{\partial\Omega} = 0. \tag{3.7}$$

Let us assume similarly to section 1 that a solution $\hat{\psi}(t,x) \in W^{1,2(2)}(Q)$ of equation (3.2) with $u(t,x) \equiv 0$ and right side $g \in L_2(Q)$ is given. Moreover, the function $\hat{\psi}(t,x)$ satisfies boundary conditions (3.3), (3.5) and the inequality

$$\left\|\hat{\psi}(0,\cdot) - \psi_0(\cdot)\right\|^2_{W_2^3(\Omega)} < \varepsilon \tag{3.8}$$

holds where $\varepsilon > 0$ is sufficiently small. The local exact controllability problem consists in constructing such control $u(t,x) \in L_2(Q)$, $\operatorname{supp} u \subset Q^\omega$, that the solution $\psi(t,x)$ of boundary value problem (3.2),(3.3),(3.5),(3.6) satisfies the condition

$$\psi(t,x)|_{t=T} = \hat{\psi}(t,x)|_{t=T}. \tag{3.9}$$

We are looking for the solution $\psi(t,x)$ in the following form

$$\psi(t,x) = w(t,x) + \hat{\psi}(t,x) \tag{3.10}$$

where w is a new unknown function. Substitution of (3.10) into (3.2) - (3.6) yields the equation for the function w :

$$\partial_t(-\Delta w(t,x)) + \Delta^2 w + B(\hat{\psi} + w, w) + B(w, \hat{\psi}) = u(t,x) \tag{3.11}$$

where

$$B(\psi,\varphi) = \partial_{x_2}((\partial_{x_1}\psi)\Delta\varphi) - \partial_{x_1}((\partial_{x_2}\psi)\Delta\varphi). \tag{3.12}$$

This also gives boundary and initial conditions

$$\left(\Delta w + \sigma\frac{\partial w}{\partial\nu}\right)\Big|_{\Sigma} = 0, \quad w|_{\Sigma} = 0, \tag{3.13}$$

$$w(t,x)|_{t=0} = w_0. \tag{3.14}$$

Here $w_0(x) = \psi_0(x) - \hat{\psi}(0,x)$. By virtue of (3.10), (3.7), (3.8) we have

$$w_0|_{\partial\Omega} = \left(\Delta w_0 + \sigma\frac{\partial w_0}{\partial\nu}\right)\Big|_{\partial\Omega} = 0, \qquad \|w_0\|^2_{W_2^3(\Omega)} < \varepsilon. \tag{3.15}$$

In Sections 2-7 the following assertion will be proved:

THEOREM 3.1. *Suppose that* $\hat{\psi} \in W^{1,2(2)}(Q)$ *satisfies (3.2) with* $u \equiv 0$, *(3.3), (3.5), and initial condition* $w_0 \in W_2^3(\Omega)$ *satisfies (3.15) with sufficiently small* $\varepsilon > 0$. *Then one can find such control* $u \in L_2(Q)$, $\operatorname{supp} u \subset (0,T) \times \omega$, *that the corresponding solution* $w \in W^{1,2(2)}(Q)$ *of problem (3.11) -(3.14) exists and satisfies the equality*

$$w(t,x)|_{t=T} = 0. \tag{3.16}$$

3.2. To prove Theorem 3.1 we use the following right inverse operator theorem:

THEOREM 3.2. *Let* X, Z *be Banach spaces and*

$$A : X \to Z \tag{3.17}$$

be a continuously differentiable mapping. Assume that for some $x_0 \in X$, *and* $z_0 \in Z$ *equality*

$$A(x_0) = z_0 \tag{3.18}$$

holds, and the derivative

$$A'(x_0) : X \to Z \tag{3.19}$$

of A *at* x_0 *is surjective operator. Then for a sufficiently small* $\varepsilon > 0$ *there exists a mapping* $M(z) : B_\varepsilon(z_0) \to X$ *defined on the ball*

$$B_\varepsilon(z_0) = \{z \in Z : ||z - z_0||_Z < \varepsilon\}$$

which satisfies conditions

$$A(M(z)) = z, \qquad z \in B_\varepsilon(z_0), \tag{3.20}$$

$$||M(z) - x_0||_X \le k\,||A(x_0) - z||_Z \text{ for all } \qquad z \in B_\varepsilon(z_0) \tag{3.21}$$

where $k > 0$ *is a certain constant.*

The Theorem 3.2 is a simple corollary of the generalization of the Implicit function theorem proved in (Alekseev V.M., Tikhomirov V.M. and Fomin S.V.(1987)).

In our case the space X consists of pairs $x = (w,u)$, and operator $A(x)$ is defined by formula (3.11):

$$A(x) = (-\partial_t \Delta w + \Delta^2 w + B(\hat{\psi} + w, w) + B(w, \hat{\psi}) - u,\ w|_{t=0}) \tag{3.22}$$

(the condition $w|_{t=T} = 0$ and boundary conditions for w are included in the definition of space X.) The space Z will be determined by set of pairs from (3.22). Set $x_0 = (0,0)$, $z_0 = (0,0)$. Evidently, equality (3.18) is fulfilled.

To check the epimorphism condition of operator (3.19) we write out the equation

$$A'(x_0)x = z.$$

In our case this equation is as follows:

$$Lw - u \equiv \partial_t(-\Delta w) + \Delta^2 w + B(\psi, w) + B(w, \psi) - u = f \qquad (3.23)$$

where $u = \chi_\omega u$, χ_ω is the characteristic function of the set ω ($\chi(x) = 1$ for $x \in \omega$; $\chi(x) = 0$ for $x \in \Omega \setminus \omega$),

$$w|_\Sigma = \left.\left(\Delta w + \sigma \frac{\partial w}{\partial \nu}\right)\right|_\Sigma = 0, \qquad (3.24)$$

$$w|_{t=0} = w_0, \qquad w|_{t=T} = 0. \qquad (3.25)$$

Note that if $x_0 = (0,0)$, $z_0 = (0,0)$ then function ψ from (3.23) coincides with $\hat{\psi}$.

Now, we define the spaces X, Z corresponding to problems (3.11)-(3.14) and (3.23)-(3.25). We start from the following lemma:

LEMMA 3.1 *Let $\omega_0 \subset\subset \omega$ be an arbitrary fixed subdomain of Ω. Then there exists a function $\beta \in C^2(\overline{\Omega})$ such that*

$$\beta(x) > 0 \; \forall x \in \Omega, \; \beta|_{\partial\Omega} = 0, \quad |\nabla \beta(x)| > 0 \quad \forall x \in \Omega \setminus \omega_0. \qquad (3.26)$$

For the proof of this lemma see (Imanuvilov O.Yu.(1995)).

Set

$$\eta(t,x) \equiv \eta^\lambda(t,x) = (e^{\lambda^2 \|\beta\|_{C(\bar{\Omega})}} - e^{\lambda\beta(x)})/((T-t)\ell(t))^3, \qquad (3.27)$$

where $\lambda > 1$ is a parameter (magnitude of λ will be fixed below), function $\beta(x)$ defined in Lemma 3.1 and function $\ell(t) \in C^\infty[0,T]$ satisfies conditions

$$\ell(t) > 0, \; \forall t \in [0,T], \; \ell(t) = t \quad \forall t \in [T/2, T].$$

Let us assume that λ such that

$$\frac{4}{3} \min_{x \in \Omega} \eta(t,x) \geq \max_{x \in \Omega} \eta(t,x). \qquad (3.28)$$

Obviously inequality (3.28) holds true for all λ sufficiently large. Finally we define the parameter λ in Lemma 3.4. Set

$$L_2(Q,\kappa) = \left\{ u(t,x) : (t,x) \in Q : \|u\|^2_{L_2(Q,\kappa)} \equiv \int_Q \kappa^2(t,x) u^2(t,x)\,dx\,dt < \infty \right\}. \qquad (3.29)$$

The weight functions κ used below are constructed by means of function (3.27). One of such weight functions is defined by the formula

$$\theta(t,x) = \begin{cases} e^{s\eta} & x \in \Omega \setminus \omega \\ 1 & x \in \omega, \end{cases} \qquad (3.30)$$

where parameter $s > 0$ will be defined in Lemma 4.3. We introduce the space

$$Y(Q) \equiv \left\{ y(t,x) \in W^{1,2(2)}(Q) : y|_\Sigma = \left(\Delta y + \sigma \frac{\partial y}{\partial \nu}\right)\Big|_\Sigma = 0, \right.$$

$$\|y\|^2_{Y(Q)} \equiv \left\| \partial_t(-\Delta y) + \Delta^2 y + B(\hat{\psi}, y) + B(y, \hat{\psi}) \right\|^2_{L_2(Q,\theta)}$$

$$\left. + \left\| e^{\frac{98}{100} s\eta} y \right\|^2_{W^{1,2(2)}(Q)} \right\}, \tag{3.31}$$

where functions θ, η are defined in (3.30), (3.27) and parameter s from (3.30). Define also

$$U_\omega(Q) = \left\{ u(t,x) \in L_2(Q) : \operatorname{supp} u \subset Q^\omega \right\}, \tag{3.32}$$

where, recall, $Q^\omega = (0,T) \times \omega$.

To apply the Theorem 3.2 in order to establish solvability of (3.11), (3.13), (3.14), (3.16) we define spaces X , Z as follows:

$$X = \{(y,u) \in Y(Q) \times U_\omega(Q) :$$

$$\|(y,u)\|^2_X = \|y\|^2_{Y(Q)} + \|u\|^2_{U_\omega(Q)} < \infty\}, \tag{3.33}$$

$$Z = L_2(Q,\theta) \times \hat{W}^3_2(\Omega) \tag{3.33'}$$

where

$$\hat{W}^3_2(\Omega) = \left\{ v(x) \in W^3_2(\Omega) : v|_{\partial\Omega} = \left(\Delta v + \sigma \frac{\partial v}{\partial \tau}\right)\Big|_{\partial\Omega} = 0 \right\}. \tag{3.33''}$$

We have

PROPOSITION 3.1 *Let the spaces X, Y be defined in (3.33), (3.33') operator $A(x)$ be defined by (3.22). Then mapping (3.17) is continuously differentiable for any point $x_0 \in X$.*

PROOF. Definition (3.31)-(3.33),(3.33') of the spaces X, Z implies directly continuity of the operator

$$(w,u) \to (\partial_t(-\Delta w) + \Delta^2 w + B(\hat{\psi}, w) + B(w, \hat{\psi}) - u,\ w|_{t=0}) \ :\ X \to Z.$$

Being linear this operator belongs to C^1. The operator B from (3.22) defined by (3.12) is bilinear. So to prove Proposition 3.1 one has to establish continuity of bilinear operator

$$B : Y(Q) \times Y(Q) \to L_2(Q,\theta). \tag{3.34}$$

Taking into account (3.12), (3.28), (3.30)-(3.32) we get by simple calculations:

$$\|B(\varphi,\psi)\|^2_{L_2(Q,\theta)} \le c \int_Q \theta(|\partial_{x_2}\Delta\psi|^2 \, |\partial_{x_1}\varphi|^2 + |\partial_{x_1}\Delta\psi|^2 \, |\partial_{x_2}\phi|^2) dx dt$$

$$\leq c \left\| e^{\frac{98}{100}s\bar{\eta}} \nabla \varphi \right\|^2_{C(\bar{Q})} \left\| e^{\frac{98}{100}s\bar{\eta}} \nabla \Delta \psi \right\|^2_{L_2(Q)} \leq c \|\psi\|^2_{Y(Q)} \|\varphi\|^2_{Y(Q)} .$$

This estimate proves continuity of operator (3.34).□

Evidently, equality (3.18) holds if A is mapping (3.22), $x_0 = (w^0, u^0) = 0$, $z_0 = 0$. So, to apply Theorem 3.2 we have to establish only that the image of operator (3.19) coincides with Z. This is reduced to the proof of problem (3.23)-(3.25) solvability for any $(f, w_0) \in Z$. Sections 3-4 are devoted to achievement of this aim.

4 Carleman Estimate for the Heat Equation

This section is devoted to solve observability problem for the operator L^*. We start from Carleman estimate for the inverse heat equation. Of course Carleman estimate for such equation is well know in the case of function with compact support (Isakov V. (1992)) or for the heat equation with zero Dirichlet or Neumann boundary conditions (Imanuvilov O.Yu.(1993), Imanuvilov O.Yu.(1994), Imanuvilov O.Yu.(1995)). But here we do not introduce boundary conditions on Σ. We set

$$\varphi(t,x) = e^{\lambda\beta(x)}/(t(T-t))^3, \quad \tilde{\varphi}(t,x) = e^{-\lambda\beta(x)}/(t(T-t))^3, \tag{4.1}$$

$$\alpha(t,x) = (e^{\lambda\beta} - e^{\lambda^2\|\beta\|_{C(\bar{\Omega})}})/(t(T-t))^3,$$

$$\tilde{\alpha}(t,x) = (e^{-\lambda\beta} - e^{\lambda^2\|\beta\|_{C(\bar{\Omega})}})/(t(T-t))^3, \tag{4.2}$$

where $\lambda > 1$ satisfies (3.28) and function β from Lemma 3.1. Note that

$$\alpha(t,x) \geq \tilde{\alpha}(t,x) \quad \forall (t,x) \in Q.$$

In the cylinder Q we consider the heat equation:

$$Gz = \partial_t z + \Delta z = f(t,x) \quad \text{in } Q, \tag{4.3}$$

Let $\omega_0 \subset\subset \omega_1 \subset\subset \omega$. We have

LEMMA 4.3 *There exists a number $\hat{\lambda} > 0$ such that for an arbitrary $\lambda > \hat{\lambda}$ there exists $s_0(\lambda) > 0$ that for any $s > s_0$ the solution $z(t,x)$ of (4.3) satisfies the Carleman estimate:*

$$\int_Q \left((s\varphi)^{-1} \left| \frac{\partial z}{\partial t} \right|^2 + s\varphi \sum_{j=1}^n \left| \frac{\partial z}{\partial x_i} \right|^2 + s^3\varphi^3 z^2 \right) e^{2s\alpha(t,x)} \, dx\, dt$$

$$\leq c \left(\int_Q f^2(t,x) e^{2s\alpha} \, dx\, dt \right.$$

$$\left. + \int_\Sigma \left(\left| \frac{\partial z}{\partial t} \frac{\partial z}{\partial \nu} \right| + s^2\varphi^2 |\nabla z||z| \right) e^{2s\alpha} d\Sigma + \int_{Q^{\omega_1}} s^3\varphi^3 z^2 e^{2s\alpha} dx dt \right), \tag{4.4}$$

where the functions $\varphi(t,x)$, $\alpha(t,x)$ are defined in (4.1), (4.2), and $c>0$ does not depend on s.

PROOF. We set $w(t,x)=e^{s\alpha}z(t,x)$, $\tilde{w}(t,x)=e^{s\tilde{\alpha}}z(t,x)$.
By (4.2) we have

$$w(T,\cdot)=\tilde{w}(T,\cdot)=w(0,\cdot)=\tilde{w}(0,\cdot)=0 \quad \text{in} \quad \Omega. \tag{4.5}$$

We define operators $P,\tilde{P}$ as the following:

$$Pw=e^{s\alpha}Ge^{-s\alpha}w, \quad \tilde{P}w=e^{s\tilde{\alpha}}Ge^{-s\tilde{\alpha}}w. \tag{4.6}$$

It follows from (4.3) that

$$Pw=e^{s\alpha}Ge^{-s\alpha}w=e^{s\alpha}g \quad \text{in} \quad Q, \tag{4.7}$$

$$\tilde{P}\tilde{w}=e^{s\tilde{\alpha}}Ge^{-s\tilde{\alpha}}\tilde{w}=e^{s\tilde{\alpha}}g \quad \text{in} \quad Q. \tag{4.8}$$

Operator P can be written explicitly as follows

$$Pw=\frac{\partial w}{\partial t}-\Delta w+2s\lambda\varphi(\nabla\beta,\nabla w)+s\lambda^2\varphi|\nabla\beta|^2w$$
$$-s^2\lambda^2\varphi^2|\nabla\beta|^2w+s\lambda\varphi w\Delta\beta-s\alpha_t w. \tag{4.9}$$

We introduce the operators L_1, L_2, $\tilde{L_1}$ and $\tilde{L_2}$ as follows

$$L_1w=-\Delta w-\lambda^2s^2\varphi^2|\nabla\beta|^2w-s\alpha_t w, \tag{4.10}$$

$$L_2w=\frac{\partial w}{\partial t}+2s\lambda\varphi(\nabla\beta,\nabla w)+2s\lambda^2\varphi|\nabla\beta|^2w, \tag{4.11}$$

$$\tilde{L_1}w=-\Delta w-\lambda^2s^2\tilde{\varphi}^2|\nabla\beta|^2w-s\tilde{\alpha}_t w, \tag{4.12}$$

$$\tilde{L_2}w=\frac{\partial w}{\partial t}-2s\lambda\tilde{\varphi}(\nabla\beta,\nabla w)+2s\lambda^2\tilde{\varphi}|\nabla\beta|^2w. \tag{4.13}$$

It follows from (4.9), (4.10) and (4.11) that

$$L_1w+L_2w=f_s \quad \text{in} \quad Q, \tag{4.14}$$

where

$$f_s(t,x)=ge^{s\alpha}-s\lambda\varphi w\Delta\beta+s\lambda^2\varphi|\nabla\beta|^2w.$$

Taking L_2-norm of both sides of (4.14), we obtain

$$||f_s||^2_{L^2(Q)}=||L_1w||^2_{L^2(Q)}+||L_2w||^2_{L^2(Q)}+2(L_1w,L_2w)_{L^2(Q)}. \tag{4.15}$$

By (4.10) and (4.11) we have the following equality:

$$(L_1w,L_2w)_{L^2(Q)}=\big(-\Delta w-\lambda^2s^2\varphi^2|\nabla\beta|^2w$$
$$-s\alpha_t w,\frac{\partial w}{\partial t}+2s\lambda^2\varphi|\nabla\beta|^2w\big)_{L^2(Q)}-\int_Q(2\lambda^3s^3\varphi^3|\nabla\beta|^2w$$

$$+2s^2\lambda\varphi\alpha_t w)(\nabla\beta,\nabla w)dxdt - \int_Q \Delta w 2s\lambda\varphi(\nabla\beta,\nabla w)dxdt. \tag{4.16}$$

Integrating by parts the first term of the right-hand-side of (4.16), we obtain

$$A_0 = \Big(-\Delta w - \lambda^2 s^2\varphi^2|\nabla\beta|^2 w - s\alpha_t w, \frac{\partial w}{\partial t} + 2s\lambda^2\varphi|\nabla\beta|^2 w\Big)_{L^2(Q)}$$

$$= \int_Q \Big((\nabla w, \nabla w_t) - \frac{\lambda^2 s^2\varphi^2}{2}|\nabla\beta|^2\frac{\partial w^2}{\partial t} - \frac{s\alpha_t}{2}\frac{\partial w^2}{\partial t} - 2s^3\varphi^3\lambda^4|\nabla\beta|^2 w^2$$

$$-2s^2\lambda^2\alpha_t\varphi|\nabla\beta|w^2 + 2s\lambda^2\varphi|\nabla\beta|^2|\nabla w|^2 + 2s\lambda^2 w(\nabla w, \nabla(\varphi|\nabla\beta|^2))\Big)dxdt$$

$$-\int_\Sigma \left(\frac{\partial w}{\partial t} + 2s\lambda^2\varphi|\nabla\beta|^2 w\right)\frac{\partial w}{\partial \nu}d\Sigma. \tag{4.17}$$

Integrating by parts in the second term of the right-hand-side of (4.16), we have

$$-\int_Q (2\lambda^3 s^3 w\varphi^3|\nabla\beta|^2(\nabla\beta,\nabla w) + 2s^2\lambda\alpha_t w\varphi(\nabla\beta,\nabla w))dxdt$$

$$= -\int_Q (\lambda^3 s^3\varphi^3|\nabla\beta|^2(\nabla\beta,\nabla w^2) + s^2\alpha_t\varphi\lambda(\nabla\beta,\nabla w^2))dxdt$$

$$= \int_Q \Big(3\lambda^4 s^3\varphi^3|\nabla\beta|^4 w^2 + w^2\varphi^3\lambda^3 s^3\sum_{i=1}^n \frac{\partial}{\partial x_i}(\beta_{x_i}|\nabla\beta|^2)$$

$$+\sum_{i=1}^n \frac{\partial}{\partial x_i}\left(\frac{s^2\lambda^2\alpha_t\varphi}{2}\frac{\partial\beta}{\partial x_i}\right)w^2\Big)dxdt$$

$$-\int_\Sigma (\lambda^3 s^3\varphi^3|\nabla\beta|^2 + s^2\alpha_t\varphi\lambda)(\nabla\beta,\nu)w^2 d\Sigma. \tag{4.18}$$

Finally, integrating by parts the third term of right-hand-side of (4.16), and taking into account (3.26) we have

$$A_1 = \int_Q -\Delta w 2s\lambda\varphi(\nabla\beta,\nabla w)dxdt = \int_Q (2s\lambda^2\varphi(\nabla\beta,\nabla w)^2$$

$$+2s\lambda\varphi\sum_{i,k=1}^n\left(\frac{\partial w}{\partial x_i}\frac{\partial\beta_{x_k}}{\partial x_i}\frac{\partial w}{\partial x_k}\right)$$

$$+2s\lambda\varphi\sum_{i,k=1}^n \frac{\partial w}{\partial x_i}\beta_{x_k}\frac{\partial^2 w}{\partial x_j\partial x_k}\Big)dxdt + \int_\Sigma 2s\lambda\varphi|\nabla\beta|\left|\frac{\partial w}{\partial\nu}\right|^2 d\Sigma$$

$$= \int_Q\Big(2s\lambda^2\varphi(\nabla\beta,\nabla w)^2 + 2s\lambda\varphi\sum_{i,k=1}^n\left(\frac{\partial w}{\partial x_i}\frac{\partial\beta_{x_k}}{\partial x_i}\frac{\partial w}{\partial x_k}\right)$$

$$+ s\lambda\varphi(\nabla\beta, \nabla|\nabla w|^2) \Big) dxdt + \int_\Sigma 2s\lambda\varphi|\nabla\beta| \left|\frac{\partial w}{\partial \nu}\right|^2 d\Sigma. \tag{4.19}$$

Integrating by parts once again, we obtain

$$A_1 = \int_Q (2s\lambda^2\varphi(\nabla\beta, \nabla w)^2 + 2s\lambda\varphi \sum_{i,k=1}^n \frac{\partial w}{\partial x_i}\frac{\partial \beta_{x_k}}{\partial x_i}\frac{\partial w}{\partial x_k}$$

$$-s\lambda^2\varphi|\nabla\beta|^2|\nabla w|^2 - |\nabla w|^2 s\lambda\varphi\Delta\beta)\, dxdt$$

$$+\int_\Sigma \left(2s\lambda\varphi|\nabla\beta| \left|\frac{\partial w}{\partial \nu}\right|^2 - s\lambda\varphi|\nabla\beta||\nabla w|^2\right) d\Sigma. \tag{4.20}$$

Now, let us transform integrals on Σ in equations (4.17) and (4.20). By virtue of (3.16) for the integral on Σ in (4.17) we have

$$\int_\Sigma \left(\frac{\partial w}{\partial t} + 2s\lambda^2\varphi|\nabla\beta|^2 w\right)(\nu, \nabla w)d\Sigma$$

$$= \int_\Sigma \left(\frac{\partial w}{\partial t} + 2s\lambda^2\varphi|\nabla\beta|^2 w\right)((\nu, \nabla z)e^{s\alpha} + s\lambda(\nu, \nabla\beta)w)\, d\Sigma; \tag{4.21}$$

On the other hand, for the integrals on Σ in (4.20) we have

$$\int_\Sigma \left(2s\lambda\varphi|\nabla\beta| \left|\frac{\partial w}{\partial \nu}\right|^2 - s\lambda\varphi|\nabla\beta||\nabla w|\right) d\Sigma$$

$$= \int_\Sigma (2s\lambda\varphi|\nabla\beta|(\nu, \nabla w)^2 - s\lambda\varphi|\nabla\beta||\nabla w|^2)d\Sigma$$

$$= \int_\Sigma s\lambda\varphi|\nabla\beta|^2 \left(2\left|\frac{\partial z}{\partial \nu}\right|^2 + 4s\lambda\varphi z \sum_{i=1}^n \nu_i \frac{\partial z}{\partial x_i}\frac{\partial \beta}{\partial x_i} + 2s^2\lambda^2\varphi^2 \left|\frac{\partial \beta}{\partial \nu}\right|^2 z^2\right.$$

$$\left. -|\nabla z|^2 - 2s\lambda\varphi(\nabla\beta, \nabla z)z - s^2\lambda^2\varphi^2|\nabla\beta|^2 z^2\right) e^{2s\alpha} d\Sigma. \tag{4.22}$$

By virtue of (4.17), (4.18) and (4.20) - (4.22) one can rewrite (4.16) as follows.

$$(L_1 w, L_2 w)_{L^2(Q)} = \int_Q (\lambda^4 s^3\varphi^3|\nabla\beta|^4 w^2 + 2s\lambda^2\varphi(\nabla\beta, \nabla w)^2)dxdt$$

$$= \int_\Sigma (2s\lambda\varphi|\nabla\beta|(\nu, \nabla w)^2 - s\lambda\varphi|\nabla\beta||\nabla w|^2)d\Sigma$$

$$= \int_\Sigma \left(s\lambda\varphi|\nabla\beta| \left(2\left|\frac{\partial z}{\partial \nu}\right|^2 + 4s\lambda\varphi z \sum_{i=1}^n \nu_i \frac{\partial z}{\partial x_i}\frac{\partial \beta}{\partial x_i} + 2s^2\lambda^2\varphi^2 \left|\frac{\partial \beta}{\partial \nu}\right|^2 z^2\right.\right.$$

$$\left. -|\nabla z|^2 - 2s\lambda\varphi(\nabla\beta, \nabla z)z - s^2\lambda^2\varphi^2|\nabla\beta|^2 z^2\right) e^{2s\alpha} d\Sigma$$

$$-\int_{\Sigma}\left(\frac{\partial w}{\partial t}+2s\lambda^2\varphi|\nabla\beta|^2 w\right)\left(s\lambda\varphi(\nu,\nabla\beta)w+\frac{\partial z}{\partial\nu}e^{s\alpha}\right)d\Sigma$$

$$-\int_{\Sigma}(\lambda^3 s^3\varphi^3|\nabla\beta|^2+s^2\alpha_t\varphi\lambda)w^2(\nabla\beta,\nu)d\Sigma+X_1, \tag{4.23}$$

where we put

$$X_1=\int_Q\Bigg(2s\lambda^2 w(\nabla w,\nabla(\varphi|\nabla\beta|^2))+\frac{1}{2}\frac{\partial}{\partial t}(\lambda^2 s^2\varphi^2|\nabla\beta|^2)w^2-\frac{s\alpha_{tt}w^2}{2}$$

$$+2s\lambda\varphi\sum_{i,k=1}^{n}\left(\frac{\partial w}{\partial x_i}\frac{\partial\beta_{x_k}}{\partial x_i}\frac{\partial w}{\partial x_k}\right)-|\nabla w|^2 s\lambda\varphi\Delta\beta$$

$$+w^2\varphi^3\lambda^3 s^3\sum_{i=1}^{n}\frac{\partial}{\partial x_i}(\beta_{x_i}|\nabla\beta|^2)-\sum_{i=1}^{n}\frac{\partial}{\partial x_i}\left(\frac{s^2\alpha_t\varphi\lambda^2}{2}\frac{\partial\beta}{\partial x_i}\right)w^2\Bigg)dxdt.$$

One can easily prove the following estimate:

$$|X_1|\le c_2\int_Q((s^3\lambda^3\varphi^3+s^2\lambda^4\varphi^3)w^2+(s\lambda\varphi+1)|\nabla w|^2)dxdt$$

$$s\ge 1,\quad \lambda\ge 1, \tag{4.24}$$

where the constant c_2 is independent on s and λ.

Similarly to (4.14) we have

$$\tilde L_1\tilde w+\tilde L_2\tilde w=\tilde f_s \quad \text{in} \quad Q, \tag{4.25}$$

where

$$\tilde f_s(t,x)=ge^{s\tilde\alpha}+s\lambda\tilde\varphi w\Delta\beta+s\lambda^2\tilde\varphi w|\nabla\beta|^2.$$

Thus,

$$||\tilde f_s||^2_{L^2(Q)}=||\tilde L_1\tilde w||^2_{L^2(Q)}+||\tilde L_2\tilde w||^2_{L^2(Q)}+2(\tilde L_1\tilde w,\tilde L_2\tilde w)_{L^2(Q)}. \tag{4.26}$$

Since $\beta(x)|_{\partial\Omega}=0$ we have

$$\tilde w|_{\Sigma}=w|_{\Sigma};\quad \tilde\varphi|_{\Sigma}=\varphi|_{\Sigma};\quad \tilde\alpha|_{\Sigma}=\alpha|_{\Sigma}. \tag{4.27}$$

By similar arguments one can obtain the analog of equality (4.23) for the scalar product $(\tilde L_1\tilde w,\tilde L_2\tilde w)_{L^2(Q)}$, and transform it using (4.27).

$$(\tilde L_1\tilde w,\tilde L_2\tilde w)_{L^2(Q)}=\int_Q(\lambda^4 s^3\tilde\varphi^3|\nabla\beta|^2\tilde w^2$$

$$+s\lambda^2\tilde\varphi|\nabla\beta|^2|\nabla\tilde w|^2+2s\lambda^2\tilde\varphi(\nabla\beta,\nabla\tilde w)^2)dxdt$$

$$-\int_{\Sigma}s\lambda\varphi|\nabla\beta|\left(2\left|\frac{\partial z}{\partial\nu}\right|^2+4s\lambda\varphi z\sum_{i=1}^{n}\nu_i\frac{\partial z}{\partial x_i}\frac{\partial\beta}{\partial x_i}+2s^2\lambda^2\varphi^2\left|\frac{\partial\beta}{\partial\nu}\right|^2 z^2\right.$$

$$-|\nabla z|^2 - 2s\lambda\varphi(\nabla\beta, \nabla z)z - s^2\lambda^2\varphi^2|\nabla\beta|^2 z^2\Big)e^{2s\alpha}d\Sigma$$

$$+\int_\Sigma \left(\frac{\partial w}{\partial t} + 2s\lambda^2\varphi|\nabla\beta|^2 w\right)\left(s\lambda\varphi(\nu, \nabla\beta)w + \frac{\partial z}{\partial \nu}e^{s\alpha}\right)d\Sigma$$

$$+\int_\Sigma (\lambda^3 s^3\varphi^3|\nabla\beta|^2 + s^2\lambda\alpha_t\varphi)w^2(\nabla\beta, \nu)d\Sigma + X_2, \tag{4.28}$$

where $|X_2|$ satisfies the estimate

$$|X_2| \le c_3\int_Q [(s^3\lambda^3\tilde\varphi^3 + s^2\lambda^4\tilde\varphi^3)\tilde w^2 + (s\lambda\tilde\varphi + 1)|\nabla\tilde w|^2]dxdt \;\; \forall\, s \ge 1, \lambda \ge 1. \tag{4.29}$$

Constant c_3 is independent of s and λ.

Hence by virtue of (4.15), (4.22), (4.24) and (4.26) we have

$$||f_s||^2_{L^2(Q)} + ||\tilde f_s||^2_{L^2(Q)} = ||\tilde L_1\tilde w||^2_{L^2(Q)} + ||L_1 w||^2_{L^2(Q)} + ||\tilde L_2\tilde w||^2_{L^2(Q)}$$

$$+||L_2 w||^2_{L^2(Q)} + 2\int_Q (\lambda^4 s^3\varphi^3|\nabla\beta|^4 w^2 + \lambda^4 s^3\tilde\varphi^3|\nabla\beta|^4\tilde w^2$$

$$+s\lambda^2\varphi|\nabla\beta|^2|\nabla w|^2 + s\lambda^2\tilde\varphi|\nabla\beta|^2|\nabla\tilde w|^2 + 2s\lambda^2\varphi(\nabla\beta, \nabla w)^2$$

$$+2s\lambda^2\tilde\varphi(\nabla\beta, \nabla\tilde w)^2)dxdt$$

$$+\int_\Sigma |\nabla\beta|\left(4\frac{\partial w}{\partial t}\frac{\partial z}{\partial \nu}e^{s\alpha} + \left(16s^2\lambda^2\varphi^2 z\sum_{i=1}^n \nu_i\frac{\partial z}{\partial x_i}\frac{\partial\beta}{\partial x_i} - 8s^2\lambda^2\varphi^2 z(\nabla\beta, \nabla z)\right)e^{2s\alpha}\right)d\Sigma$$

$$+X_1 + X_2. \tag{4.30}$$

Hence, taking parameter $\lambda > 0$ sufficiently large in (4.30), by virtue of (4.24) and (4.29) we obtain: There exists $s_0(\lambda) > 0$ such that

$$||\tilde L_1\tilde w||^2_{L^2(Q)} + ||L_1 w||^2_{L^2(Q)} + ||L_2 w||^2_{L^2(Q)} + ||\tilde L_2\tilde w||^2_{L^2(Q)}$$

$$+\int_Q (\lambda^4 s^3\varphi^3 w^2 + \lambda^4 s^3\tilde\varphi^3\tilde w^2 + s\lambda^2\varphi|\nabla w|^2 + s\lambda^2\tilde\varphi|\nabla\tilde w|^2)dxdt$$

$$\le c_4\Big(\int_{Q^{\omega_0}} (\lambda^4 s^3\varphi^3 w^2 + \lambda^4 s^3\tilde\varphi^3\tilde w^2 + s\lambda^2\varphi|\nabla w|^2 + s\lambda^2\tilde\varphi|\nabla\tilde w|^2)dxdt + ||ge^{s\alpha}||^2_{L^2(Q)}$$

$$+||ge^{s\tilde\alpha}||^2_{L^2(Q)} + \int_\Sigma \left(\left|\frac{\partial w}{\partial t}\frac{\partial z}{\partial \nu}\right|e^{s\alpha} + s^2\lambda^2\varphi^2|\nabla z||z|e^{2s\alpha}\right)d\Sigma\Big) \quad \forall s \ge s_0. \tag{4.31}$$

Thus, from (4.10) - (4.13), (4.31) we have

$$\int_Q\left\{\frac{1}{s\varphi}\left(\frac{\partial w}{\partial t}\right)^2 + \frac{1}{s\tilde\varphi}\left(\frac{\partial\tilde w}{\partial t}\right)^2\right.$$

$$\left.+s\lambda^2\varphi|\nabla w|^2 + s\lambda^2\tilde\varphi|\nabla\tilde w|^2 + \lambda^4 s^3\varphi^3 w^2 + \lambda^4 s^3\tilde\varphi^3\tilde w^2\right\}dxdt$$

$$\leq c_5\Big(\int_{Q^{\omega_0}}(\lambda^4s^3\varphi^3w^2+\lambda^4s^3\tilde{\varphi}^3\tilde{w}^2+s\lambda^2\varphi|\nabla w|^2+s\lambda^2\tilde{\varphi}|\nabla\tilde{w}|^2)dxdt+||ge^{s\alpha}||^2_{L^2(Q)}$$

$$+||ge^{s\tilde{\alpha}}||^2_{L^2(Q)}+\int_\Sigma\left(\left|\frac{\partial w}{\partial t}\frac{\partial z}{\partial\nu}\right|e^{s\alpha}+s^2\lambda^2\varphi^2|\nabla z||z|e^{2s\alpha}\right)d\Sigma\Big)\quad\forall s\geq s_0. \tag{4.32}$$

Replacing w by $e^{s\alpha}z$ and $\tilde{w}$ by $e^{s\tilde{\alpha}}z$ respectively in (4.32), we get

$$\int_Q\Bigg\{\left(\frac{1}{s\varphi}\left(\frac{\partial z}{\partial t}\right)^2+s\lambda^2\varphi|\nabla z|^2+s^3\lambda^4\varphi^3z^2\right)e^{2s\alpha}$$

$$+\left(\frac{1}{s\tilde{\varphi}}\left(\frac{\partial z}{\partial t}\right)^2+s\lambda^2\tilde{\varphi}|\nabla z|^2+s^3\lambda^4\tilde{\varphi}^3z^2\right)e^{2s\tilde{\alpha}}\Bigg\}dxdt$$

$$\leq c_6(\lambda)(\int_{Q^{\omega_0}}(\lambda^4s^3\varphi^3z^2e^{2s\alpha}+\lambda^4s^3\tilde{\varphi}^3z^2e^{2s\tilde{\alpha}}+s\lambda^2\varphi|\nabla z|^2e^{2s\alpha}$$

$$+s\lambda^2\tilde{\varphi}|\nabla z|^2e^{2s\tilde{\alpha}})dxdt+||ge^{s\alpha}||^2_{L^2(Q)}+||ge^{s\tilde{\alpha}}||^2_{L^2(Q)}$$

$$+\int_\Sigma\left(\left|\frac{\partial w}{\partial t}\frac{\partial z}{\partial\nu}\right|e^{s\alpha}+s^2\lambda^2\varphi^2|\nabla z||z|e^{2s\alpha}\right)d\Sigma\Big)\quad\forall s\geq s_1. \tag{4.33}$$

Let us consider the function $\rho(x)\in C_0^\infty(w)$, $\rho(x)\equiv 1$ in ω_0. We multiply the equation (4.7) by $s\lambda^2\varphi ze^{2s\alpha}$ scalarly in $L^2(Q)$. Integrating by parts with respect to t and x, applying the Cauchy-Bunyakovskii inequality, we obtain

$$\int_{Q^{\omega_0}}s\lambda^2\varphi|\nabla z|^2e^{2s\alpha}dxdt\leq c_7(\int_{Q^{\omega_1}}s^3\lambda^4\varphi^3z^2e^{2s\alpha}dxdt+||ge^{s\alpha}||^2_{L^2(Q)}), \tag{4.34}$$

where the constant c_7 is independent of s.

Similarly

$$\int_{Q^{\omega_0}}s\lambda^2\tilde{\varphi}|\nabla z|^2e^{2s\tilde{\alpha}}dxdt\leq c_8(\int_{Q^{\omega_1}}s^3\lambda^4\tilde{\varphi}^3z^2e^{2s\tilde{\alpha}}dxdt+||ge^{s\tilde{\alpha}}||^2_{L^2(Q)}), \tag{4.35}$$

where the constant c_8 is independent of s.

By virtue of (4.3), (4.33), (4.34) and (4.35) we have

$$\int_Q\Bigg(\left(\frac{1}{s\varphi}\left(\frac{\partial z}{\partial t}\right)^2+s\lambda^2\varphi|\nabla z|^2+s^3\lambda^4\varphi^3z^2\right)e^{2s\alpha}$$

$$+\left(\frac{1}{s\tilde{\varphi}}\left(\frac{\partial z}{\partial t}\right)^2+s\lambda^2\tilde{\varphi}|\nabla z|^2+s^3\lambda^4\tilde{\varphi}^3z^2\right)e^{2s\tilde{\alpha}}\Bigg)dxdt$$

$$\leq c_9\Big(\int_{Q^{\omega_1}}(\lambda^4s^3\varphi^3z^2e^{2s\alpha}+\lambda^4s^3\tilde{\varphi}^3z^2e^{2s\tilde{\alpha}})dxdt+||ge^{s\alpha}||^2_{L^2(Q)}+||ge^{s\tilde{\alpha}}||^2_{L^2(Q)}$$

$$+\int_\Sigma\left(\left|\frac{\partial w}{\partial t}\frac{\partial z}{\partial\nu}\right|e^{s\alpha}+s^2\lambda^2\varphi^2|\nabla z||z|e^{2s\alpha}\right)d\Sigma\Big)\quad\forall s\geq s_0. \tag{4.36}$$

We observe that for all $\lambda > 0$ there exist constants $c_{10}(\lambda) > 0, c_{11}(\lambda), c_{12}(\lambda) > 0, c_{13}(\lambda)$ such that the following inequalities hold

$$c_{10}(\lambda)|\varphi| \le |\tilde{\varphi}| \le c_{11}(\lambda)|\varphi|, \; c_{12}(\lambda)\frac{1}{|\varphi|} \le \frac{1}{|\tilde{\varphi}|} \le c_{11}(\lambda)\frac{1}{|\varphi|} \;\forall (t,x) \in Q. \tag{4.37}$$

By (4.36), (4.37) we finally obtain (4.4).□

Let us consider the Dirichlet boundary value problem for the Laplace operator

$$\Delta\psi = z \text{ in } \Omega, \quad \psi|_{\partial\Omega} = 0. \tag{4.38}$$

We have

LEMMA 4.2 *There exists a number $\hat{\lambda} > 0$ such that for an arbitrary $\lambda > \hat{\lambda}$ there exists $s_0(\lambda)$ such that for each $s \ge s_0(\lambda)$ the solutions of problem (4.38) satisfy the following inequality*

$$\int_\Omega \left(\frac{1}{s\varphi}\left(\sum_{i,j=1}^n \left|\frac{\partial^2\psi}{\partial x_i \partial x_j}\right|^2\right) + s\varphi|\nabla\psi|^2 + s^3\varphi^3\psi^2\right) e^{2s\alpha}dx + \int_\Sigma s\varphi\left(\frac{\partial\psi}{\partial\nu}\right)^2 e^{2s\alpha}d\Sigma$$

$$\le c_1\Big(\int_\Omega |z|^2 e^{2s\alpha}dx + \int_{\omega_1} s^3\varphi^3\psi^2 e^{2s\alpha}dxdt\Big), \tag{4.39}$$

For the proof of this Lemma see for example in (Fursikov and Imanuvilov (1996b)).

Now we consider the parabolic equation

$$L^*p \equiv \partial_t(\Delta p(t,x)) + \Delta^2 p + B_2^*(\psi,p) + B_1^*(p,\psi) = f, \tag{4.40}$$

$$p|_\Sigma = \left(\Delta p + \sigma\frac{\partial p}{\partial\nu}\right)\Big|_\Sigma = 0, \tag{4.41}$$

$$p(T,\cdot) = p_0, \tag{4.42}$$

where $B_1^*(\cdot,\psi)$, $B_2^*(\psi,\cdot)$ are operators adjoint formally to linear operators $B(\cdot,\psi)$, $B(\psi,\cdot)$ respectively. By definition (3.12) of operator the $B(\psi,\varphi)$ we have

$$B_1^*(h,\psi) = \partial_{x_1}(\Delta\psi\partial_{x_2}h) - \partial_{x_2}(\Delta\psi\partial_{x_1}h), \tag{4.43}$$

$$B_2^*(\psi,h) = \Delta(\partial_{x_1}h\partial_{x_2}\psi - \partial_{x_2}h\partial_{x_1}\psi). \tag{4.44}$$

The following Lemma can be easily proved by the standard energy methods.
LEMMA 4.3 *Let $f \in L^2(Q)$, $p_0 \in W_2^3(\Omega)$ and $p_0|_{\partial\Omega} = (\Delta p_0 + \sigma\frac{\partial p_0}{\partial\nu})|_{\partial\Omega} = 0$. Then there exists the unique solution of problem (4.40)-(4.42) which satisfy the estimate*

$$||p||_{W^{1,2(2)}(Q)} \le c(||f||_{L^2(Q)} + ||p_0||_{W_2^3(\Omega)}). \tag{4.45}$$

We have

LEMMA 4.4 *There exists $s>0$ such that the solutions of the problem (4.40)- (4.42) satisfy the estimate*

$$\int_Q\left(\frac{1}{(T-t)^{12}}|\nabla p|^2+\frac{1}{(T-t)^{18}}|p|^2\right)e^{2s\overline{\alpha}}dxdt\le C(s)\Big(\int_Q f^2e^{2s\overline{\alpha}}dxdt$$
$$+\int_{Q_{\omega_1}}\left(\frac{1}{(T-t)^{18}}p^2+\frac{1}{(T-t)^{12}}|\Delta p|^2\right)e^{2s\overline{\alpha}}dxdt. \qquad (4.46)$$

PROOF. Applying the Carleman estimates (4.4), (4.39) to equation (4.40) for all s sufficiently large we obtain

$$\int_Q\Big(\frac{1}{s\varphi}\left|\frac{\partial p}{\partial t}\right|^2+s\varphi|\nabla\Delta p|^2+s^3\varphi^3|\Delta p|^2+s^4\varphi^4|\nabla p|^2+s^6\varphi^6p^2\Big)e^{2s\alpha}dxdt+$$
$$\int_\Sigma\left(\left|\frac{\partial p_t}{\partial\nu}\right|^2+s^4\varphi^4\left|\frac{\partial p}{\partial\nu}\right|^2\right)e^{2s\alpha}d\Sigma\le C\Big(\int_{Q_1^\omega}(s^3\varphi^3|\Delta p|^2+s^6\varphi^6|p|^2)e^{2s\alpha}dxdt$$
$$+\int_\Sigma\left(\left|\frac{\partial p}{\partial t}\frac{\partial\Delta p}{\partial\nu}\right|+s^2\varphi^2|\nabla\Delta p|\left|\frac{\partial p}{\partial\nu}\right|\right)e^{2s\alpha}d\Sigma\Big). \qquad (4.47)$$

Applying the Cauchy-Bynyakovskii inequality to the last integral in the right hand side of (4.47) we have

$$\int_Q\left(\frac{1}{s\varphi}\left|\frac{\partial p}{\partial t}\right|^2+s\varphi|\nabla\Delta p|^2+s^3\varphi^3|\Delta p|^2+s^4\varphi^4|\nabla p|^2+s^6\varphi^6p^2\right)e^{2s\alpha}dxdt$$

$$\le C\Big(\int_{Q^{\omega_1}}(s^3\varphi^3|\Delta p|^2+s^6\varphi^6|p|^2)e^{2s\alpha}dxdt+\int_\Sigma\left(\left|\frac{\partial^2 p}{\partial t\partial\nu}\right|^2+|\nabla\Delta p|^2\right)e^{2s\alpha}d\Sigma\Big). \qquad (4.48)$$

Let $x_0\in\partial\Omega$. Set $\overline{\alpha}(t)=\alpha(x_0,t)$, $q(t,x)=e^{s\overline{\alpha}(t)}p(t,x)$, $r(t,x)=e^{\overline{\alpha}(t)}f(t,x)$. The function $q(t,x)$ satisfy equations

$$\partial_t(\Delta q(t,x))-l_1q+\Delta^2q+B_2^*(\psi,q)+B_1^*(q,\psi)=r, \qquad (4.49)$$

$$q|_\Sigma=\left(\Delta q+\sigma\frac{\partial q}{\partial\nu}\right)\Big|_\Sigma=0, \qquad (4.50)$$

$$q(0,\cdot)=q(T,\cdot)=0, \qquad (4.51)$$

where $l_1=\frac{\partial}{\partial t}e^{s\overline{\alpha}}$. By Lemma 4.3 the solution of problem (4.49)-(4.51) satisfy the estimate

$$\|q\|_{W^{1,2(2)}(Q)}\le c(\|r\|_{L^2(Q)}+\|l_1\Delta q\|_{L^2(Q)}). \qquad (4.52)$$

From (4.52) by (4.48) and the Sobolev imbedding theorem we have

$$\|\Delta q\|_{L^2(0,T:W_2^{\frac{1}{2}}(\partial\Omega))}+\left\|\frac{\partial^2 q}{\partial\nu\partial t}\right\|_{L^2(\Sigma)}\le c(\|s\varphi^{\frac{4}{3}}\Delta q\|_{L^2(Q)}+\|r\|_{L^2(Q)}). \quad (4.53)$$

The inequalities (4.48) and (4.54) yield:

$$\int_Q \left(\frac{1}{s\varphi} \left| \frac{\partial p}{\partial t} \right|^2 + s\varphi |\nabla \Delta p|^2 + s^3\varphi^3 |\Delta p|^2 + s^4\varphi^4 |\nabla p|^2 + s^6\varphi^6 p^2 \right) e^{2s\alpha} dx dt$$

$$\leq C(\int_{Q^{\omega_1}} (s^3\varphi^3 |\Delta p|^2 + s^6\varphi^6 |p|^2) e^{2s\alpha} dx dt + s \int_Q f^2 e^{2s\alpha} dx dt). \qquad (4.54)$$

Thus the statement of the our Lemma follows from (4.54) and estimate (4.45).□

5 Exact Controllability of the Linearized Navier–Stokes System

In this section we will prove an existence theorem for the exact controllability problem (2.23)-(2.25).In the previous section we proved an estimate (3.46) which solves observability problem for the operator (3.40). The following Lemma convert observability result into controllability one.

LEMMA 5.1 *Let* $\hat{\psi} \in W^{1,2(2)}(Q)$, $f \in L_2(Q, e^{2s\eta})$,$w \in W_2^3(\Omega)$ *and satisfy (2.24). Then there exists solution of problem (2.23)-(2.25)* $w \in W_2^{1,2(0)}(Q) \cap L^2(Q, e^{2s\eta}/(T-t)^{18})$, $u \in L^2(0,T : (W_2^2(\Omega))^*)$, $supp\, u \in Q^{\omega_0}$ *which satisfy the estimate*

$$||(w,u)||_{W_2^{1,2(0)}(Q) \cap L^2(Q, e^{s\eta}/(T-t)^{18}) \times L^2(Q, e^{2s\eta})} \leq c(||w_0||_{W_2^1(\Omega)} + ||f||_{L^2(Q, e^{2s\eta})}). \qquad (5.1)$$

The proof of Lemma 5.1 is similar to the proof of Theorem 2.1 from (Fursikov and Imanuvilov (1996b)).

Let us consider the initial boundary value problem for the linearized Navier-Stokes system with the slip boundary conditions

$$Lw = -\frac{\partial \Delta w}{\partial t} + \Delta^2 w + B(w, \hat{\psi}) + B(\hat{\psi}, w) = f \text{ in } Q, \qquad (5.2)$$

$$w|_{\partial\Omega} = \left(\Delta w + \sigma \frac{\partial w}{\partial \nu} \right) \Big|_{\partial\Omega} = 0, \qquad (5.3)$$

$$w(0, \cdot) = w_0. \qquad (5.4)$$

We have

LEMMA 5.2 *Let* $\hat{\psi} \in W^{1,2(2)}(Q)$, $w_0 \in W_2^3(\Omega)$, *and satisfy (2.24). Then for an arbitrary* $f \in L^2(0,T; (W_2^s(\Omega))^*)$ *there exists the unique solution* $w \in W^{1,2(2-s)}(Q)$ *of the problem (5.2)- (5.4) which satisfies the estimate*

$$||w||_{W^{1,2(2-s)}(Q)} \leq c(||w_0||_{W_2^3(\Omega)} + ||f||_{L^2(Q)}), \qquad (5.5)$$

where $s = 0, 1, 2$.

This lemma can be easily proved by the standard energy method.

The main result of this section is the following lemma.

LEMMA 5.3 *Let* $\hat{\psi} \in W^{1,2(2)}(Q)$, $f \in L_2(Q,\theta)$, $w_0 \in W_2^3(\Omega)$ *and satisfy (2.24). Then there exists solution of problem (2.23)-(2.25)* $w \in W^{1,2(0)}(Q) \cap L^2(Q, e^{2s\eta}/(T-t)^{18})$, $u \in$ *which satisfy the estimate*

$$\|(w,u)\|_{Y(Q)\times L^2(Q)} \le c(\|w_0\|_{W_2^3(\Omega)} + \|f\|_{L^2(Q,\theta)}). \tag{5.6}$$

PROOF. Note, that without loss of generality we can assume that $w_0 \equiv 0$. Indeed, let $w_0 \not\equiv 0$. For this case we are looking for solution of problem (2.23)-(2.25) in the form

$$w(t,x) = \overline{w}(t,x) + \ell_2(t)\tilde{w}(t,x),$$

where $\tilde{w}(t,x)$ is a solution of the boundary value problem (5.2)-(5.4) with initial datum $w_0, f \equiv 0$, and $\ell_2(t)$ is a function which satisfy the following property:

$$\ell_2(t) = 1 \ \forall\, t \in [0, T/4], \quad \ell_2(t) = 0 \ \forall\, t \in [3T/4, T]. \tag{5.7}$$

The function $\overline{w}$ is a solution of exact controllability problem

$$L\overline{w} = L(\ell_2\tilde{w}) + f + u \text{ in } Q,$$
$$\overline{w}|_{\partial\Omega} = \left(\Delta\overline{w} + \sigma\frac{\partial\overline{w}}{\partial\nu}\right)\Big|_{\partial\Omega} = 0,$$
$$\overline{w}(0,\cdot) = 0.$$

From Lemma 5.2 we get $L(\ell_3 w) \in L^2(Q)$. By virtue of (5.7) $L(\ell_3 w) \in L^2(Q,\theta)$. Hence we reduced our original problem to problem (2.23)-(2.25) with initial datum $w_0 \equiv 0, f \in L^2(Q,\theta)$.

Now let us assume that $f \in L^2(Q, e^{s\eta})$. By virtue of Lemma 5.1 there exists solution of the problem (2.23)-(2.25) (w,u) which satisfies the estimate (5.1). Let $x_0 \in \partial\Omega$. Set $\overline{\eta(t)} = \eta(t,x_0)$. Denote $q(t,x) = e^{s\overline{\eta(t)}}w(t,x)$. Then the function q satisfy the system of equations

$$Lq = \overline{\eta}_t\Delta q + (u+f)e^{s\overline{\eta}} \text{ in } Q, \tag{5.8}$$

$$q|_{\partial\Omega} = \left(\Delta q + \sigma\frac{\partial q}{\partial\nu}\right)\Big|_{\partial\Omega} = 0, \tag{5.9}$$

$$q(0,\cdot) = 0. \tag{5.10}$$

Multiplying equation (5.8) by q scalarly in $L^2(\Omega)$ and integrating by parts we obtain

$$\frac{1}{2}\frac{d}{dt}\|\nabla q\|^2_{L^2(\Omega)} + \|\Delta q\|^2_{L^2(\Omega)} \le \frac{1}{2}\|\Delta q\|^2_{L^2(\Omega)}$$
$$+c(\|\nabla q\|^2_{L^2(\Omega)} + \|(u+f)e^{s\overline{\eta}}\|^2_{L^2(\Omega)} + \|\overline{\eta}_t q\|^2_{L^2(\Omega)}). \tag{5.11}$$

Then (5.11), (5.5) and Gronwall's inequality imply

$$\|q\|_{C(0,T;W_2^1(\Omega))} + \|q\|_{L^2(0,T;W_2^2(\Omega))} \le c. \tag{5.12}$$

Using (5.12) we can estimate $\frac{\partial q}{\partial t}$

$$\left\| \frac{\partial q}{\partial t} \right\|_{L^2(Q)} \le c\|q\|_{L^2(Q)} \le c.$$

Thus finally we obtain

$$\|q\|_{W^{1,2(0)}(Q)} \le c. \tag{5.13}$$

Let $\rho_1 \in C^\infty(\Omega)$, $\rho_1(x) = 1$ for $x \in \Omega \setminus \omega_1$; $\rho_1(x) = 1$ $x \in \omega_1$. Denote $q_1(t,x) = \rho_1 we^{\frac{99s\overline{\eta}}{100}}$, $f_1(t,x) = \rho_1 fe^{\frac{99s\overline{\eta}}{100}}$. By (5.2),(5.4) function q_1 satisfy the system of equations

$$Lq_1 = \overline{\eta}_t \Delta q_1 + u_1 + f_1 \text{ in } Q, \tag{5.14}$$

$$q_1|_{\partial\Omega} = \left(\Delta q_1 + \sigma\frac{\partial q_1}{\partial \nu}\right)\bigg|_{\partial\Omega} = 0, \tag{5.15}$$

$$q_1(0,\cdot) = 0, \tag{5.16}$$

where $u_1 \in L^2(Q)$, $\operatorname{supp} u_1 \subset Q^{\omega_1}$. Then by virtue of Lemma 5.2

$$q_1 \in W^{1,2(1)}(Q). \tag{5.17}$$

Now we apply this trick again by introducing the function $q_2(t,x) = \rho_1^4 we^{\frac{98s\overline{\eta}}{100}}$, $f_1(t,x) = \rho_1 fe^{\frac{99s\overline{\eta}}{100}}$. This function satisfy the system

$$Lq_2 = \overline{\eta}_t \Delta q_2 + u_2 + f_2 \text{ in } Q, \tag{5.18}$$

$$q_2|_{\partial\Omega} = \left(\Delta q_2 + \sigma\frac{\partial q_2}{\partial \nu}\right)\bigg|_{\partial\Omega} = 0, \tag{5.19}$$

$$q_2(0,\cdot) = 0, \tag{5.20}$$

where right hand side of (5.17) belongs to the space $L^2(Q)$ by virtue of (5.16). Thus applying the Lemma 5.2 we got the statement of our lemma.

Now let $f(t,x) \in L^2(Q,\theta)$ be an arbitrary function. We can write it in the form

$$f(t,x) = \chi_\omega f(t,x) + (1-\chi_\omega)f(t,x).$$

By definition of the space $L^2(Q,\theta)$ function $(1-\chi_\omega)f$ belongs to $L^2(Q,e^{s\eta})$. Hence there exists a solution $(w,u) \in Y(Q) \times U_\omega(Q)$ of problem (2.23)-(2.25) for initial data $(w_0,(1-\chi_\omega)f)$. Obviously the pair $(w, u-\chi_\omega f)$ is the solution of (2.23)-(2.25) for initial date (w_0, f). □

References

Alekseev V.M., Tikhomirov V.M. and Fomin S.V.(1987): Optimal control. Consultants Bureau, New York

Coron J.-M. (1993): Contrôlabilité exacte frontiere de l'équation d'Euler des fluids parfaits incompresibles bidimensionnesls. C.R.Acad. Sci., Paris. T. 317. Série I. 271-276

Coron J.-M. (1996): On the Controllability of the 2-D incompressible Navier-Stokes Equations with the Navier-Slip boundary conditions. ESIAM: Control, Optimization and Calculus of Variations, **1**, 35-75

Coron J.-M. (1996): On the controllability of 2-D incompressible perfect fluids. J. Math. Pures et Appl. **75**, 155–188

Coron J.-M., Fursikov A.V. (1996): Global exact controllability of the 2D Navier-Stokes equations on manifold without boundary. Russian Journal Of Math. Physics, **4**, 3, 1-20

Fursikov A.V. (1995): Exact boundary zero controllability of three dimensional Navier-Stokes equations. J. of Dynamical and Control Syst., **1**, 3, 325-350

Fursikov A.V., Imanuvilov O.Yu. (1994): On exact boundary zero-controllability of two-dimensional Navier-Stokes equations, Acta Applicandae Mathematicae, **37**, 67-76

Fursikov A.V., Imanuvilov O.Yu. (1995): On controllability of certain systems simulating a fluid flow. in Flow Control, IMA vol. Math. Appl., **68**, Ed. by M. D. Gunzburger, Springer-Verberg, New York, 148-184

Fursikov A.V., Imanuvilov O.Yu. (1996): Local exact controllability for 2-D Navier-Stokes equations. Math. Sbornik, **59**, 8 (in Russian)

Fursikov A.V., Imanuvilov O.Yu. (1996): Controllability of evolution equations. Lecture Notes Series **34**, Recearch Institute of Mathematics Global Analysis Research Center

Fursikov A.V., Imanuvilov O.Yu. (1996): Local exact controllability of the Navier-Stokes equations. C.R. Acad. Sci. Paris, **323**, Série I, 275-280

Fursikov A.V., Imanuvilov O.Yu. (1997): Local exact controllability of the Boussinesq equation. SIAM J. Cont. Optim., (to appear)

Imanuvilov O.Yu. (1993): Exact boundary controllability of the parabolic equation. Russian Math. Surveys **48**, 211-212

Imanuvilov O.Yu. (1994): Exact controllability of the semilinear parabolic equation. Vestnik R.U.D.N. ser. Math. **1** 109–116.(in Russian)

Imanuvilov O.Yu. (1995): Boundary controllability of parabolic equations. Russian Acad. Sci. Sb. Math., **186**, 6, 109–132

Isakov V. (1992): Carleman type estimates in an anisotropic case and applications. J. Diff. Equ. **8**, 193–206

Lions J.-L. (1990): Are there connections between turbulence and controllability? 9e Conférence internationale de l'INRIA. Antibes. 12-15 juin 1990

Lions J.-L. and Magenes E. (1968): Problemes aux limites nonhomogenes et applications v.1. Dunod, Paris

Nonexistence of Global Solutions to Nonlinear Wave Equations

Varga Kalantarov

Hacettepe University, Faculty of Science
Department of Mathematics, 06532 Beytepe
Ankara, Turkey

Abstract. Sufficient conditions for global nonexistence of solutions to Cauchy problem for a class of second order nonlinear differential-operator equations and applications to some nonlinear wave equations are given.

1 Introduction

The purpose of this article is to find sufficient conditions on global nonexistence of solutions to some class of nonlinear wave equations.There are many investigations on global nonexistence of solutions to initial and initial-boundary value problems for nonlinear evolutionary PDE's. Various examples of parabolic equations where solutions blow up in a finite time are given in Ladyzhenskaya, Solonnikov, Ural'tseva 1968. Whereas in the articles of Keller 1957, Kaplan 1963, Fujita 1969, Glassey 1973 , second order parabolic and hyperbolic equations of the following form are considered

$$u_t + Lu = f(u), \tag{1}$$

$$u_{tt} + Lu = f(u), \tag{2}$$

where L is a Laplacian or a self-adjoint uniformly elliptic operator. In these articles, sufficient conditions for blowing up of solutions of initial -boundary value problems for equations like (1) and (2) are given. Their methods are based on positiveness of the Green's function or positiveness of the first eigenvalue of the elliptic operator L .

In the articles Levine 1974a ,Levine 1973 it is developed a very power full method (the so called *concavity method*) for finding conditions of global nonexistence for the solutions to the Cauchy problem for differential-operator equations of the form

$$Pu_{tt} + Au = F(u), \tag{3}$$

$$Pu_t + Au = F(u), \tag{4}$$

In the equations (3), (4) P and A are linear symmetric operators, with P positive and A nonnegative. The operator $F(u)$ is a nonlinear potential operator satisfying the condition:

$$(F(u), u)) - bG(u) \geq 0,$$

where b is a positive number greater than 2, $G(u)$ is the functional, whose gradient is the operator $F(u)$, and $(\cdot, \cdot)$ is the inner product of the Hilbert space H,on which the equations are considered.

In the articles Knops et al. 1974, Levine 1974b, Levine 1974c, Straughan 1975, Levine Paine 1974, using the concavity method of H.A.Levine, global nonexistence theorems are proven for differential-operator equations with the nonlinear principle part, for second order differential-operator equations with a dissipative term, for linear parabolic and hyperbolic equation with nonlinear boundary conditions and for various equations and systems of equations from the continuum mechanics.

The idea of the concavity method is based on a construction of some positive functional $\Psi(t) = \psi(u(t))$,which is defined in terms of the local solution of the problem (the local solvability of the problem is therefore required) and proving that the function $\Psi(t)$ satisfies the inequality (5) given in the following statement:

Lemma 1.(see Levine 1974a, Levine 1973) Let $\Psi(t)$ be a positive, twice differentiable function, which satisfies, for $t > 0$, the inequality

$$\Psi''(t)\Psi(t) - (1+\alpha)\left[\Psi'(t)\right]^2 \geq 0 \tag{5}$$

with some $\alpha > 0$. If $\Psi(0) > 0$ and $\Psi'(0) > 0$, then there exists a time $t_1 \leq \frac{\Psi(0)}{\alpha\Psi'(0)}$ such that $\Psi(t) \to +\infty$ as $t \to t_1$.

Thanks to this approach the authors were able not only to cover many problems considered in Fujita 1969, Glassey 1973, Kaplan 1963, Keller 1957, but also a wide class of new problems, to which the methods mentioned in the beginning are not applicable.

However as it is noted in Levine 1974a and Levine 1973, in the framework of the concavity method the condition of symmetry and non negativity of the operator A are essential. The *generalized concavity method* suggested in Kalantarov, Ladyzhenskaya 1977 allows one to consider equations of type (3) and (4), where the operator A is not necessarily symmetric and nonnegative. This method, generalizing the concavity method, relies on the following lemma:

Lemma 2. (Kalantarov, Ladyzhenskaya 1977) Suppose that a positive, twice-differentiable function $\Psi(t)$ satisfies for $t > 0$ the inequality

$$\Psi''(t)\Psi(t) - (1+\alpha)\left[\Psi'(t)\right]^2 \geq -2C_1\Psi'(t)\Psi(t) - C_2\left[\Psi(t)\right]^2, \tag{6}$$

where $\alpha > 0, C_1, C_2 \geq 0$.

If $\Psi(0) > 0$, $\Psi'(0) > -\gamma_2 \cdot \alpha^{-1} \cdot \Psi(0)$, and $C_1 + C_2 > 0$, then $\Psi(t) \to +\infty$ as

$$t \to t_1 \leq \frac{1}{2\sqrt{C_1^2 + \alpha C_2}} \cdot \ln \frac{\gamma_1\Psi(0) + \alpha\Psi'(0)}{\gamma_2\Psi(0) + \alpha\Psi'(0)},$$

where $\gamma_1 = -C_1 + \sqrt{C_1^2 + \alpha C_2}, \gamma_2 = -C_1 - \sqrt{C_1^2 + \alpha C_2}$.

In this case, the local solvability of the corresponding problem is also assumed. In order to find sufficient conditions for global nonexistence, it is necessary to construct a positive functional $\Psi(t) = \psi(u(t))$, and prove that it satisfies the inequality (6). Let us note that, as in the case of the concavity method a good choice of such a functional is one of the essential ingredients of the proof.

Later on, the concavity method as well as the generalized concavity method was used in many articles on global behavior of solutions to various nonlinear evolutionary equations. (see Kalantarov 1983, Kalantarov 1987, Turitsyn 1993, Palais 1988)

2 Abstract wave equation

Let D be a dense linear subspace of a real Hilbert space H with the inner product $(\cdot,\cdot)$ and the corresponding norm $\|\cdot\|$. We will denote by u a vector-function with domain $[0,T)$ and range D. Suppose that P, Q and A are symmetric linear operators defined on D such that the operators A, Q are nonnegative and the operator P is positive definite. Let $B(\cdot,\cdot)$ be a nonlinear operator defined on $D \times D$ and $F(\cdot,\cdot)$ be a nonlinear operator defined on $[0,T) \times D$, we also assume that for each fixed t the operator $F(t,u)$ is the Frechet differential of some nonlinear functional $G(t,u)$:

$$\frac{d}{d\tau} G(t, u(\tau)) = (F(t, u(\tau)), u_\tau(\tau)) \tag{7}$$

Moreover, $G(t,u)$ smoothly depends on t so that for $u(t) \in C^1(0,T;D)$ we have:

$$\frac{d}{dt} G(t, u(t)) = (F(t, u(t)), u_t(t)) + G_t(t, u(t)) \tag{8}$$

In the Hilbert space H consider the following Cauchy problem :

$$Pu_{tt} + Qu_t + Au = B(u, u_t) + F(t, u), \tag{9}$$

$$u(0) = u_0, u_t(0) = u_1, \tag{10}$$

For the sake of simplicity we assume that $u(t)$ is a strong solution of (9),(10) that is all terms in (9) are elements of $L_2(0,T;H)$ and $u(.), u_t(.) \in C(0,T;H)$. Our main result is the following theorem:

Theorem 1. Suppose that the operators P, Q, A, B and F satisfy all the conditions mentioned above and let $u(t)$ be a strong solution of the equation (9). In addition assume that

$$(F(t,u), u) \geq 2(1+2\alpha) G(t,u), \forall u \in D, \tag{11}$$

for some $\alpha > 0$,

$$G_t(t,u) \geq M_1 G(t,u), \forall u \in D, \tag{12}$$

for some $M_1 > 0$,

$$|(B(u,v),u)| \le d_1(Au,u) + d_2(Pv,v) + M_2(Pu,u), \forall u,v \in D, \tag{13}$$

for some $d_1 \in (0,2\alpha]$,$d_2 \in (0,2\alpha)$ and $M_2 > 0$,

$$|(B(u,v),v)| \le \frac{1}{2} M_1 \left[(Au,u) + (Pv,v)\right], \forall u,v \in D. \tag{14}$$

Assume also that the initial elements satisfy:

$$(Pu_0,u_0) + (Qu_0,u_0) > 0,$$

$$2(Pu_1,u_0) + 2(Qu_1,u_0) > -\gamma_2 \alpha_0^{-1} \left[(Pu_0,u_0) + (Qu_0,u_0)\right],$$

and

$$-\frac{1}{2}(Pu_1,u_1) - \frac{1}{2}(Au_0,u_0) + G(0,u(0)) \ge 0. \tag{15}$$

Then

$$\lim_{t\to t_1}(Pu(t),u(t)) + \int_0^t (Qu(s),u(s))ds = \infty,$$

for some t_1 satisfying,

$$t_1 \le \frac{1}{2\sqrt{(1+\alpha_0)^2 + \alpha_0(2M_2+1+\alpha_0)}} \times$$

$$\times \ln \frac{\gamma_1\left[(Pu_0,u_0) + (Qu_0,u_0)\right] + \alpha_0\left[(Pu_1,u_0) + (Qu_0,u_0)\right]}{\gamma_2\left[(Pu_0,u_0) + (Qu_0,u_0)\right] + \alpha_0\left[(Pu_1,u_0) + (Qu_0,u_0)\right]},$$

where $\alpha_0 = \alpha - \frac{d_2}{2}$, $\gamma_1 = -(1+\alpha_0) + \sqrt{(1+\alpha_0)^2 + \alpha_0(2M_2+1+\alpha_0)}$,
$\gamma_2 = -(1+\alpha_0) - \sqrt{(1+\alpha_0)^2 + \alpha_0(2M_2+1+\alpha_0)}$

Proof. Consider the following function:

$$\Psi(t) = (Pu(t),u(t)) + \int_0^t (Qu(s),u(s))ds + (Qu_0,u_0).$$

It is clear that

$$\Psi'(t) = 2(Pu_t(t),u(t)) + 2\int_0^t (Qu_s(s),u(s))ds + (Qu_0,u_0) \tag{16}$$

and

$$\Psi''(t) = 2(Pu_{tt}(t) + Qu_t(t),u(t)) + 2(P(u_t(t),u_t(t)).$$

Using the equation (9) we get from the last relation:

$$\Psi''(t) = -2(Au,u) + 2(B(u,u_t),u) + 2(F(t,u),u) + 2(Pu_t,u_t)$$

Due to condition (11) we have:

$$\Psi''(t) \geq -2(Au,u) + 2(Pu_t,u_t) + 4(1+2\alpha)G(t,u) + 2(B(u,u_t),u) =$$
$$4(1+2\alpha)\left[-\frac{1}{2}(Au,u) - \frac{1}{2}(Pu_t,u_t) + G(t,u)\right] + 4\alpha(Au,u)+$$
$$4(1+\alpha)(Pu_t,u_t) + 4(1+\alpha)(Pu_t,u_t) + 2(B(u,u_t),u). \qquad (17)$$

Using the notation

$$E(t) = -\frac{1}{2}(Au,u) - \frac{1}{2}(Pu_t,u_t) + G(t,u)$$

and the condition (13) we get from (17):

$$\Psi''(t) \geq 4(1+2\alpha)E(t) + (4\alpha - 2d_1)(Au,u)+$$
$$+ (4(1+\alpha) - 2d_2)(Pu_t,u_t) - 2M_2(Pu,u). \qquad (18)$$

Since $d_1 \leq 2\alpha$, the inequality (18) implies:

$$\Psi''(t) \geq 4(1+2\alpha)E(t) + 4(1+\alpha_0)(Pu_t,u_t) - 2M_2(Pu,u), \qquad (19)$$

where $\alpha_0 = \alpha - \frac{d_2}{2}$. Multiplication of the equation (9) by u_t gives us the following equality:

$$\frac{d}{dt}(\frac{1}{2}(Pu_t,u_t) + \frac{1}{2}(Au,u) - G(t,u)) + (Qu_t,u_t) = (B(u,u_t),u_t) - G_t(t,u).$$

Here we have used (8). So, thanks to (12) we have:

$$\frac{d}{dt}E(t) = (Qu_t,u_t) - (B(u,u_t),u_t) + G_t(t,u) \geq$$
$$\geq (Qu_t,u_t) - (B(u,u_t),u_t) + M_1G(t,u).$$

Using the condition (14) we can easily get from the last inequality:

$$\frac{d}{dt}E(t) \geq (Qu_t,u_t) + M_1E(t). \qquad (20)$$

Integrating the inequality (20) we obtain:

$$E(t) \geq E(0)\cdot\exp(M_1t) + \int_0^t (Qu_s(s),u_s(s))ds.$$

From (15) it follows that $E(0)$ is nonnegative. Therefore, the last inequality implies:

$$E(t) \geq \int_0^t (Qu_s(s),u_s(s))ds \qquad (21)$$

It follows then from (19) and (21) that

$$\Psi''(t) \geq 4(1+2\alpha)\int_0^t (Qu_s(s), u_s(s))ds + 4(1+\alpha_0)(Pu_t(t), u_t(t)) - 2M_2(Pu(t), u(t)) \tag{22}$$

Finally using the inequality (22) and the relation (16) we get

$$\Psi''\Psi - (1+\alpha_0)\left[\Psi'\right]^2 \geq$$

$$4(1+\alpha_0)\left\{\left[(Pu_t, u_t) + \int_0^t (Qu_s, u_s)ds\right]\left[(Pu, u) + \int_0^t (Qu, u)ds + (Qu_0, u_0)\right] - \right.$$

$$\left.\left[(Pu_t, u) + \int_0^t (Qu_s, u)ds + \frac{1}{2}(Qu_0, u_0)\right]^2\right\} - 2M_2(Pu, u)\Psi =$$

$$4(1+\alpha_0)\left\{\left[(Pu_t, u_t) + \int_0^t (Qu_s, u_s)ds\right]\cdot\left[(Pu, u) + \int_0^t (Qu, u)ds\right] - \right.$$

$$\left.\left[(Pu_t, u) + \int_0^t (Qu_s, u)ds\right]^2\right\} - 2M_2\Psi^2 + (Qu_0, u_0)\left[(Pu_t, u_t) + \int_0^t (Qu_s, u_s)ds\right] -$$

$$4(1+\alpha_0)(Qu_0, u_0)\left[(Pu_t(t), u(t)) + \int_0^t (Qu_s(s), u(s))ds\right] - (1+\alpha_0)(Qu_0, u_0)^2. \tag{23}$$

It follows from the Cauchy-Schwarz inequality that the expression $\left[(Pu, u) + \int_0^t (Qu, u)ds\right]\cdot\left[(Pu_t, u_t) + \int_0^t (Qu_s, u_s)ds\right] - \left[(Pu_t, u) + \int_0^t (Qu_s, u)ds\right]^2$ is non-negative. Therefore, the inequality (23) implies :

$$\Psi''(t)\Psi(t) - (1+\alpha_0)\left[\Psi'(t)\right]^2 \geq$$

$$-4(1+\alpha_0)(Qu_0, u_0)\left[(Pu_t(t), u(t)) + \int_0^t (Qu_s(s), u(s))ds\right] - 2M_2\left[\Psi\right]^2 -$$

$$-(1+\alpha_0)(Qu_0, u_0)^2 \geq -2(1+\alpha_0)\Psi(t)\Psi'(t) - (2M_2 + 1 + \alpha_0)\left[\Psi(t)\right]^2.$$

Thus we have proved that the function $\Psi(t)$ satisfies the inequality (6) with $C_1 = 1+\alpha_0$ and $C_2 = 2M_2+1+\alpha_0$. So the conclusion of the theorem follows from the Lemma 2, since all the conditions of this lemma are satisfied.

Suppose that $B(u,v) = B_1u + B_2v$, where B_1 and B_2 are linear operators.In this case following Kalantarov, Ladyzhenskaya 1977 under some restrictions on the operators $B_i,\ i = 1,2.$ we can show that the condition (12) can be replaced by the weaker condition:

$$G_t(t,u) \geq 0. \tag{24}$$

Let $u(t)$ be a solution of the equation (9), then the function $v(t) = u(t)e^{-mt}$ satisfies the equation:

$$Pv_{tt} + \hat{Q}\, v_t + \hat{A}\, v = \hat{B}(v, v_t) + \hat{F}\,(t,v), \tag{25}$$

where $\hat{Q} = 2mP + Q$, $\hat{A} = m^2P + mQ + A$, $\hat{B}(v,v_t) = (B_1 + mB_2)v + B_2v_t$ and $\hat{F}\,(t,v) = e^{-mt}F(t,e^{mt}v)$. It is clear that the inequality (11) remains valid for the operator $\hat{F}\,(t,v)$ and its potential $\hat{G}\,(t,v) = e^{-2mt}G(t,e^{mt}v)$, and

$$\hat{G}_t\,(t,v) = e^{-2mt}G_t(t,e^{mt}v) - 2m\,\hat{G}\,(t,v) + m(\hat{F}\,(t,v),v).$$

Due to the condition (11) it follows from the last equality that if G satisfies (24) then

$$\hat{G}_t\,(t,v) \geq 4\alpha m\,\hat{G}\,(t,v).$$

Taking $m = \frac{M_1}{4\alpha}$, we satisfy the condition (12).

Suppose now that B_1 and B_2 are linear symmetric or skew symmetric operators, which satisfy the inequalities:

$$|(B_iu,v)| \leq b_i(Au,u) + c_i(Pv,v) + D_i(Pu,u), i = 1,2., \forall u,v \in D, \tag{26}$$

and

$$|(B_2u,u)| \leq D_3(Pu,u), \forall u \in D, \tag{27}$$

where $b_i, c_i,\ i = 1,2$ and $D_i,\ i = 1,2,3$, are some nonnegative numbers.

Using the conditions (26) and (27) we can easily show that

$$\left|(\hat{B}(u,v),u)\right| \leq (b_1+b_2)(Au,u) + c_2(Pv,v) + (c_1+D_1+D_2+mD_3)(Pu,u),$$

$$\left|(\hat{B}(u,v),v)\right| \leq (b_1+mb_2)(Au,u) + (c_1+mc_2+D_3)(Pv,v) + (D_1+mD_2)(Pu,u).$$

Finally it is easy to see that, if

i) $b_1 + b_2 \leq 2\alpha$, $c_2 < 2\alpha$,

ii) $b_1 + \frac{M_1}{4\alpha} b_2 \leq \frac{1}{2} M_1$, $c_1 + \frac{M_1}{4\alpha} c_2 + D_3 \leq \frac{1}{2} M_1$,

iii) $D_1 + \frac{M_1}{4\alpha} D_2 \leq \frac{M_1^3}{32\alpha^2}$,

then the operator $\hat{B}(u,v)$ satisfies the conditions

$$| (\hat{B}(u,v), u) | \leq d_1(\hat{A} u, u) + d_2(Pv, v) + M_2(Pu, u), \tag{28}$$

$$| (\hat{B}(u,v), v) | \leq \frac{1}{2} M_1 \left[(\hat{A} u, u) + (Pv, v) \right], \tag{29}$$

with $d_1 = b_1 + b_2$, $d_2 = c_2$, $M_2 = c_1 + D_1 + D_2 + mD_3$.

Thus we can apply the Theorem 1 and see that true

Theorem 2. Let $v(t)$ be a strong solution of the equation (25), satisfying the initial conditions

$$v(0) = v_0, v_t(0) = v_1, \tag{30}$$

and suppose that:

1.the operators B_i, $i = 1,2$, satisfy (26),(27), where the parameters b_i, c_i, $i = 1,2$ and D_i, $i = 1,2,3$, satisfy i)-iii) ,

2. $(Pv_0, v_0) + (\hat{Q} v_0, v_0) > 0$,

$$2(Pv_1, v_0) + 2(\hat{Q} v_1, v_0) > -\gamma_2 \alpha_0^{-1} \left[(Pv_0, v_0) + (\hat{Q} v_0, v_0) \right],$$

$$-\frac{1}{2}(Pv_1, v_1) - \frac{1}{2}(\hat{A} v_0, v_0) + \hat{G}(0, v_0) \geq 0.$$

Then

$$\lim_{t \to t_0} (Pv(t), v(t)) + \int_0^t (\hat{Q} v(s), v(s)) ds = \infty,$$

for some

$$t_0 \leq \frac{1}{2\sqrt{(1+\alpha_0)^2 + \alpha_0(2M_2 + 1 + \alpha_0)}} \ln \frac{\gamma_1 \hat{\Psi}(0) + \alpha_0 \hat{\Psi}'(0)}{\gamma_2 \hat{\Psi}(0) + \alpha_0 \hat{\Psi}'(0)},$$

where

$\alpha_0 = \alpha - \frac{c_2}{2}$, $\gamma_1 = -(1+\alpha_0) + \sqrt{(1+\alpha_0)^2 + \alpha_0(2M_2 + 1 + \alpha_0)}$,

$\gamma_2 = -(1+\alpha_0) - \sqrt{(1+\alpha_0)^2 + \alpha_0(2M_2 + 1 + \alpha_0)}$

and

$$\hat{\Psi}(t) = (P(v(t), v(t)) + \int_0^t (\hat{Q} v(s), v(s)) ds + (\hat{Q} v_0, v_0).$$

2.1 Applications

In this section we shall consider some concrete problems.Using Theorem 1, or Theorem 2 one can find sufficient conditions on the data guarantiying global nonexistence of solutions to corresponding problems.

Semilinear second order hyperbolic equation : Suppose that $u(x,t)$ is a solution of the following initial-boundary value problem

$$u_{tt} - u_{xx} + r_1 u + r_2 u_t = |u|^p u, x \in (0,1), t > 0,$$

$$u(x,0) = u_0(x), x \in (0,1),$$

$$u(0,t) = u(1,t) = 0, t > 0,$$

where $p > 0$ and r_1, r_2 are arbitrary constants.

This problem is a special case of the problem (9),(10) with $H = L_2(0,1)$, $D = H^2(0,1) \cap H_0^1(0,1)$, $P = I$, $A = -\frac{d^2}{dx^2}$, $B(u, u_t) = B_1 u + B_2 u_t$, where $B_1 = r_1 I$, $B_2 = r_2 I$, $F(u) = |u|^p u$.The operator $F(u) = |u|^p u$ is the gradient of the functional $G(u) = \frac{1}{p+2} \int_0^1 |u|^{p+2} dx$. Therefore the condition (11) is satisfied with $\alpha = \frac{p}{4}$. Due to the Cauchy inequality we have

$$|(B_i u, v)| = |r_i| \, |(u,v)| \le |r_i| (\int_0^1 u^2 dx)^{\frac{1}{2}} (\int_0^1 v^2 dx)^{\frac{1}{2}} =$$

$$|r_i| \, (Pu, u)^{\frac{1}{2}} (Pv, v)^{\frac{1}{2}} \le \frac{p}{4}(Pv, v) + \frac{r_i^2}{p}(Pu, u).$$

Thus the conditions (26) and (27) satisfy with $c_1 = c_2 = \frac{p}{4}$, $b_1 = b_2 = 0$, $D_i = \frac{r_i^2}{p}, i = 1,2$, and $D_3 = |r_2|$. It is easy to see that the condition i) is satisfied and the conditions ii),iii) are satisfied with $M_1 \ge \max\{p + 4|r_2|, z_0\}$, where z_0 is a positive root of the equation $z^3 - 2p|r_1|z - 2p^2|r_1| = 0$. So using Theorem 2 we can find sufficient conditions of global nonexistence.

Hyperbolic thermoelasticity equations Consider the initial boundary value problem for the nonlinear thermoelasticity equations:

$$w_{tt} - w_{xx} + k\theta_{xt} = f_1(w, \theta), x \in (0,1), t > 0, \tag{31}$$

$$\theta_{tt} - \theta_{xx} + k w_{xt} = f_2(w, \theta), x \in (0,1), t > 0, \tag{32}$$

satisfying the following initial and boundary conditions:

$$w(x,0) = w_0(x), w_t(x,0) = w_1(x), \theta(x,0) = \theta_0(x), \theta_t(x,0) = \theta_1(x), \tag{33}$$

$$w(0,t) = w(1,t) = 0, \theta(0,t) = \theta(1,t) = 0, t > 0, \tag{34}$$

where k is a given parameter, $w(x,t)$ and $\theta(x,t)$ are unknown functions. The components of the vector-function $f(p,y) = \begin{pmatrix} f_1(p,y) \\ f_2(p,y) \end{pmatrix}$ are given sufficiently smooth functions defined on R^2 and there exists the differentiable function $G(p,y) : R^2 \to R^1$, such that for arbitrary differentiable functions $p(t), y(t)$ the following equality holds:

$$\frac{d}{dt} G(p(t), y(t)) = f_1(p,y)p'(t) + f_2(p,y)y'(t).$$

Moreover we assume that there exists a positive number l such that

$$f_1(p,y)p + f_2(p,y)y \geq (2+l)G(p,y),$$

for each $\{p,y\} \in R^2$

The problem (31)-(34) can be written in the form(9),(10), where $u = \begin{pmatrix} w \\ \theta \end{pmatrix}$ is the unknown vector-function,

$$P = \begin{pmatrix} I & 0 \\ 0 & I \end{pmatrix},$$

is a symmetric, positive-definite operator acting in $H = L_2(0,1) \times L_2(0,1) \equiv [L_2(0,1)]^2$.

$$A = \begin{pmatrix} -\frac{d^2}{dx^2} & 0 \\ 0 & -\frac{d^2}{dx^2} \end{pmatrix}$$

is a symmetric positive definite operator from $D = \left[H^2(0,1) \cap H_0^1(0,1)\right]^2$ into $H = [L_2(0,1)]^2$. The operator $B_1 = 0$. The operator

$$B_2 = \begin{pmatrix} 0 & k\frac{d}{dx} \\ k\frac{d}{dx} & 0 \end{pmatrix},$$

is skew symmetric, and for each $u = \begin{pmatrix} w \\ \theta \end{pmatrix}$ and $v = \begin{pmatrix} z \\ \sigma \end{pmatrix}$ from D we have

$$(B_2 u, u) = k \int_0^1 [\theta_x(x,t)w(x,t) + w_x(x,t)\theta(x,t)]\, dx =$$

$$= k \int_0^1 [-\theta(x,t)w_x(x,t) + w_x(x,t)\theta(x,t)]\, dx = 0, \tag{35}$$

$$|(B_2 u, v)| = |k \int_0^1 [\theta_x(x,t)z(x,t) + w_x(x,t)\sigma(x,t)]\, dx| \leq$$

$$\frac{l}{2} \int_0^1 \left[w_x^2(x,t) + \theta_x^2(x,t)\right] dx + \frac{k^2}{2l} \int_0^1 \left[z^2(x,t) + \sigma^2(x,t)\right] dx =$$

$$= \frac{l}{2}(Au, u) + \frac{k^2}{2l}(Pv, v).$$

It is not difficult to see that the operator

$$F(u) = \begin{pmatrix} f_1(w, \theta) \\ f_2(w, \theta) \end{pmatrix}$$

satisfies the condition (11) with $\alpha = \frac{l}{4}$. Now it is clear that if $|k| < l$, then the Theorem 2 is applicable.

Nonlinear pseudohyperbolic equation Let the function $u(x, t)$ be the solution of the following problem:

$$-\Delta u_{tt} - \Delta u - \Delta u_t + \sum_i b_i u_{tx_i} + c(u, u_t) = e^{2t}u^5, x \in \Omega, t > 0, \qquad (36)$$

$$u(x, 0) = u_0(x), u_t(x, 0) = u_1(x), x \in \Omega; u(x, t) = 0, x \in \partial\Omega, t > 0, \qquad (37)$$

where , $b_i, i = 1, ..., n$, are arbitrary constants and $c(.,.) : R^2 \to R^1$ is a continuous function satisfying the condition:

$$|c(s, v)| \leq C_1 |s| + C_2 |v|, \forall s, v \in R^1,$$

where the nonnegative numbers C_1 and C_2 satisfy the inequality $C_1 + C_2 \leq \lambda_1$, and λ_1 is the first eigenvalue of the operator $-\Delta$ under the homogeneous Dirichlet boundary condition.

The problem (36), (37) can be written in the form(9),(10) with $P = -\Delta : D \to L_2(\Omega) \equiv H$, $Q = P$, $A = P$, $B(u, u_t) = \sum_i b_i u_{tx_i} + c(u, u_t)$, $F(t, u) = e^{2t}u^5$. In this case the potential of $F(t, u)$ is $G(t, u) = \frac{1}{6}e^{2t}\int_\Omega u^6 dx$. So the condition (11) holds with $\alpha = 1$, and (12) holds with $M_1 = 2$. Thanks to the Friedrichs inequality we can get

$$|(B(u, v), u)| = \left| \int_\Omega (\sum_i b_i v_{x_i} u + c(u, v)u) dx \right| \leq$$

$$\leq \frac{b}{\lambda_1^{\frac{1}{2}}} \|\nabla u\| \|\nabla v\| + C_1 \|u\|^2 + C_2 |(u, v)| \leq \frac{d_2}{2} \|\nabla v\|^2 + \frac{b^2}{2d_2\lambda_1} \|\nabla u\|^2 + C_1\lambda_1^{-1} \|\nabla u\|^2 +$$

$$\frac{d_2}{2} \|\nabla v\|^2 + \frac{C_2^2}{2d_2\lambda_1^2} \|\nabla u\|^2 = d_2(Pv, v) + M_2(Pu, u$$

where $d_2 < 2$ is some positive number , $b = \sqrt{\sum b_i^2}$, $M_2 = \frac{b^2\lambda_1 + 2C_1 d_2 + C_2^2}{2d_2\lambda_1^2}$. It is clear that

$$|(B(u,v),v)| = \left| \int_\Omega c(u,v) v dx \right| \leq$$

$$C_1 \int_\Omega |u|\,|v|\,dx + C_2 \int_\Omega |v|^2\,dx \leq \frac{C_1}{2}\lambda_1^{-1}\|\nabla u\|^2 + (\frac{C_1}{2} + C_2)\lambda_1^{-1}\|\nabla v\|^2 \leq$$

$$(C_1 + C_2)\lambda_1^{-1}\left[(Pv,v) + (Au,u)\right].$$

Thus the conditions (13) and (14) are also satisfied and we can apply Theorem 1.

References

Fujita H., *On the blowing up of solutions of the Cauchy problem for* $u_t = \Delta u + u^{1+\alpha}$ J. Faculty. Sci., Univ. Tokyo, 13, 109-124, 1969.

Glassey R.T., *Blow-up theorems of nonlinear wave equations,* Math.Z., 132, 183-203, 1973.

Kalantarov V.K., Ladyzhenskaya O.A. , *The occurrence of collapse for quasilinear equations of parabolic and hyperbolic type,* J. Soviet Math. ,10, 53- 70.(1978). Translated from Zap. Nauch. Sem. LOMI ,69,77-102, (1977)

Kalantarov V.K., *Collapse of the solutions of parabolic and hyperbolic equations with nonlinear boundary conditions,* J.Sov.Math. 27(1984), 2601-2606.Translated from Zap.Nauchn.Sem.LOMI, 127(1983), 75-83.

Kalantarov V.K., *On the global behavior of solutions of the Cauchy problem for second order differential-operator equations,* Izv.A.N. Az.SSR, ser. fiz. tech. mat. nauk., 1987, 7, no.6 ,36-43

Kaplan S., *On the growth of solutions of quasilinear parabolic equations,* Comm. Pure Appl. Math. ,16, 305-330 (1963).

Keller J.B., *On solutions of nonlinear wave equations,* Comm.Pure Appl.Math.,10, 523-530, 1957.

Knops R.J. ,Levine H.A. ,Pane L.E., *Nonexistence ,instability and growth theorems for solutions of a class of abstract nonlinear equations with applications to nonlinear elastodynamics,* Arch. Rational Mech. Anal. 1974, 55, 52-72.

Ladyzhenskaya O.A. ,Solonnikov V.A. ,Ural'tseva N.N. ,*Linear and Quasilinear Equations of Parabolic Types,* A.M.S., Providence, Rhode Island (1968).

Levine H.A., *Instability and nonexistence of global solutions to nonlinear wave equations of the form* $Pu_{tt} = -Au + F(u)$, Trans. Am. Math. Soc. ,192, 1-21, (1974)

Levine H.A., *Some nonexistence and instability theorems for formally parabolic equations of the form* $Pu_t = -Au + F(u)$, Arch. Rational Mech. Anal. 1973, 51, 371-386.

Levine H.A., *Some additional remarks on the nonexistence of global solutions to nonlinear wave equations,* SIAM J. Math. Anal. 1974, 5, 138-146.

Levine H.A., *A note on a nonexistence theorem for some nonlinear wave equations,* SIAM J. Math. Anal. 1974, 5, 644-648.

Levine H.A., Paine L.E., *Nonexistence theorems for the heat equations with nonlinear boundary conditions and for porous medium equation backward in time,* J. Diff. Eq. ,1974, 16, 319-334.

Palais B., *Blowup for nonlinear equations using a comparison principle in Fourier space,* Comm.Pure Appl.Math.,151, 163-196, 1988

Straughan B., *Further global nonexistence theorems for abstract nonlinear wave equations,* Proc. Amer. Math. Soc. ,1975, 48, 381-390.

Turitsyn S.K., *On a Toda lattice model with a transversal degree of freedom. Sufficient criterion of blow-up in the continuum limit,* Physics Letter A 173(1993) 267-269.

Applications of Direct Numerical Simulation to Complex Turbulent Flows

Sedat Biringen and Robert S. Reichert

Department of Aerospace Engineering Sciences
Campus Box 429
University of Colorado, Boulder, Colorado 80309-0429, USA

Abstract. The technique of direct numerical simulation (DNS) is discussed with emphasis on application to complex turbulent flows. Several specific DNS schemes for time-integration of the governing three-dimensional Navier-Stokes equations are briefly described. The application of the methods reviewed here involve first the simulation of a spherical particle in a flat-plate boundary layer providing insight into the processes of bypass transition. Secondly, a turbulent square duct flow simulation has yielded a detailed understanding of the origins of secondary flow in a streamwise corner in addition to allowing direct assessment of turbulence models for such complex turbulent flows. Finally, the simulation of a supersonic turbulent wall-shear layer has allowed the evaluation of compressibility effects and has revealed the structure of compressible turbulence close to a wall.

1 Introduction

In this paper we outline the direct numerical simulation approach for the computation/prediction of turbulent flows. Our goal is to provide examples from our group's work with sufficient complexity to have engineering relevance. For a comprehensive critique of direct numerical simulation (DNS) and large-eddy simulation (LES) methods, the reader is referred to the recent articles by Reynolds (1990) and Hussaini and Speziale (1990).

The DNS approach can be placed within the hierarchy of turbulent flow computation methods which can be broadly classified as follows:

Methods Based on the Reynolds-Averaged Navier-Stokes Equations (RANS). These methods solve a form of the time-averaged Navier-Stokes equations and can obtain only the mean (averaged) quantities. A phenomenological model (closure) is required for all the scales of motion. Generally, no constitutive relation is available for turbulent flows, and the constants and empirical functions required to obtain closure are flow dependent. Consequently, modeling information based on simple flows may not work for complex (engineering) flows. Computational resource demands can be significant, especially if multi-equation (stress-equation) models are used for closure in complex geometries. Although RANS methods (closure formulae) are not universal and require calibration for each flow, they provide estimates for engineering problems at high Reynolds numbers, high Mach numbers, and when shock-boundary layer interactions are present. They are the current means used for industrial and design purposes.

Large-Eddy Simulation Approach (LES). In this approach, the full Navier-Stokes equations are numerically integrated, and it is possible to obtain solutions for both the mean and instantaneous quantities. The Navier-Stokes equations are locally averaged over space to filter the small scales of motion. The filtered equations describe the dynamics of the large scales which are responsible for most of the turbulent transport and carry most of the turbulence energy. The small eddies are assumed to be more isotropic and are modeled by a subgrid scale (SGS) closure model. The LES method is less prone to inaccuracies due to closure approximations and can have sufficient resolution at moderately high Reynolds numbers. However, SGS modeling is currently a very active research area with the main problems being proper representation of backscatter (reverse energy cascade) and effects of numerical accuracy. Comprehensive accounts of the LES method and SGS models are given by Lesieur and Metais (1996), Piomelli (1994), and Ghosal (1995). The reader is also referred to the article by Rogallo and Moin (1984).

Direct Numerical Simulation Approach (DNS). This method offers a third alternative for calculation of turbulent flows. In this method, the full, three-dimensional, time-dependent Navier-Stokes equations are integrated, and the solution provides both the mean and fluctuating quantities. No averaging is applied and no modeling assumptions are required as all the scales of the turbulence are resolved. This method has the highest computer resource demands and requires high-order space and time accurate discretization of the governing equations to minimize dispersion and dissipation errors in the numerical scheme. The main disadvantage of the method is the Reynolds number limitation stemming from the requirement that in order to resolve all the scales of the motion, the number of grid points in one direction is proportional to L/η, where L is the length scale for the largest scales (on the order of the computational domain) and η is the Kolmogorov length scale which represents the smallest, dissipative scales. The ratio L/η is proportional to $Re^{3/4}$, so that for one order of magnitude increase in Reynolds number, resolution requirements increase by more than two orders of magnitude. Recalling that current computers can provide sufficient resolution up to Reynolds number on the order of 10^4, aerodynamically significant Reynolds numbers on the order of 10^8 are certainly beyond the present day hardware capabilities.

In spite of these limitations, DNS has been a very useful tool to simulate transitional flows (see e.g. Biringen and Laurien (1990) and Kleiser and Zang (1991) for overviews) as well as turbulent flows at low Reynolds numbers and has proven to be a very significant adjunct to experimental studies. Perhaps the most significant contribution of DNS to date has been to provide databases used to evaluate and develop closure models for RANS and LES methods (see e.g. Kasagi and Shikazono (1995)). In addition, low Reynolds number limitations may not be very significant because the fundamental physics remains the same once the flow has reached the fully developed turbulence stage (Reynolds (1990)). In addition, flows in turbomachinery have lower Reynolds numbers (on the order of 10^4–10^5) than external, aerodynamic flows so that DNS can be useful both as a research tool to probe turbulence physics and to compute some flows of engineering relevance with current supercomputer capabilities.

Because both LES and DNS methods require high spatial accuracy, spectral methods (Orszag (1972), Gottlieb and Orszag (1977), Canuto, Hussaini, Quarteroni, and Zang (1988)) are implemented, especially in rectangular domains. Spectral methods minimize truncation errors but are prone to aliasing errors. Solutions can be de-aliased at the cost of extra computational effort by the 3/2 rule on a uniform mesh as Fourier transforms are used for de-aliasing. Consequently, high-order finite-difference methods (Rai and Moin (1991), Huser (1992)) are considered attractive alternatives, especially in complex geometries and nonuniform graded mesh systems where the application of Fourier transforms are difficult. In most DNS simulations, including the incompressible flow simulations that we present in this paper, time-advancement is done by semi-implicit methods using a time-splitting procedure, and mass conservation is satisfied to machine zero on a staggered mesh (Harlow and Welch (1965)). Linear algebraic systems resulting from the elliptic pressure Poisson equation and the elliptic Helmholtz type momentum equations are very efficiently solved by the tensor product method (see e.g. Huser and Biringen (1993)). In the compressible flow simulations, explicit high-order finite difference methods have found wide application. In our work, we have implemented the Two-Four method (Gottlieb and Turkel (1976)) which is a variant of the explicit MacCormack method (MacCormack (1969)). More detail is available on computational methods for LES and DNS in Ferziger and Perić (1996) and Härtel (1996).

In what follows, we present three simulations as examples of the DNS work conducted by our group. The first two are incompressible flow simulations of complex turbulent and transitional flows, and the third simulation investigates the effects of compressibility on the structure of turbulence at high Mach numbers.

2 Survey of Results

2.1 Complex Turbulent Flows

Bradshaw (1975) defines complex turbulent flows as flows other than simple shear flows containing a dominant mean rate of strain. In contrast, in complex turbulent flows, extra rates of strain contribute significantly to turbulent transport and consequently all the components of the Reynolds stress tensor attain significant magnitudes. These flows are the most difficult to model using RANS techniques (Moin et al. (1994)). Examples of complex turbulent flows are interacting shear layers, secondary flows, high angle of attack airfoils, and separated flows. In what follows, we discuss two examples that contain such effects.

DNS of a Spherical Particle in a Boundary Layer. The effects of particles on boundary layer transition have received considerable attention as an example of bypass transition which is representative of the transition process observed in engineering applications. In this type of transition, the flow bypasses the linear development phase and suddenly evolves into incipient turbulence. Such transition processes can be observed in underwater vehicles moving through a particulate environment. Individual particulates entering the boundary layer produce additional fluid motion through shedding of vortices and/or impacting the surface to create an isolated, three-dimensional roughness element. Saiki (1995) and Saiki and Biringen (1997) numerically investigated the mechanisms by which an elevated, stationary sphere (Fig. 1) can induce transition in a boundary layer.

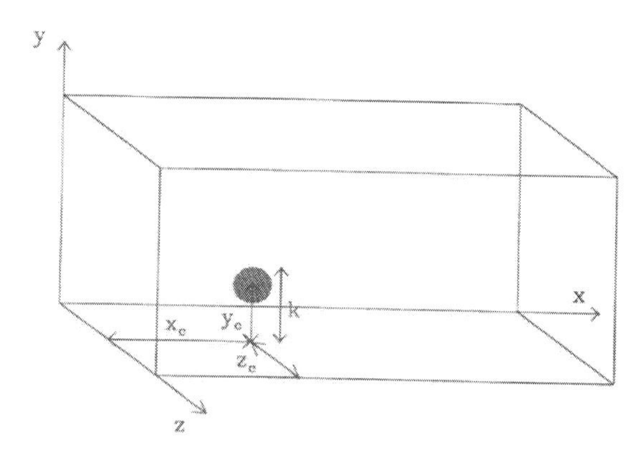

Fig. 1. Geometry and coordinate system of a sphere in a flat-plate boundary layer.

As a precursor case, numerical simulations were performed to investigate the development of an isolated disturbance induced by two pairs of counter-rotating streamwise vortices in a spatially developing boundary layer. A $241 \times 71 \times 33$ mesh was employed in the x, y, and z directions, respectively, on a $141 \times 75 \times 50$ domain, nondimensionalized by displacement thickness, δ^*. To allow for spatial evolution of disturbances, a buffer domain technique (Streett and Macaraeg (1989)) was used at the outflow boundary. No-slip was imposed on the bottom of the domain, while the spanwise direction was periodic. The Reynolds number specified at the inflow was $Re_{\delta^*_o} = 950$, and the vortex pairs were centered around $z = 25$, with maximum inflow perturbation amplitudes of $v_{max} = 5.8 \times 10^{-3}$ and $w_{max} = 1.86 \times 10^{-2}$. The case delineates the two-part behavior of the three-dimensional disturbance.

This two-part evolution of the disturbance as it travels downstream is shown in streamwise perturbation velocity contours in x-y planes at $z = 25$ in Fig. 2. The u' contours are dominated by a high amplitude transient structure appearing as a negative region followed by a positive region, which develop in accordance with the vertical motion of the fluid. Upstream of the transient, the contours are similar to those which develop due to the presence of a Tollmien-Schlichting (T-S) wave. The wave part of the disturbance travels at the phase speed of the T-S wave lagging behind the transient part. The two-part structure associated with the isolated disturbance was determined in order to identify such disturbances in the subsequent sphere computations.

The subcritical and supercritical behavior of the wake of a sphere in a boundary layer was examined by simulating two cases of different inflow displacement boundary layer Reynolds numbers, $Re_{\delta_o^*} = 500$ and $Re_{\delta_o^*} = 750$. A mesh of $306 \times 71 \times 33$ was defined on a domain of dimensions $137.36 \times 75 \times 2\pi$. The length of the outflow buffer domain was 62.64 nondimensional lengths, comprising 116 grid points. The sphere cases employed the same boundary conditions used for the isolated disturbance case. The sphere was placed well within the boundary layer at a nondimensional downstream position from the inflow of $x_c = 31.14$. In both computations, the diameter of the sphere was $d^* = 2.297\ mm$ (roughly, $d^* = \frac{1}{3}\delta_p$; δ_p is the dimensional laminar boundary layer thickness at the sphere position, x_c). The sphere was elevated 0.3 nondimensional units above the plate. The boundary of the sphere was imposed by addition of a virtual boundary forcing term (Goldstein, Handler, and Sirovich (1993)) applied on 368 points with feedback coefficients set to $\alpha_f = -500$ and $\beta_f = -4$. Additional mesh points in the streamwise direction were clustered in the vicinity of the sphere. The body of the sphere was introduced by "turning on" the forcing term at $t = 0$ within a Blasius boundary layer.

Like the isolated disturbance case, the subcritical case exhibited a clear two-part behavior. To investigate the prominent frequencies/wavenumbers in the flow field, the streamwise velocity field was Fourier transformed in time and in the spanwise direction to obtain Fourier amplitudes in the frequency/wavenumber space (Fig. 3). Here, we denote these modes as (f, k_z) pairs, where f is nondimensional frequency ($\omega = 2\pi f$) and k_z is spanwise wavenumber. The highest amplitude modes represent the streamwise trailing vortices in the near vicinity of the sphere and the wake further downstream. The amplitude of the $(0.2, 0)$ mode exhibits a large increase throughout the trailing and hairpin vortex region followed by a dramatic decrease downstream. This mode corresponds to the shedding frequency of the hairpin vortices. A second dominant frequency, $(0.017, 0)$, also appears in Fig. 3. This lower frequency mode dominates the vortex shedding mode further down-

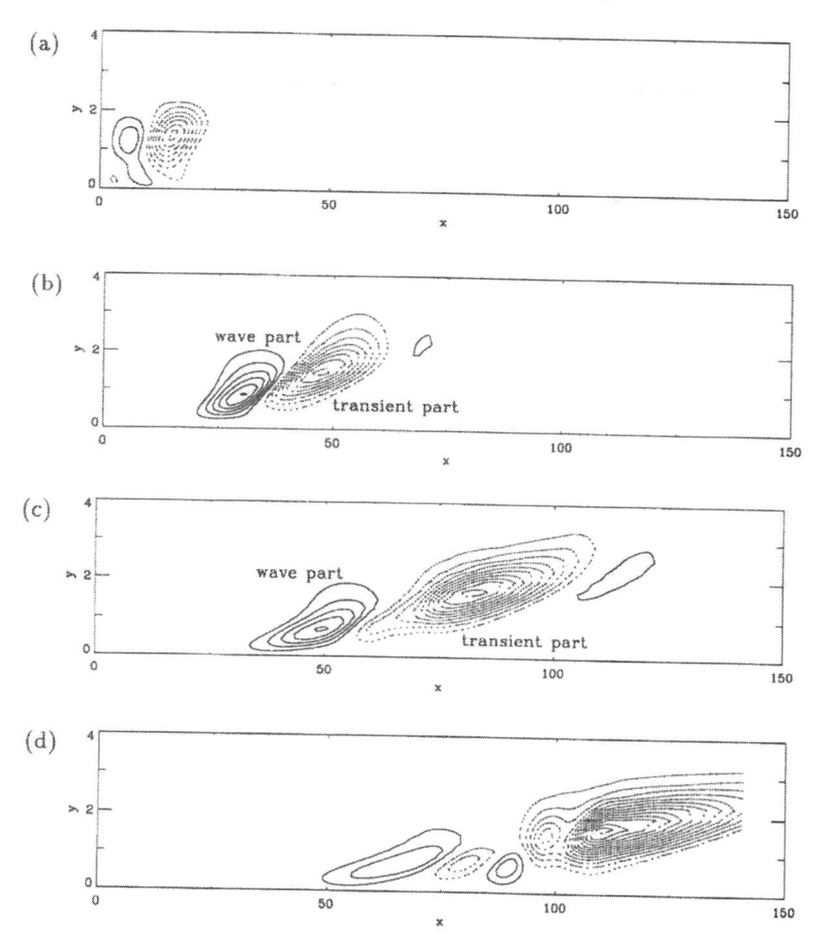

Fig. 2. Contours of streamwise perturbation velocity in the x-y plane at $z = 25$: (a) $t = 32.7$, (b) $t = 85.1$, (c) $t = 137.44$, and (d) $t = 189.8$. Note that negative quantities are denoted by the dotted contours and the vertical direction is magnified.

stream, and the frequency of this signal, $f_l = 0.017$ (or $\omega_l = 0.107$), is not a subharmonic of the shedding frequency.

The low frequency signal appears to be related to the most unstable Tollmien-Schlichting (T-S) wave predicted by linear stability theory for the given Reynolds number range of the computational domain. This is confirmed by consultation of the stability chart obtained from solutions of the Orr-Sommerfeld equation for the boundary layer correlated with a properly scaled frequency and Reynolds number, based upon local displacement thickness at the sphere position, $Re_{\delta_p^*} = 544$. For reference, the neutral stability curve for boundary layers, as determined by Jordinson (1970), is reproduced in Fig. 4. In this figure, the region of interest for the current study is de-

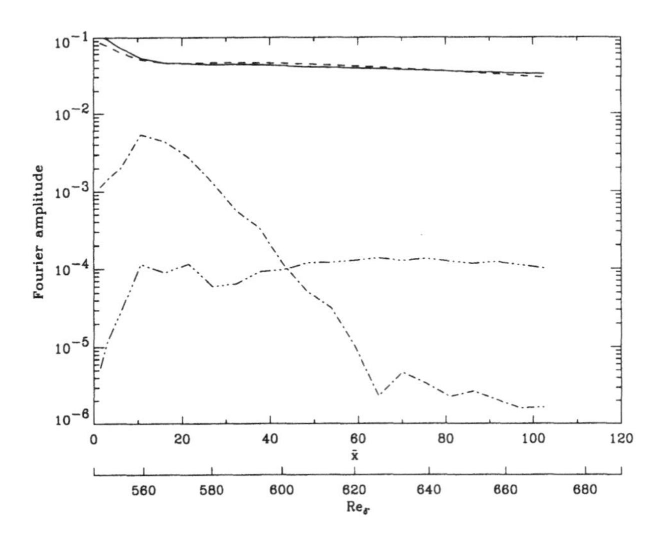

Fig. 3. Subcritical case: Streamwise distribution of Fourier amplitudes of dominant frequency/wavenumber pairs, (f, k_z). ——— , $(0,1)$; — — — , $(0,2)$; –·–·–, $(0.2, 0)$; –···–···, $(0.017, 0)$. The second horizontal axis indicates the behavior of these modes with respect to the local boundary layer displacement thickness Reynolds number, Re_{δ^*}.

noted by a line drawn from the Reynolds number at the sphere position, $Re_{\delta^*_p} = 544$, to the Reynolds number at the end of the computational domain, $Re_{\delta^*} = 673$, for the low frequency, $\omega_l = 0.107$. According to Fig. 4, a wave at this frequency experiences growth in the streamwise direction over the majority of the domain followed further downstream by amplitude decay. This behavior is reflected in the streamwise distribution of the low frequency wave amplitude (with some modulation due to the wake of the sphere) shown in Fig. 3. The second horizontal axis included in Fig. 3 shows the direct correspondence of the $\overline{x}$ coordinate (where $\overline{x}$ is the x coordinate measured from the downstream edge of the sphere) with the local Reynolds number for comparison with Fig. 4. The stability chart further reveals that ω_l falls near the frequency of the most unstable wave for the Reynolds number at the location of the sphere, $Re_{\delta^*_p} = 544$ (marked by the asterisk in Fig. 4).

In Fig. 5, instantaneous contours of streamwise and normal perturbation velocities in the $\overline{x}$-y planes at $z = 2\pi$ detail the evolution of the sphere's disturbance. The two-part characteristics of the development are similar to those observed for an isolated, three-dimensional disturbance induced by two pairs of streamwise vortices, as discussed at the beginning of this section. A transient portion leaves the computational domain with a wave portion trailing behind inducing a decaying, low frequency T-S wave in the flow field.

For the supercritical case, the location and dimension of the sphere remained the same, but Reynolds number was increased to a supercritical value.

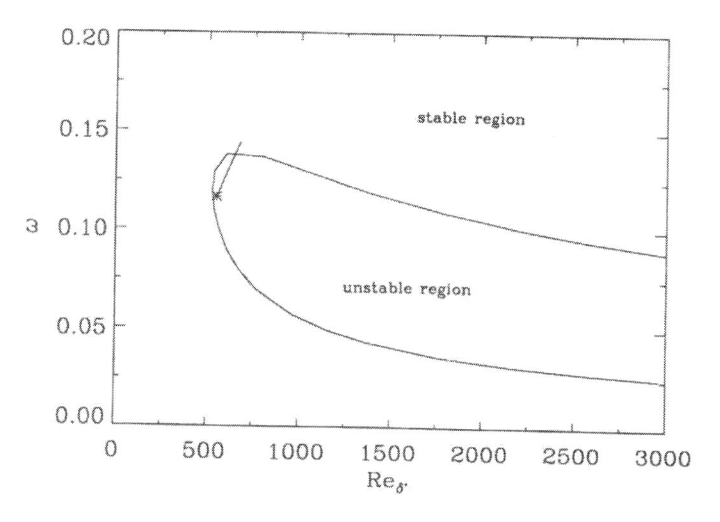

Fig. 4. Neutral curve (Jordinson, 1970) predicted by linear stability theory for boundary layers; frequency nondimensionalized by δ^* (ω) vs. Reynolds number (Re_{δ^*}). The region spanned in the subcritical case from the position of the sphere to the end of the domain with respect to the low frequency is denoted by the solid line. The position of the sphere is indicated by $*$.

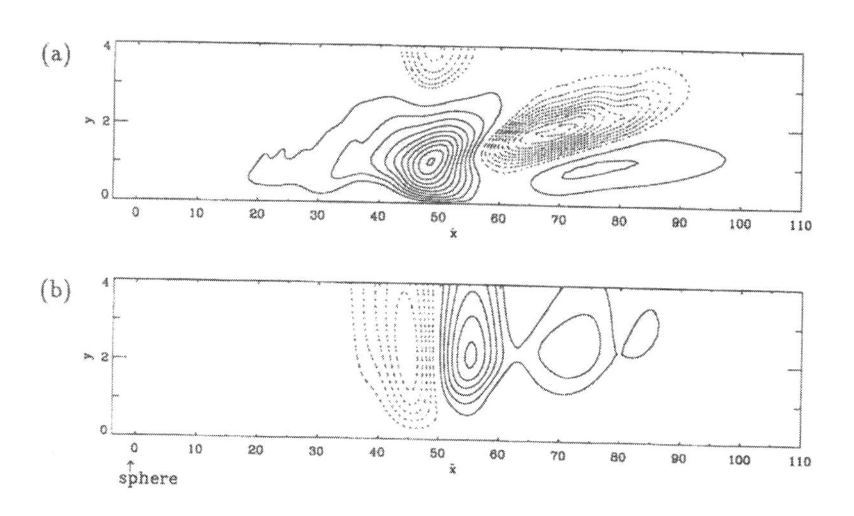

Fig. 5. Subcritical case: Contours of the disturbance at $t = 104.35$. (a) Streamwise and (b) normal perturbation velocity contours in the $\overline{x}$-y plane at $z = 2\pi$. Dotted contours denote negative values.

The mesh resolution was $306 \times 71 \times 65$ on a domain size of $137.36 \times 75 \times 4\pi$. This case resulted in the development of complex fluid motion akin to turbulence. The available machine resolution enabled the simulation of an incipient turbulent state characterized by the formation of a turbulent wedge, but was insufficient for the simulation of a statistically steady, fully developed turbulent flow. Consequently, this high Reynolds number simulation captured only the salient characteristics of transitional behavior and early turbulence in this complex flow field.

Instantaneous streamwise velocity contours reveal the time evolution of the planform of the turbulent wedge structure in $\bar{x}$-z planes at $y = 1.3$ (Fig. 6). The breakdown appears to originate at the upstream part of the structure in a manner similar to the breakdown of an isolated disturbance, and as the disturbance moves downstream, the complexity of the flow field influences the downstream portion of the disturbance. The increase in the streamwise length of the spot is evident; the downstream portion propagates faster than the upstream portion. Note that the speed of the downstream part closely corresponds to the local mean velocities in accordance with the development of the transient part of an isolated, three-dimensional disturbance. The upstream portion moves at a velocity closer to that of the most unstable T-S wave predicted by linear stability analysis for the Reynolds numbers used in this case.

The experiments of Blackwelder, Browand, Fisher, and Tanaguichi (1992) observed the production of a turbulent spot when an isolated sphere entered the boundary layer. The sphere hit the wall and then rolled along it creating a wake as the spot quickly moved downstream. The results suggest that the turbulent spot observed by Blackwelder, Browand, Fisher, and Tanaguichi (1992) evolved from an isolated three-dimensional disturbance generated by the interaction of the sphere and the boundary layer. Comparison of mean streamwise velocity and root-mean-square velocity fluctuation profiles with experiments provides evidence that the computational turbulent spot is representative of low Reynolds number turbulence (see Saiki (1995) and Saiki and Biringen (1997) for details).

DNS of Turbulent Flow in a Square Duct. Turbulent flow along a streamwise corner has application to flow in the root region of a lifting section, complex flow in turbomachinery and heat exchangers, flow in ducts with noncircular cross section, and flow in open channels and rivers. These complex turbulent flows have two inhomogeneous directions and are characterized by the existence of secondary flows of the second kind (as classified by Prandtl (1926)), which are secondary mean flows created by the turbulent motion. Although this type of secondary flow is relatively weak (about 2-3 percent of the streamwise bulk velocity), its effects on wall shear stress distribution, heat transfer rates, and transport of passive tracers are quite significant (Demuren (1990)). Secondary flow of the second kind is most frequently studied

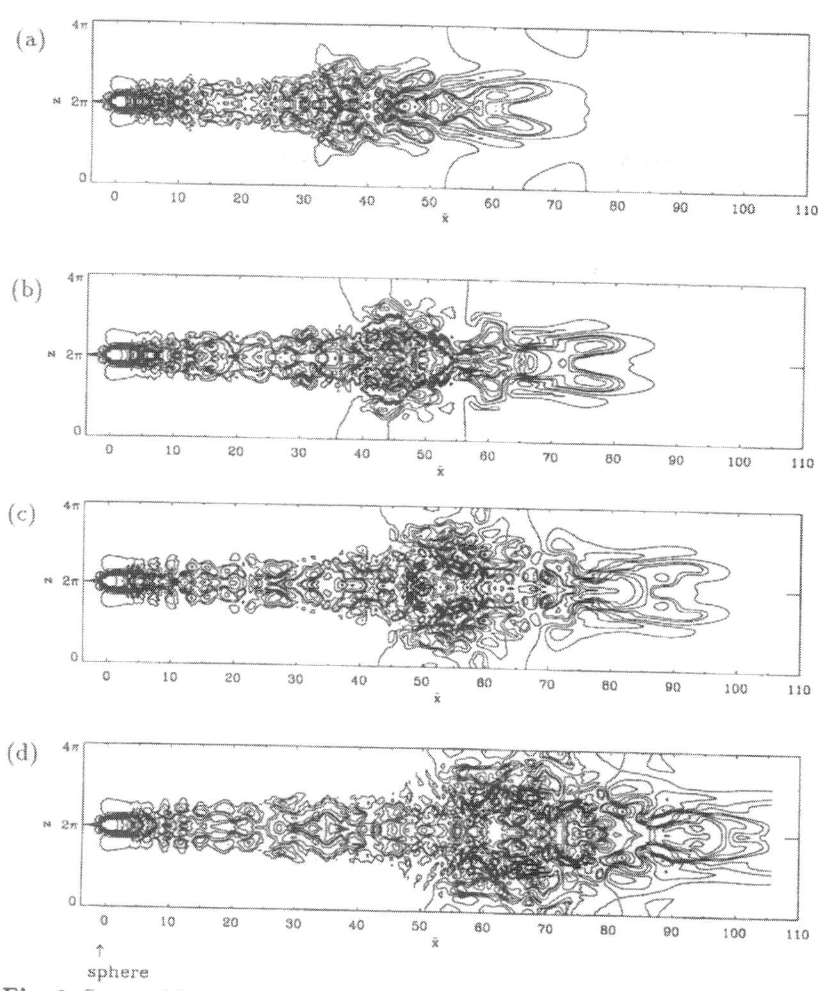

Fig. 6. Supercritical case: Time evolution of streamwise velocity contours in the $\bar{x}$-z plane at $y = 1.3$. (a) $t = 98.74$, (b) $t = 116.69$, (c) $t = 134.64$, and (d) $t = 152.6$.

by considering turbulent flow in a square duct because of its relatively simple geometry.

Huser and Biringen (1993) performed a DNS of incompressible square duct flow (Fig. 7) at a Reynolds number $Re_\tau = 600$, in the well-established turbulent regime. No-slip boundary conditions were imposed on each wall, while the streamwise direction was periodic. The domain was $1 \times 1 \times 6.4$ in the z, y, and x directions, respectively, resolved with a $100 \times 100 \times 96$ grid. The results provided a database from which turbulence statistics were obtained. This, in turn, gave a detailed description of the corner influence on turbulence statistics and on the origin of secondary flows of the second kind.

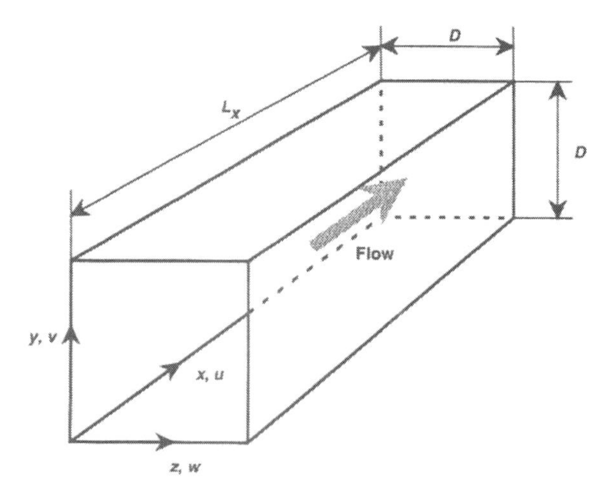

Fig. 7. Geometry and coordinate system of turbulent square duct flow.

The square duct mean flow is displayed in Fig. 8. Long term statistics were obtained by averaging the flow field in the homogeneous direction, x, and in time over the time it would take a particle at the channel center to travel a distance equal to $330D$, where D is the duct width. The distributions were then ensemble-averaged over the eight similar octants formed when the wall and corner bisectors are drawn. The mean secondary flow, which consists of weak streamwise vortices, is apparent in the $\overline{v}, \overline{w}$ vectors plotted in Fig. 8.

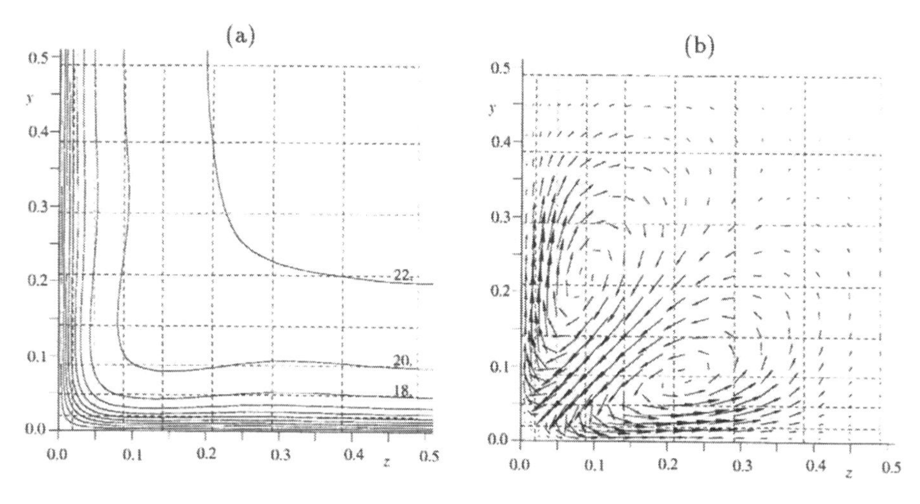

Fig. 8. Ensemble-averaged mean velocities: (a)$\overline{u}$ contours, increment=2; (b) $\overline{v}, \overline{w}$ velocity vectors.

The mean streamwise vorticity equation yields insights into the origin of the secondary flow and is given by (Perkins (1970)):

$$\overline{v}\frac{\partial \overline{\omega_x}}{\partial y} + \overline{w}\frac{\partial \overline{\omega_x}}{\partial z} = \frac{\partial^2}{\partial y \partial z}\left(\overline{v'^2} - \overline{w'^2}\right) + \left(\frac{\partial^2}{\partial z^2} - \frac{\partial^2}{\partial y^2}\right)\overline{v'w'} + \frac{1}{Re_\tau}\nabla^2\overline{\omega_x}, \quad (1)$$

where $\overline{\omega_x} = \partial \overline{w}/\partial y - \partial \overline{v}/\partial z$. Equation (1) demonstrates the importance of the secondary Reynolds stresses on the mean streamwise vorticity production. The y-distributions of the four terms in Eq. (1) are plotted in Fig. 9 for two z locations. The balance indicates small deviations from zero. At $z = 0.1$ the distributions reach peak values near the wall where the maximum production of streamwise vorticity takes place (Madabhushi and Vanka (1991) and Gavrilakis (1992)). Figure 9(a) presents the distribution of these terms near the vertical wall at $z = 0.015$; the convection has a minimum at $y \approx 0.09$ where the vorticity is negative. Normal stress gradients decrease the magnitude of the vorticity here in agreement with the results of Gavrilakis (1992). Figure 9(b,c) demonstrate how the negative vorticity is produced *below* the corner bisector: for $y^+ < 18$ (Fig. 9(c)), negative vorticity is produced by the secondary normal stress and for $y^+ > 18$, negative vorticity is produced by the secondary shear stress by transporting it outward from the wall.

Figure 10 is an illustrative cross section of the instantaneous turbulence structures which contribute to the production of shear stresses. The figure displays streamwise velocity contours and secondary velocity vectors in the y, z plane. An example of an ejection which is not influenced by the side walls appears at $z = 0.5$ near the horizontal wall. A characteristic mushroom-like shape is depicted in Fig. 10(a), with corresponding counter-rotating vortices shown in Fig. 10(b).

Near the corner, several ejections can be observed. The strongest ejection, whose mushroom shape is evident in Fig. 10(a), starts from the vertical wall at around $y = 0.25$. The velocity vectors display two counter-rotating vortices, one below and one above this ejection structure. The lower vortex is most pronounced, and it appears that this vortex interacts with another ejection structure from the horizontal wall. Figure 10(a) indicates that an ejection event is starting at around $z = 0.25$ from the horizontal wall, and the stem of this ejection stretches out toward the vertical wall and joins the lower vortex belonging to the stronger ejection from the vertical wall.

To obtain a more precise view of how these ejection structure interactions contribute to production of the secondary flow, Huser and Biringen (1993) also investigated the behavior of the pressure-velocity correlations. It was observed that the $\overline{v'^2}$ gradient, $\partial \overline{v'^2}/\partial y$, is reduced near the corner along the horizontal wall, which becomes an important contributor to the source of streamwise vorticity governed by (1). The reduction may be explained by the redistribution of $\overline{v'^2}$-energy to $\overline{w'^2}$-energy through the pressure-strain (or velocity-pressure gradient) correlations. In Fig. 11, the diagonal terms in the pressure-strain tensor are plotted as a function of y at two z locations. Near

(a)

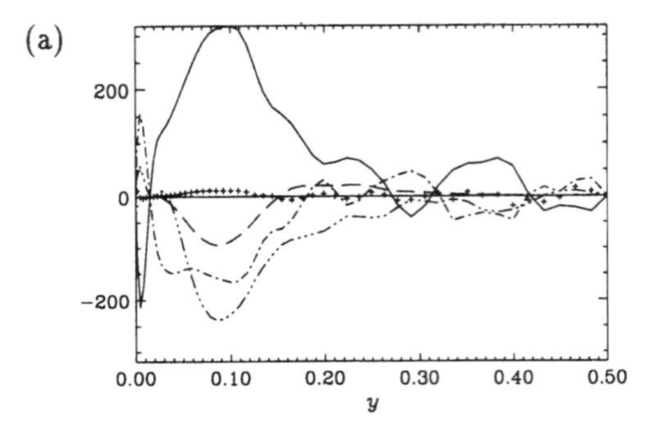

(b)

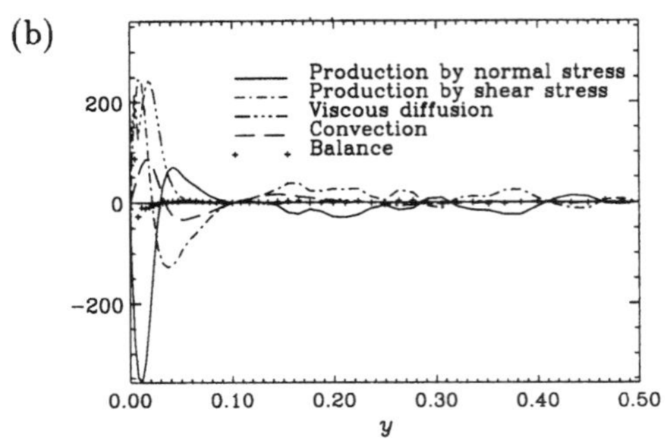

(c)

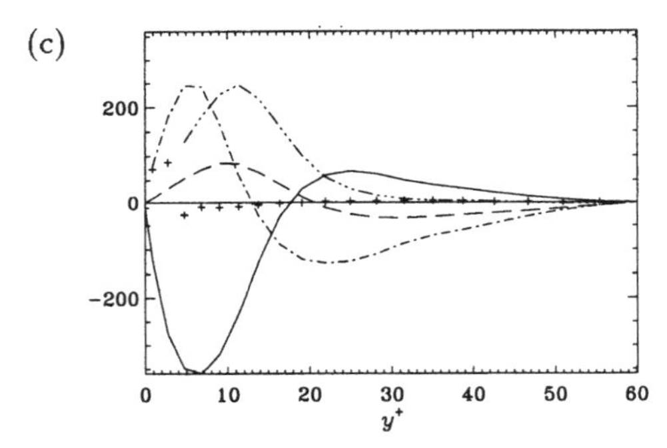

Fig. 9. Ensemble-averaged $\overline{\omega}_x$ budget distribution at two z locations: (a) $z = 0.015$, (b) $z = 0.1$, (c) $z = 0.1$ plotted versus y^+ coordinate.

the corner, Φ_{22} becomes a loss to the $\overline{v'^2}$ budget for $y < 0.05$. The loss is distributed only to the $\overline{w'^2}$ budget for $0.02 < y < 0.05$. This "corner" effect can be explained by the interactions between ejections arising from both walls consecutively. The upward motion in an ejection from the horizontal wall is bent toward the vertical wall when an ejection has taken place there. The ejection from the vertical wall will cause a downdraft to the corner caused by the streamwise eddy near the corner. This downdraft will influence the ejection from the horizontal wall and bend its ejection stem toward the vertical wall. Because of this bending, energy is transferred from the v' to

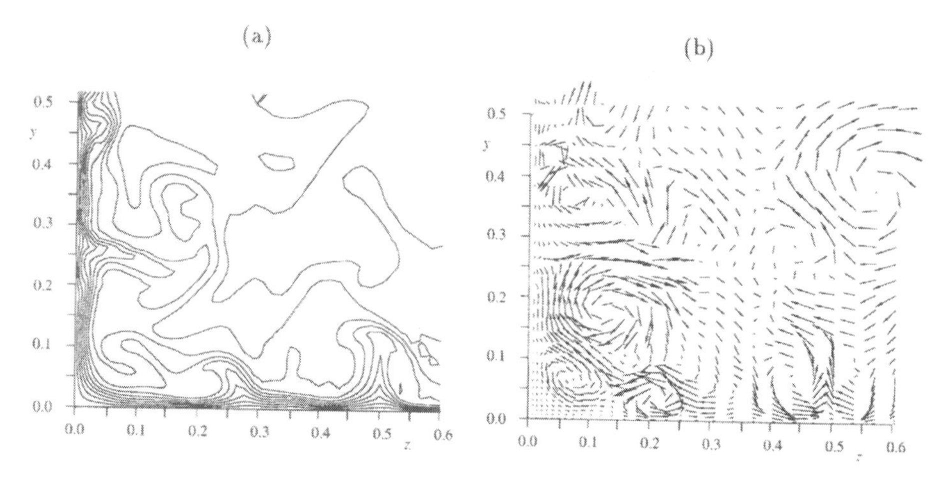

Fig. 10. Instantaneous velocities: (a)u contours, increment=2; (b) v, w velocity vectors.

the w' component. It should be stressed that the underlying reason for the interaction of ejections giving a mean secondary flow is the lack of bursts from the corner, causing the ejections to be "locked" at locations away from the corner.

The DNS database from the turbulent duct flow is useful for direct evaluation of turbulence models. Huser, Biringen, and Hatay (1994) performed such an evaluation of the nonlinear $k-\epsilon$ model of Speziale (1987) in an attempt to uncover why such models fail to predict the strength of secondary flows of the second kind. Specifically, the focus was on the Speziale model's capability to represent the anisotropy induced by the intersecting walls of the square duct.

The formation of the secondary flow is directly associated with the generation of streamwise vorticity $\overline{\omega_x}$, as described by (1). The model of Speziale (1987) may be used to approximate the Reynolds stresses which appear in the source terms of (1). The model requires knowledge of mean flow quantities, mean strain rate, turbulent kinetic energy, and turbulent dissipation rate. Huser, Biringen, and Hatay (1994) used the DNS database to provide these quantities for computation of the model's Reynolds stress approximations. Finally, the source terms of (1) were computed using these approximations and directly from the DNS for comparison.

Figure 12 displays the source terms for the secondary flow as they appear in the streamwise vorticity equation (1). Although some anisotropy close to the corners is captured by the model, the distributions and the magnitudes are not in agreement.

Clearly, significant differences exist between the model behavior and the DNS results. For example, there is a difference of two orders of magnitude

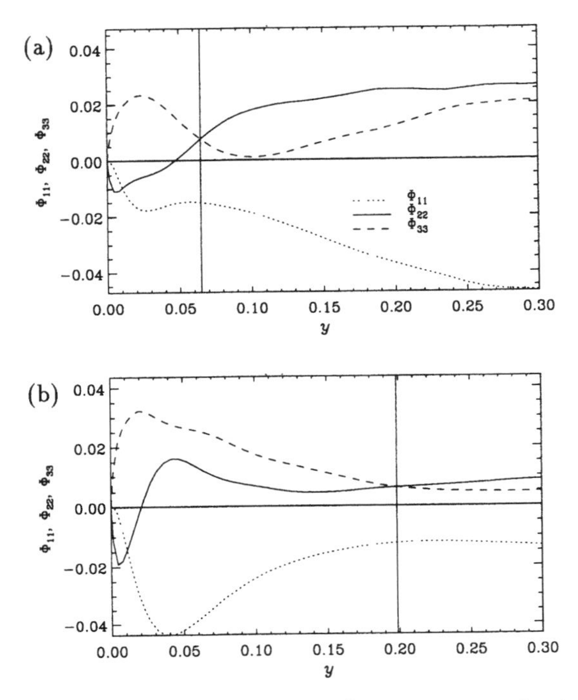

Fig. 11. Diagonal terms in the pressure-strain tensor at two z locations: (a) $z = 0.064$, (b) $z = 0.198$. The vertical line indicates the corner bisector.

between the model predictions and DNS data for the source terms depicted in Fig. 12. This is caused by grossly negative values predicted by the model for the secondary normal stresses which leads to an unrealistic partition of the turbulent kinetic energy between the primary and the secondary flows and to a weak secondary flow pattern.

Proper Orthogonal Decomposition. The method of proper orthogonal decomposition (POD) was applied to a database derived from the DNS of turbulent flow in the square duct (Reichert, Hatay, Biringen, and Huser (1994)). The main focus was to assess the usefulness and applicability of the POD method for this complex turbulent flow, which has two inhomogeneous directions and is characterized by both primary and secondary mean flows.

The proper orthogonal decomposition method of Lumley (1970) is an attempt to systematically identify coherent structures in turbulent flows. Berkooz, Holmes, and Lumley (1993) discuss some of the method's important aspects. The method is analytically based and uses statistics from experiment or numerical simulation to obtain coherent structures with maximal energy content. Identification of the energy-maximizing modes can aid in data com-

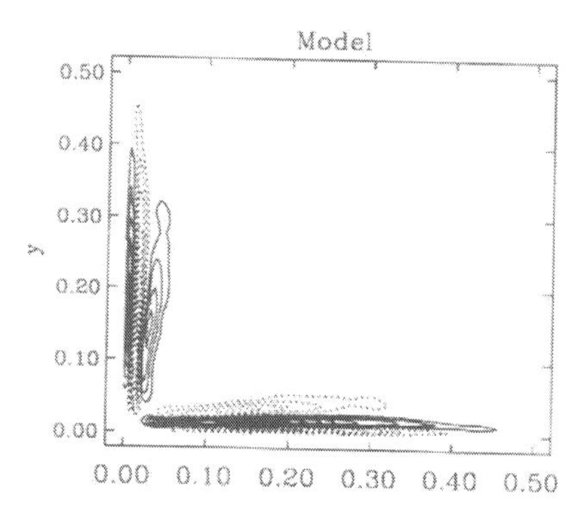

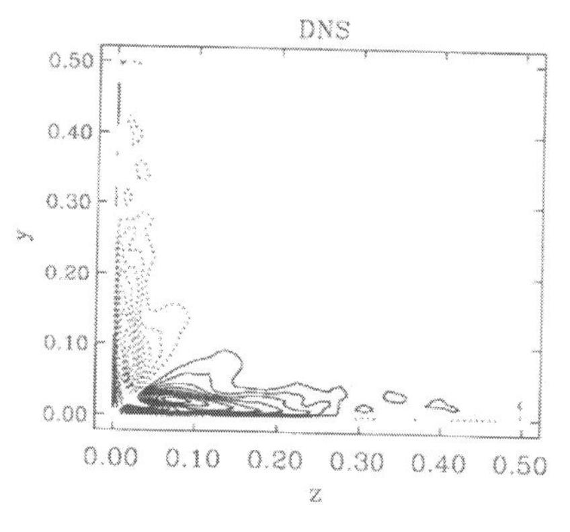

Fig. 12. Comparison of the source term in the $\overline{\omega}_x$ equation, $-[\partial^2(\boldsymbol{R}_{22} - \boldsymbol{R}_{33})/\partial y \partial z] + (\partial^2 \boldsymbol{R}_{23}/\partial y^2) - (\partial^2 \boldsymbol{R}_{23}/\partial z^2)$, from DNS and model of Speziale (1987). The model–predicted values lie between -5000 and +5000 in the upper plot, whereas the range of DNS values in the lower plot is between -250 and 250.

pression if a small number of such modes capture most of the turbulent energy.

The POD method searches for a complete, orthonormal set of coherent structures $\phi_i^n(y,z)$ such that an instantaneous fluctuating velocity field $u_i'(y,z)$ may be written as

$$u_i'(y,z) = \sum_{n=1}^{N} a^n \phi_i^n(y,z). \tag{2}$$

The constraint for the set's uniqueness is that the modes should have maximal energy content. Such modes are found from the solution of the following integral eigenvalue problem:

$$\lambda^n \phi_i^n(y,z) = \int_y \int_z R_{ij}(y,z,y',z') \phi_j^n(y',z') \, dz' \, dy'. \tag{3}$$

Here, $R_{ij}(y,z,y',z')$ is the symmetric two-point correlation tensor defined as

$$R_{ij}(y,z,y',z') = \overline{u'_i(y,z)u'_j(y',z')}, \tag{4}$$

where the overbar denotes mean over some homogeneous set. For the duct DNS, averaging is in time, the streamwise x direction, and over the eight similar octants of the duct cross section. Since the POD modes form a complete set, reconstruction of any flow property, instantaneous or mean, is possible. Reynolds stress $\overline{u'_i u'_j}(y,z)$ reconstruction, for instance, is given by

$$\overline{u'_i u'_j}(y,z) = \sum_{n=1}^{N} \lambda^n \phi_i^n(y,z)\phi_j^n(y,z). \tag{5}$$

Figure 13 shows an example reconstruction of $\overline{u'v'}$ in the turbulent square duct, exhibiting surprisingly fast convergence. Figure 13(a) is the total $\overline{u'v'}$ extracted from the two-point correlation tensor. Only about 40 modes are necessary to reproduce the time-averaged stress, and it is seen that 100 mode additions give essentially converged contours. Though slower, kinetic energy and instantaneous velocity field reconstruction also converge in about 100 modes. Examination of the POD eigenvalue spectrum reveals that the turbulent square duct flow has no single dominant eigenfunction. Rather, the flow is characterized by structures with a gradual cascading of energies and scales. The largest mode contains only about 12% of the total energy, but the first 10 modes capture over 50% of the total energy. This suggests that a set of, say, 100 modes can act as a truncated, but adequate, description for this flow, leading to over an order of magnitude compression in the DNS database (Reichert, Hatay, Biringen, and Huser (1994)).

2.2 Compressible Turbulent Wall-Shear Layer

Measuring difficulties in compressible wall-shear layers prohibit a detailed experimental description of compressible turbulence, especially in the near-wall region. Typically, measurements are limited to the region above 0.1 boundary layer thickness, excluding the near-wall region (Kistler (1959), Spina, Donovan, and Smits (1991)). The lack of reliable data in the near-wall region, where most of the turbulence production and dissipation take place significantly retards the development and validation of turbulence models. Hatay (1994) and Hatay and Biringen (1995) performed direct numerical simulation to provide a detailed description of compressible turbulence in supersonic wall-shear layers as an adjunct to experimental observations. The geometry of the simulation is shown in Fig. 14.

Two simulations of Hatay (1994) and Hatay and Biringen (1995) are described here: the first one, Case-**Q1**, was performed at $M = 2.5$ and $Re_{\delta^*} = 1000$ (based on displacement thickness, δ^*) at a resolution of $258 \times 128 \times 130$

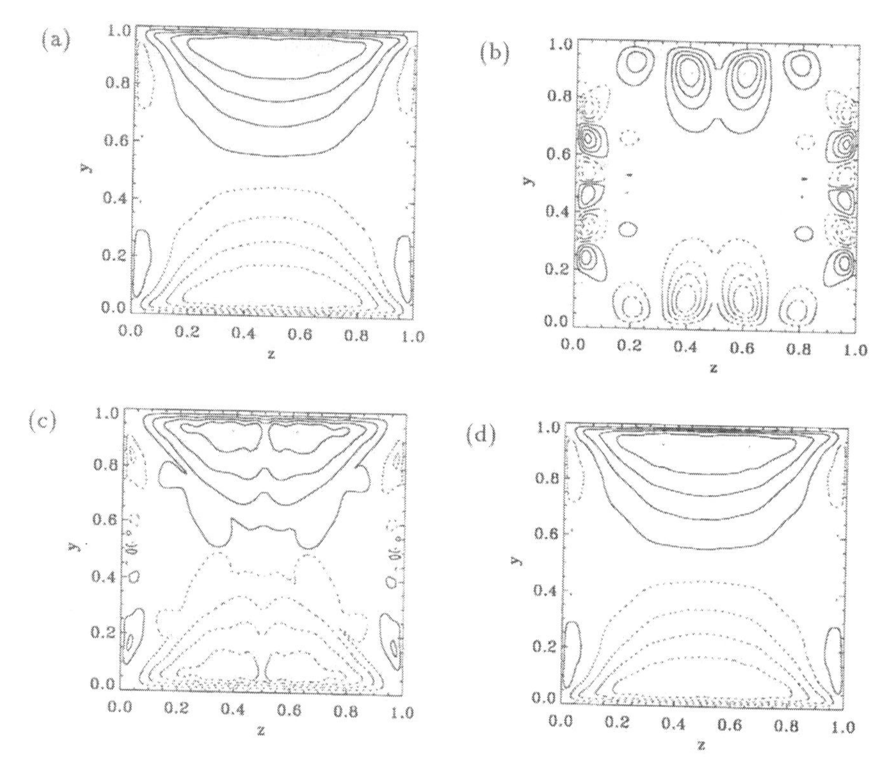

Fig. 13. Reynolds stress $\overline{u'v'}$ reconstruction for full octant decomposition, mapped into full duct cross section: (a) field from DNS, (b) 1 mode, (c) 10 modes, (d) 100 modes. Dashed contours denote negative values.

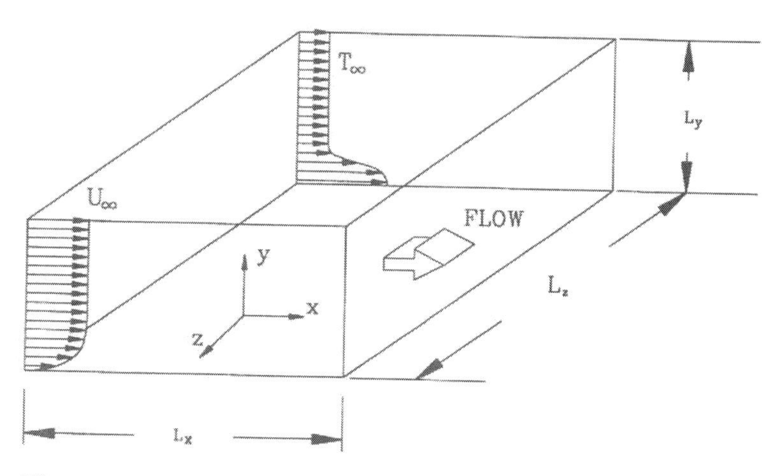

Fig. 14. Geometry and coordinate system for compressible wall-shear layer over a flat plate.

in the streamwise, normal, and spanwise directions, respectively. The second simulation, Case-**Q2**, was a continuation of Case-**Q1** at a resolution of $386 \times 128 \times 258$. Both simulations used a domain $125\delta^* \times 75\delta^* \times 32\delta^*$. The bottom boundary was a no-slip, adiabatic wall, while along the top boundary, zero-gradient (shear-free) conditions were imposed on all variables. Periodicity was employed in both the spanwise and streamwise directions; forcing terms were added to the governing equations to allow a streamwise periodic model of the spatially growing wall-shear layer. After the turbulence level was sustained in the simulations, statistics were collected for a total time of $T = 187.2$ in Case-**Q1** over 31 instantaneous data sets and $T = 55.5$ in Case-**Q2** over 48 instantaneous data sets.

Morkovin (1962) compiled a hypothesis stating that the direct effects of density fluctuations on turbulence behavior are small if the root-mean-square density fluctuation is small compared with the absolute (mean) density. Originally the hypothesis was advanced to explain time-averaged behavior and was thought to apply to wall-shear layers with free stream Mach number less than five. As acoustic fluctuations do not assume a dominant role within the limits of the Morkovin hypothesis, the only apparent compressibility effect is due to the mean density gradient. In this regard, the hypothesis is a generalization of the "Reynolds Analogy" which applies to heated, incompressible wall-shear layers.

The Morkovin hypothesis (Morkovin (1962)), Strong Reynolds Analogy (Bradshaw (1977)), is based on the similarity of density and temperature fluctuation fields to velocity fluctuations,

$$\frac{\rho'}{\overline{\rho}} \approx \frac{T'}{\overline{T}} \approx (\gamma - 1)M^2 \frac{u'}{U_\infty}, \tag{6}$$

under the assumption that pressure fluctuations are negligible, velocity fluctuations are small, and Mach number is not larger than four or five for boundary layer flows. The Strong Reynolds Analogy implies that temperature and density fluctuations are caused by the transport of mean-temperature or mean-density fields by the velocity fluctuations. Total velocity fluctuation, q, is defined as:

$$q^2 = \frac{1}{2}\left(u'u' + v'v' + w'w'\right). \tag{7}$$

In the simulations of Hatay (1994) and Hatay and Biringen (1995), the pressure fluctuations are several orders of magnitude smaller than the velocity fluctuations. The validity of the Strong Reynolds Analogy is checked directly in Fig. 15. The density and temperature profiles coalesce as a consequence of an almost uniform pressure field. The velocity fluctuations fall within 20% of the analogy given in (6).

Figure 16 compares the two-point correlations in the $(\Delta x^+, \Delta z^+)$ plane with one (computational) probe fixed at different $y_a^+ = y_b^+$ locations. These distributions are best interpreted as the horizontal cross sections of a typical event in the wall-shear layer. The significant change of sign in $-\boldsymbol{R_{uv}}$ with

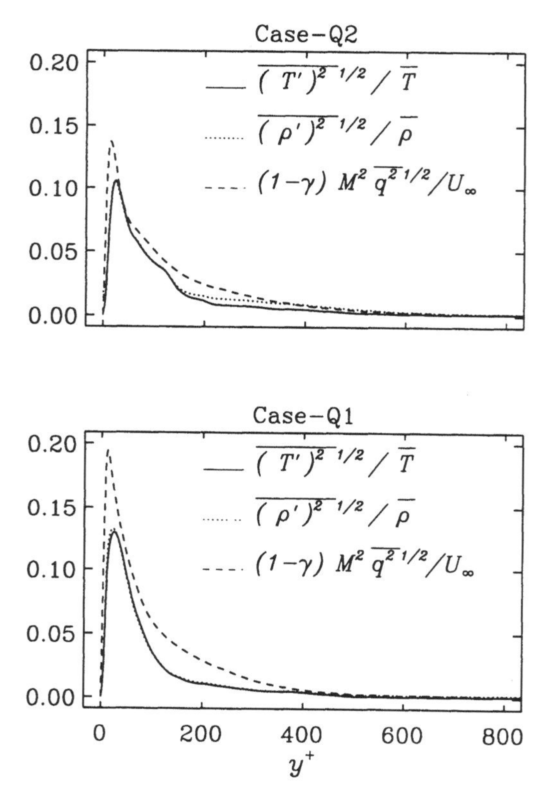

Fig. 15. Comparison of density, temperature, and (total) velocity fluctuations.

spanwise separation (Δz^+) implies a high degree of organization of the large-scale eddies similar to in incompressible turbulent wall-shear layers (Kovasnay (1970)). This consistent behavior is the result of large eddies having a characteristic (or preferred) structure and orientation. Incompressible studies by Kim, Moin, and Moser (1987) and Head and Bandyopadhyay (1981) showed that during a typical burst event in the near-wall region, the low-momentum fluid is carried away from the wall region producing positive Reynolds-shear-stress. Complementing the strong bursts (on a statistically averaged sense), high-momentum fluid is slowed down and brought to the near-wall region by a weak sweep event with negative contributions to $-\boldsymbol{R_{uv}}$. Figure 16 echoes the same argument for wall-shear layers at supersonic velocities. Moreover, very strong similarity of the statistical distributions in the inner and outer parts of the turbulent wall-shear layer is observed. This suggests that burst/sweep structures are not confined to the near-wall region but also extend or correlate with the outer layer.

Figure 17 compares the two-point correlations in the $(\Delta x^+, y_a^+)$ plane by fixing one probe at different y_b^+ locations. $\boldsymbol{R_{uu}}$ distributions are not sym-

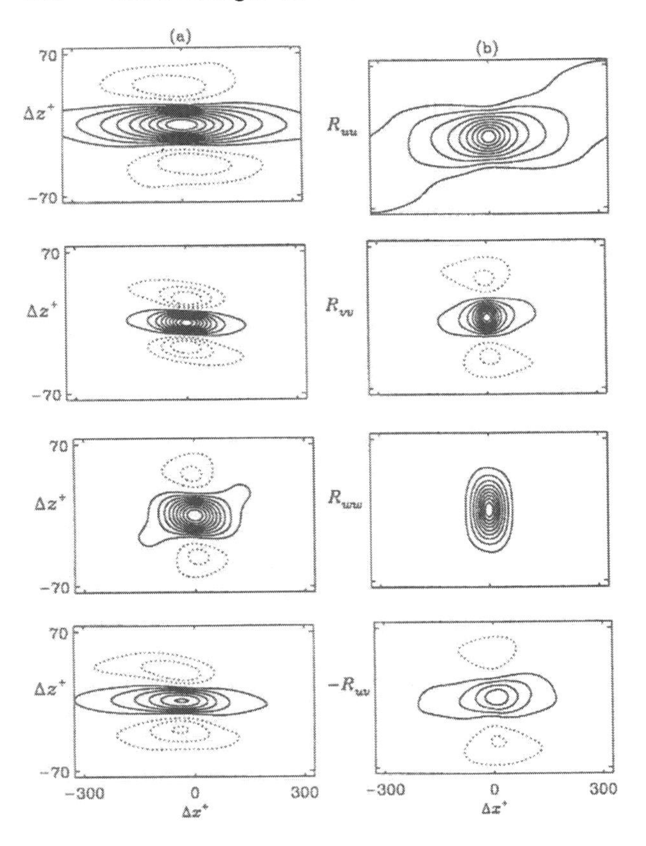

Fig. 16. Two-point correlations of the velocity fluctuations with streamwise and spanwise separations, (a) $\boldsymbol{R}(\Delta x^+, \Delta z^+, y_a^+ = 5.98, y_b^+ = 5.98)$, (b) $\boldsymbol{R}(\Delta x^+, \Delta z^+, y_a^+ = 34.60, y_b^+ = 34.60)$. Contour levels with constant increments of 0.1 between -0.5 and 1.0 are shown. Zero contour level is suppressed. Dotted contours denote negative values.

metric around the zero-streamwise-separation line, $\Delta x^+ = 0$, but are inclined towards positive separations. The same phenomenon was also shown in the space-time isocorrelation distributions in the high-Reynolds number experiments of Robinson (1986) and Spina, Donovan, and Smits (1991) at $M = 3$. Despite the one order of magnitude difference in the Reynolds numbers, the resemblance of the present results in Fig. 17 with the experimental results is very striking and indicates that the basic structures of turbulence are relatively free from Reynolds number effects.

3 Concluding Comments

The studies reviewed in this paper demonstrate the usefulness of direct numerical simulations applied to complex transitional and turbulent flows.

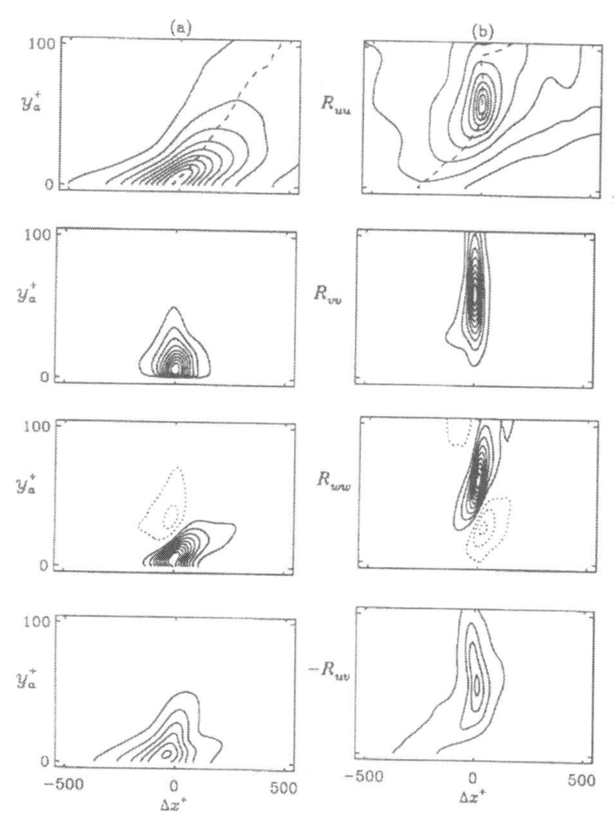

Fig. 17. Two-point correlations of the velocity fluctuations with streamwise and wall-normal separations, (a) $\boldsymbol{R}(\Delta x^+, \Delta z^+ = 0, y_a^+, y_b^+ = 5.98)$, (b) $\boldsymbol{R}(\Delta x^+, \Delta z^+ = 0, y_a^+, y_b^+ = 59.75)$. Contour levels with constant increments of 0.1 between -0.5 and 1.0 are shown. Zero contour level is suppressed. Dotted contours denote negative values.

These studies were aimed primarily at detailed analysis of the underlying physics of transition and turbulence in the particular geometries. However, the problems considered are of sufficient complexity to be representative of simple engineering flows. The current trends of ever greater supercomputer speed with increased memory provide promise for direct simulation of even larger and more complex geometries. Consequently, it is apparent that DNS will be useful not only as a tool for examining flow physics, but also as a means of computing flows for increasingly complex engineering applications.

4 Acknowledgements

The authors gratefully acknowledge the support of ONR and NASA through grants DOD-N00014-94-0923, DOD-N00014-95-1-0419, and NAG-1-1161.

References

Berkooz G., Holmes P., Lumley J.L. (1993): The Proper Orthogonal Decomposition in the Analysis of Turbulent Flows. Annual Review of Fluid Mechanics. **25**, 539–575

Biringen S., Laurien E. (1990): Nonlinear Structures of Transition in Wall-Bounded Flows. Applied Numerical Mathematics. **7**, 129–150

Blackwelder R.F., Browand F.K., Fisher C., Tanaguichi P. (1992): Initiation of Turbulent Spots in a Laminar Boundary Layer by Rigid Particulates. Bulletin of the American Physical Society. **37**, 1812

Bradshaw P. (1975): Calculation Methods for Complex Flows. In *Prediction Methods for Turbulent Flows.* VKI LS 76

Bradshaw P. (1977): Compressible Turbulent Shear Layers. Annual Review of Fluid Mechanics. **9**, 33–54

Canuto C., Hussaini M.Y., Quarteroni A., Zang T.A. (1988): *Spectral Methods in Fluid Dynamics.* (Springer-Verlag, New York)

Demuren A.O. (1990): Calculation of Turbulence-Driven Secondary Motion in Ducts with Arbitrary Cross-Section. AIAA Paper 90-0245

Ferziger J.H., Perić M. (1996): *Computational Methods for Fluid Dynamics.* (Springer, Germany)

Gavrilakis S. (1992): Numerical Simulation of Low Reynolds Number Turbulent Flow Through a Straight Square Duct. Journal of Fluid Mechanics. **244**, 101–129

Ghosal S. (1995): Analysis of Discretization Errors in LES. Annual Research Briefs, CTR

Goldstein D., Handler R., Sirovich L. (1993): Modeling a No-Slip Flow Boundary with an External Force Field. Journal of Computational Physics. **105**, 354–366

Gottlieb D., Orszag S.A. (1977): *Numerical Analysis of Spectral Methods: Theory and Applications.* (SIAM-CBMS, Philadelphia)

Gottlieb D., Turkel E. (1976): Dissipative Two-Four Methods for Time-Dependent Problems. Mathematics of Computation. **30**(136), 703–723

Harlow F.H., Welch J.E. (1965): Numerical Calculation of Time-Dependent Viscous Incompressible Flow of Fluid with Free Surface. Physics of Fluids. **8**, 2182–2189

Härtel C. (1996): Turbulent Flows: Direct Simulation and Large-Eddy Simulation. In *Handbook of Computational Fluid Mechanics* (Ed. R. Peyret). (Academic Press, Great Britain)

Hatay F.F. (1994): Direct Numerical Simulation of Transitional and Turbulent Wall-Shear Layers. Ph.D. Thesis, Department of Aerospace Engineering Sciences, University of Colorado, Boulder, Colorado

Hatay F.F., Biringen S. (1995): Direct Numerical Simulation of a Compressible Turbulent Boundary Layer. AIAA Paper 95-0581

Head M.R., Bandyopadhyay P. (1981): New Aspects of Turbulent Boundary-Layer Structure. Journal of Fluid Mechanics. **107**, 297–338

Huser A. (1992): Direct Numerical Simulation of Turbulent Flow in a Square Duct. Ph.D. Thesis, Department of Aerospace Engineering Sciences, University of Colorado, Boulder, Colorado

Huser A., Biringen S. (1993): Direct Numerical Simulation of Turbulent Flow in a Square Duct. Journal of Fluid Mechanics. **257**, 65–95

Huser A., Biringen S., Hatay F.F. (1994): Direct Simulation of Turbulent Flow in a Square Duct: Reynolds-Stress Budgets. Physics of Fluids. **6**(9), 3144–3152

Hussaini M.Y., Speziale C.G. (1990): The Potential and Limitations of Direct and Large Eddy Simulations. Comment 2. Lecture Notes in Physics. **357**, 354–368

Jordinson R. (1970): The Flat Plate Boundary Layer. Part 1. Numerical Integration of the Orr-Sommerfeld Equation. Journal of Fluid Mechanics. **43**, 801–811

Kasagi N., Shikazono N. (1995): Contribution of Direct Numerical Simulation to Understanding and Modeling Turbulent Transport. Proceedings of the Royal Society of London A. **451**, 257

Kim J., Moin P., Moser R. (1987): Turbulence Statistics in Fully Developed Channel Flow at Low Reynolds Number. Journal of Fluid Mechanics. **177**, 133–166

Kistler A.L. (1959): Fluctuation Measurements in a Supersonic Turbulent Boundary Layer. Physics of Fluids. **2**, 290–296

Kleiser L., Zang T.A. (1991): Numerical Simulation of Transition in Wall-Bounded Shear Flows. Annual Review of Fluid Mechanics. **23**, 495

Kovasnay L.S.G. (1970): The Turbulent Boundary Layer. Annual Review of Fluid Mechanics. **2**, 95–112

Lesieur M., Metais O. (1996): New Trends in Large-Eddy Simulations of Turbulence. Annual Review of Fluid Mechanics. **28**, 45

Lumley J.L. (1970): *Stochastic Tools in Turbulence.* (Academic Press, New York)

MacCormack R.W. (1969): The Effect of Viscosity in Hypervelocity Impact Cratering. AIAA Paper 69-354

Madabhushi R.K., Vanka S.P. (1991): Large Eddy Simulation of Turbulence-Driven Secondary Flow in a Square Duct. Physics of Fluids A. **3**, 2734–2745

Moin P., Carati D., Lund T., Ghosal S., Akselvoll K. (1994): Developments and Applications of Dynamic Models for Large-Eddy Simulation of Complex Flows. In *Application of Direct and Large Eddy Simulation to Transition and Turbulence.* Paper 1, AGARD CP 551

Morkovin M.V. (1962): Effects of Compressibility on Turbulent Flows in *Mechanique de la Turbulence* (Centre National de la Recherche Scientifique, Paris) 367

Orszag S.A. (1972): Comparison of Pseudospectral and Spectral Approximations. Studies in Applied Mathematics. **51**, 253

Perkins H.J. (1970): The Formation of Streamwise Vorticity in Turbulent Flow. Journal of Fluid Mechanics. **44**, 721–740

Piomelli U. (1994): Large Eddy Simulation of Turbulent Flows. TAM Report No. 767, UILU-ENG-945-6023. University of Illinois, Champagne, Illinois

Prandtl L. (1926): Uber die Ausgebildete Turbulenz. NACA Technical Memorandum. **435**, 62

Rai M.M., Moin P. (1991): Direct Numerical Simulation of Flow Using Finite Difference Schemes. Journal of Computational Physics. **96**, 15–53

Reichert R.S., Hatay F.F., Biringen S., Huser A. (1994): Proper Orthogonal Decomposition Applied to Turbulent Flow in a Square Duct. Physics of Fluids. **6**(9), 3086–3092

Reynolds W.C. (1990): The Potential and Limitations of Direct and Large Eddy Simulations. Lecture Notes in Physics. **357**, 313–343

Robinson S.K. (1986): Space-Time Correlation Measurements in a Compressible Turbulent Boundary Layer. AIAA Paper 86-1130

Rogallo R.S., Moin P. (1984): Numerical Simulation of Turbulent Flows. Annual Review of Fluid Mechanics. **16**, 99–137

Saiki E.M. (1995): Spatial Numerical Simulation of Boundary Layer Transition: Effects of a Spherical Particle. Ph.D. Thesis, Department of Aerospace Engineering Sciences, University of Colorado, Boulder, Colorado

Saiki E.M., Biringen S. (1997): Spatial Numerical Simulation of Boundary Layer Transition: Effects of a Spherical Particle. Submitted to Journal of Fluid Mechanics.

Speziale C.G. (1987): On Nonlinear k-l and k-ϵ Models of Turbulence. Journal of Fluid Mechanics. **178**, 459–475

Spina E.F., Donovan J.F., Smits A.J. (1991): On the Structure of High-Reynolds Number Supersonic Turbulent Boundary Layers. Journal of Fluid Mechanics. **222**, 293–327

Streett C.L., Macaraeg M.G. (1989): Spectral Multi-Domain for Large-Scale Fluid Dynamic Simulations. Applied Numerical Mathematics. **6**, 123–139

Computational Aspects of Three-Dimensional Vortex Element Methods: Applications with Vortex Rings

Celalettin Ruhi Kaykayoğlu

İstanbul University, Faculty of Engineering
Department of Mechanical Engineering, Avcılar 34850, İstanbul, Turkey

Abstract. The field of three-dimensional computational Vortex Elements Methods (VEMs) has matured considerably. VEMs are a means of simulating time dependent, incompressible fluid flow in which the flow is represented by a collection of vorticity elements. The vorticity elements concentrate in the flow region of interest where the vorticity is concentrated. This feature makes the method very attractive for complex flow simulations. In this paper a brief introduction to three-dimensional vortex methods for flows of incompressible nature are provided with a special emphasis on the Vortex Particle Method so called Vortons Method. The paper reports results for the simulation of various kinds of vortex ring interactions. The new results and evidence on the possibility of simulating turbulent flows in the inertial range by using the vorton representation of the vorticity field are also presented.

1 Introduction

In incompressible flows, the evolution of the vorticity field determines the velocity field. Hence, once the vorticity field is given the velocity field can be evaluated. Over the last two decades, considerable progress has been made towards developing three dimensional vortex dynamics algorithms so as to simulate complex flow fields. Researchers can refer to the excellent reviews in the literature provided by Widnall (1975), Saffman and Baker (1979), Leonard (1980, 1985), Sarpkaya (1989), Lewis (1991), Shariff and Leonard (1992), Saffman (1993), Meiburg (1995).

The capability of the Vortex Element Methods in simulating a wide range of turbulent shear flows at high Reynolds numbers and homogeneous turbulent flows is one of the current challenges. Computational methods based upon vorticity dynamics have been proven to be very powerful in simulating two–dimensional turbulence (Chorin, 1994). The question whether VEMs can describe the Kolmogorov's $k^{-5/3}$ energy cascade in three–dimensional turbulence and consequently the possibility of simulating turbulent flows with them needs verifications. Lesieur et al. (1995) have argued that the vortex methods fail to describe Kolmogorov's $k^{-5/3}$ energy cascade in three–dimensional isotropic turbulence, instead they provide a k^{-1} type energy cascade. On the other hand, Kiya and Ishii (1991) have shown that the inviscid interaction of vortex filaments can provide the Kolmogorov spectrum. Recently,

Kaykayoğlu et al. (1995) have provided results about the possibility of representing turbulent flows with vortex elements.

The area of multiple vortex ring interaction has been and is currently an area of active research. A vortex ring may be regarded as a simple eddy of turbulence which is commonly found in various complex flows such as: in the initial region of a circular jet, in turbulent boundary layers in the form of a curved filament of concentrated vorticity which evolves into a vortex ring as a result of self induction effect and in the outer region of a turbulent boundary layer. The ring vortices are known as turbulence generators due to their complex interaction dynamics. The transition from a laminar ring to a turbulent one through the complex stages of ring instabilities is one of the ways of obtaining turbulent vortex rings (Shariff and Leonard, 1992). Knio and Ghoniem (1990) have provided a detailed study of the transition process from a laminar ring to a turbulent one by using vortex element approach. Moreover, vortex ring interactions play an important role in sound generation (Kambe et al., 1989). The understanding of organized vortex structures and their interaction is one of the key questions in turbulence (Saffman, 1993).

It is well known that the solution of the Navier–Stokes equations can be approximated by using Vortex Element Methods (Chorin, 1989). The most widely used vortex elements are: *vortex blobs, vortex filaments, vortex sticks or vortons* (Leonard, 1985, Meiburg, 1995). Since VEMs can provide approximate solutions of Navier–Stokes which also describe turbulent flow fields, we can visualize turbulence as the interaction of various kinds of vortex phenomenon. The Lagrangian VEMs provide an efficient representation of vorticity confined flows and the possible representation of the energy cascade with the evolution of the vortex elements seems to be very promising.

The main goal of the present study is to provide details on the cascade of flow energy during unsteady evolution of vortex rings by using vorton elements. This goal is emphasized so that the possibility of simulating turbulent flows by using Vortex Element Methods will be opened to discussion. VEM calculations and their related models may provide a new and helpful picture of the energy cascade process in the inertial range of the turbulent flow energy spectrum. The realization of the inertial range "-5/3 Kolmogorov Law" for the energy spectrum was investigated for different vortex rings interaction mechanisms. A three–dimensional Vortex–In–Cell approach combined with the vorton representation of the vortex rings provides the unsteady evolution of the vortex ring dynamics. The computational approach is inviscid although the projection of the flow properties to grids acts to diffuse vorticity between vortons (Zawadzki and Aref 1991). The author intends to investigate the nature of the three-dimensional energy spectrum evolution in parallel to the flow field dynamics. The detailed character of the flow properties are not reported here since there have been already extensive papers such as (Kida et al., 1991) on the interaction of vortex rings.

The plan of this paper is as follows. In section II, a brief discussion of the computational aspects of the VEMs will be discussed. Section III discusses the concepts of vorton energy and the nature of the three–dimensional energy spectra of vorton rings. After the basic definitions, instantaneous interaction patterns and energy spectra are presented for three different cases. The interaction of two vortex rings starting in a side by side arrangement, the coaxial interaction of two vortex rings starting one behind the other and the motion of four vortex rings along a parallel axes are the three flow cases investigated in the current study. Finally, the concluding remarks are presented.

2 Mathematical Modeling

2.1 Governing Equations

The equations of motion for a vortex system in an inviscid, incompressible fluid can be written as

$$\frac{D\boldsymbol{\omega}}{Dt} = \frac{\partial \boldsymbol{\omega}}{\partial t} + (\boldsymbol{u} \cdot \boldsymbol{\nabla})\,\boldsymbol{\omega} = (\boldsymbol{\omega} \cdot \boldsymbol{\nabla})\,\boldsymbol{u} \tag{1}$$

$$\boldsymbol{\omega} = \boldsymbol{\nabla} \times \boldsymbol{u} \tag{2}$$

$$\boldsymbol{\nabla} \cdot \boldsymbol{u} = 0 \tag{3}$$

where $\boldsymbol{u}(\boldsymbol{x},t)$ is the velocity field, $\boldsymbol{\omega}(\boldsymbol{x},t)$ is the vorticity field. Most of the three–dimensional VEMs use the vorticity–velocity formulation of the incompressible Navier–Stokes equations in the form of Vorticity Transport Equation (1). The velocity field is calculated either by the Poisson Equation

$$\nabla^2 \boldsymbol{u} = -\boldsymbol{\nabla} \times \boldsymbol{\omega} \tag{4}$$

or by applying the Biot–Savart Law

$$\boldsymbol{u}(\boldsymbol{x}) = -\frac{1}{4\pi} \int \frac{(\boldsymbol{x} - \boldsymbol{x}') \times \boldsymbol{\omega}(\boldsymbol{x}')}{|\boldsymbol{x} - \boldsymbol{x}'|^3} d\boldsymbol{x}' \tag{5}$$

A vortical flow field can be represented by sufficiently large number of vortex elements that carry vorticity. The time dependent flow patterns are then obtained by tracing the vortex elements. The change in the vorticity strength of the elements are calculated by using the appropriate form of the vorticity stretching term in Equation (1). Different methods of mimicking the viscous diffusion of vorticity can be applied (Winckelmans and Leonard 1993). The fluid velocity at the locations of these elements can be evaluated by using either the Equation (4) or Equation (3). The Vortex Element Methods differ in various ways. The important properties which characterize the VEMs are as follows:

- discretization of the spatial vorticity field,

- procedures in tracing the discrete vortex elements,
- computational procedures for the calculation of velocity field required to move the vortex elements,
- modeling of the vorticity stretching term,
- modeling of the viscous effects.

Below, the brief presentation of the three different Vortex Element Methods is provided. Readers are advised to refer to the excellent reviews cited above for further details about VEMs.

2.2 Vortex Filament Method, VFM

In many three-dimensional flows, both the vorticity and viscous effects are concentrated in tube like regions which are known as vortex filaments. In Vortex Filament Method, VFM, the initial vorticity field is discretized into N vortex filaments. A single vortex filament is discretized by: the vortex filament strength, Γ_i, core radius, σ_i, and a space curve $r_i(\xi, t)$ on which the filament located (ξ denotes a parameter along the filament curvature). The overall vorticity field is represented by the superposition of the individual vortex filaments (Leonard, 1985),

$$\boldsymbol{\omega}(\boldsymbol{x}, t) = \sum_{i=1}^{N} \Gamma_i \int \frac{1}{\sigma^3{}_i} p\left(\frac{|\boldsymbol{x} - r_i(\xi, t)|}{\sigma_i}\right) \frac{\partial r_i}{\partial \xi} d\xi \tag{6}$$

where $p(\cdot)$ is a smoothing function used in order to desingularize the Biot–Savart integral. Each vortex filament is discretized by a finite number of points. Then the vortex filaments are moved by the induced velocities at these points. The initial strength of the vortex filament, Γ_i, is updated proportional to the change in its length as the flow evolves. Leonard (1980, 1985) and Meiburg (1995) have reported the details of the method. The simulations of three dimensional shear layers with VFM were well demonstrated by the works of Ashurst and Meiburg (1988) and Inoue(1993).

2.3 Vortex Blob Method, VBM

In an inviscid fluid of constant density, the evolution of a vorticity field can be represented by a collection of vortex blobs with overlapping cores (Winckelmans and Leonard 1993). A vortex blob is a vortex element which is introduced to represent a region of distributed vorticity inside a flow field. A single vortex blob α is characterized by the position of its center $\boldsymbol{x}^\alpha$, the vorticity $\boldsymbol{\omega}^\alpha(\boldsymbol{x}^\alpha)$, the volume $d\boldsymbol{x}^\alpha$, and the cut–off radius, σ_α. The strength of the vortex blob $\boldsymbol{\gamma}^\alpha$ is defined by $\boldsymbol{\omega}^\alpha(\boldsymbol{x}^\alpha)\, d\boldsymbol{x}^\alpha$. Then the vorticity field around the vortex blob is defined as

$$\boldsymbol{\omega}^\alpha(\boldsymbol{x}) = \frac{1}{\sigma^3{}_\alpha} p\left(|\boldsymbol{x} - \boldsymbol{x}^\alpha| / \sigma_\alpha\right) \gamma^\alpha \tag{7}$$

where $p(\cdot)$ is an appropriate smoothing function. The velocity induced at the center $\boldsymbol{x}^\alpha$ is computed by the Biot–Savart Law, while the change in blob's strength is calculated by using the inviscid vorticity transport equation (1). The vortex blob position $\boldsymbol{x}^\alpha$ and the strength γ^α are updated by using appropriate schemes. The accuracy of the method depends on several parameters which should be monitored carefully during the simulation process. These parameters are as follows,

- the dimension of the vortex blob, $d\boldsymbol{x}^\alpha$,
- the smoothing function, p,
- the cut off radius, σ_α
- time step for advancing the vortex blobs,
- number of vortex blobs for simulating the desired vorticity region,
- the accurate representation of the spatial resolution of the blobs,
- the control of the increase of vorticity inside the flow field.

The use of a vortex blob method was well demonstrated by several researchers. Winckelmans and Leonard (1993) used the approach to simulate vortex ring interactions. On the other hand, Kiya and Ishii (1991) made use of the method to study vortex interactions and also turbulent energy cascade. Kiya et al. (1994) investigated the unsteady evolution of axisymmetric jets with and without active forcing by using the VBM.

2.4 Vortex Particle Method or Vorton Method, VM

Novikov (1983) introduced the use of three–dimensional singularities so called vortons so as to simulate three–dimensional vortex dynamics. Later, Aksman, Novikov and Orszag (1985) applied the method to study the problem of the mutual penetration of two or four vorton rings by using very small number of vorton singularities. The vorticity field is represented by vortex particles, so called vortons (or vortex sticks) (Winckelmans and Leonard 1993) in the present study. A position and the strength vectors are associated with each vorton. Hence the vorticity field can be written as

$$\boldsymbol{\omega}(\boldsymbol{x},t) = \sum_{k=1}^{N} (\boldsymbol{\omega}_k vol_k)(t)\,(\delta(\boldsymbol{x} - \boldsymbol{x}_k(t)) \tag{8}$$

where vol_k denotes the volume element represented by vorton k and ω_k is the strength and direction of vorticity in this volume element. Eq. (8) can be expressed as

$$\boldsymbol{\omega}(\boldsymbol{x},t) = \sum_{k=1}^{N} \boldsymbol{\gamma_k}(t)\,(\delta(\boldsymbol{x} - \boldsymbol{x}_k(t)) \tag{9}$$

The evolution equations for the vorton position and strength vector are usually taken as (Winckelmans and Leonard 1993),

$$\frac{d}{dt}\boldsymbol{x}_k = \boldsymbol{u}_k\left(\boldsymbol{x}_k(t)\right) \tag{10}$$

$$\frac{d}{dt}\gamma_k = \gamma_k \cdot \left(\nabla \boldsymbol{u}_k\left(\boldsymbol{x}_k(t)\right)\right) \tag{11}$$

Saffman and Meiron (1986) have shown that the equations (10) and (11) does form the weak solution of the three-dimensional vorticity equation (1). However, later, Winckelmans and Leonard (1987) have shown that vortons constitute a weak solution to the vorticity transport equation if the formulation

$$\frac{d}{dt}\gamma_k = \gamma_k \cdot \left(\nabla \boldsymbol{u}_k\left(\boldsymbol{x}_k(t)\right)\right)^T \tag{12}$$

is used instead of Equation (11). Now, equations (10) and (12) provide the transpose scheme as it was outlined in detail by Rehbach (1978) and Winckelmans and Leonard (1993).

Traditionally, Vortex Element Method calculations have been performed by using the Biot–Savart Law. Such an approach requires operations proportional to N_v, where N_v is the number of vortex elements. The computations become slow and very costly when many vortex elements appear inside the flow field. Christiansen (1973) introduced the well known Vortex–In–Cell (VIC) method for two–dimensional flows and this method reduced the operation count to approximately $M \log M$, where M is the number of cells used. In two–dimensional applications, VIC is proven to be very powerful, accurate and faster than the traditional method. The three–dimensional evolution of one and two inviscid buoyant bubbles are studied by using the Vortex–In–Cell (VIC) method by Brechet and Ferrante (1989). Zawadzki and Aref (1991) have utilized the same approach to investigate the flow and mixing characteristics during the head on collision of two vortex rings. In the present study, the VIC approach was used to determine the velocity field and the vorticity stretching values which were than back mapped to the moving vortons. Hence, the current approach is a hybrid Lagrangian-Eulerian one. The velocity field is calculated by solving the Poisson Equation on a computational grid. The local velocity of the vorton is then calculated by interpolation of the grid velocities. The vorticity, $\boldsymbol{\omega}(\boldsymbol{x}, t)$, at grid points is determined using the three–dimensional extension of the two–dimensional VIC approach of Christiansen, (1973).

3 Applications with Vortex Rings

3.1 Vortex Ring Representation

Knio and Ghoniem (1990) studied various spatial discretization schemes of vortical flow fields by using vortex elements. In this paper, only one kind of discretization scheme is used for the representation of the vortex rings by using vortons. First, the vortex ring is divided into n_p, $(p = 1, ...M)$

segments in the azimuthal direction. Then, at each cross section the vortex core radius, R, is divided into n_r, $(r = 1, ...N)$ concentric rings. For $r = 1$, i.e. the center of the vortex core, a single vorton is placed. Vortons are then placed on concentric rings: on each ring eight vortons are placed. The number of concentric rings are chosen as $N = 3$ and kept constant throughout the computations. By this approach, the vortex core is represented by seventeen –17– vortons. Figure 1a shows the details of the vorticity field discretization scheme. Figure 1b shows the isometric view of a vortex ring of finite core constructed by 5040 vortons. The velocity field of a vortex ring is presented in Figure 1c at the symmetry plane.

As it was outlined by Knio and Ghoneim (1990), the evolution of vortex particles depends on the initial placement of vortex elements within a vortex ring cross section. Since the main objective of the author is to study the long term deformation of a vortex core which results in random distribution of vortex particles providing the three–dimensional energy cascade inside the rings, no efforts are given to study the details of the discretization schemes.

3.2 The Nature of Three-Dimensional Energy Spectrum of a Vortex Ring

The vortex ring provides a very rich three–dimensional energy spectrum (Leonard, 1985). A ring with zero cross section has two distinct energy cascade regions: the $1/k$ behavior for large k and k^2 behavior for small k. The peak energy corresponds to a length scale equal to the radius of the ring. Aksman et al. (1985) have discussed and obtained the expressions for the energy of a vorton system. They have defined a new form of energy so called interaction energy of vortons since the self energy of a vorton is infinite. Due to three–dimensional vorticity stretching, the interaction energy can be transformed into self energy of vortons. They believe that such kind of inviscid dissipation of energy can support the realization of the turbulent inertial range "-5/3 law" for the energy spectrum. We hope to show this with more evidences and support in the present study. Aksman et al. (1985) have argued that even four vortons are enough for a reasonable description of a vortex ring. However as it will be shown below, the energy spectrum of a thin vortex ring strongly depends on the number of vortons placed tangentially to the ring circle. Moreover, the flow energy extends to high wavenumbers if there are more vortons inside the flow field to represent the flow kinematics. Figure 2 shows the three–dimensional spectra of a thin vortex ring (zero cross section) where the amount of the vortons used in representation of the vortex ring is used as a parameter.

All the cases provides power law of energy decay towards high wavenumbers. However, for the case N_v equals to 24, this decay is very short and followed by a plateau. As the amounts of vortons are increased, the spectrum begins to follow $1/k$ type behavior at high wave numbers. For low wavenumber range, the k^2 type behavior is observed. The second series of tests were performed

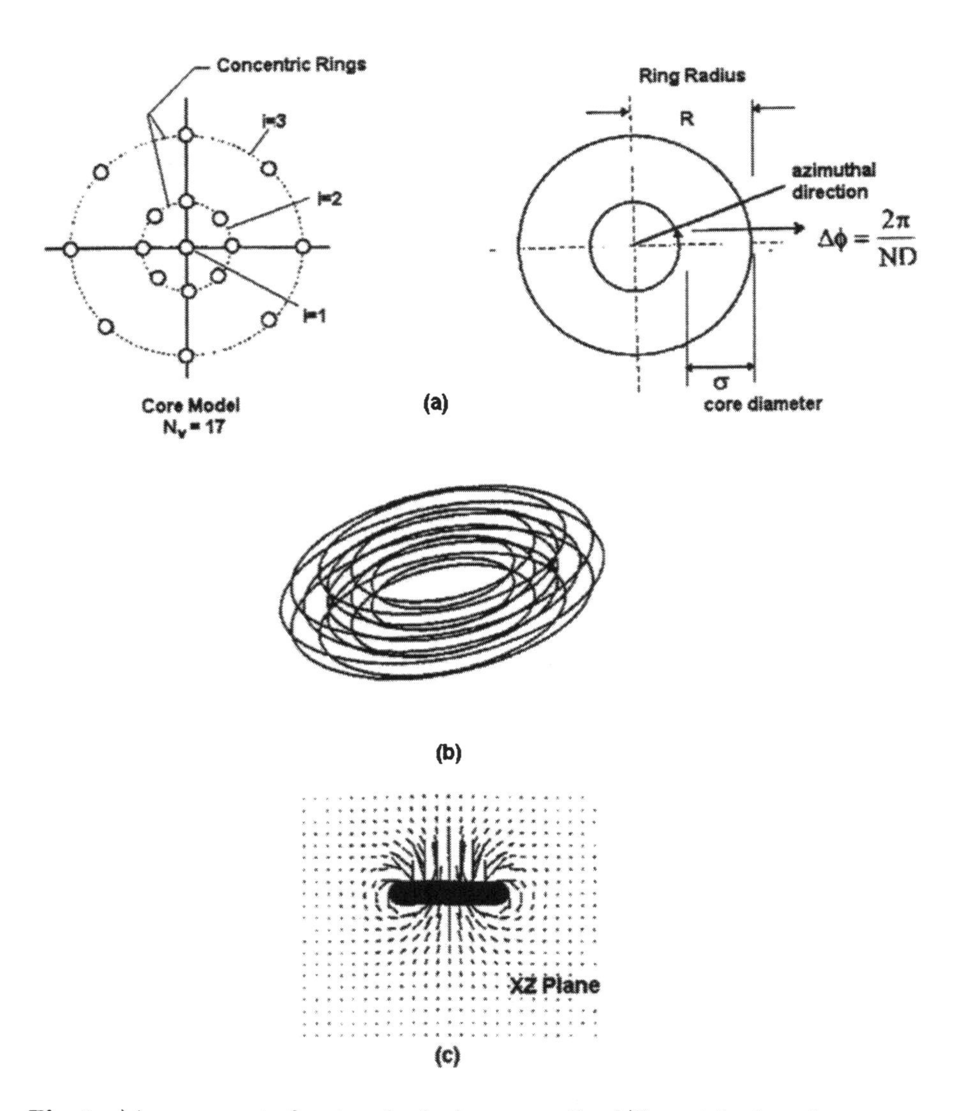

Fig. 1. a)Arrangement of vortons in ring's cross section b)Isometric view of a vorton ring c)Velocity field due to a vorton ring at the symmetry plane

by using a vortex rings with a finite core. Figure 3 show the initial energy spectrum of a vortex ring for the Gaussian type vorticity distribution. The slight hump at the energy on the right of the main peak corresponds to the core radius. The first peaks corresponds to the main radius of the vortex ring.

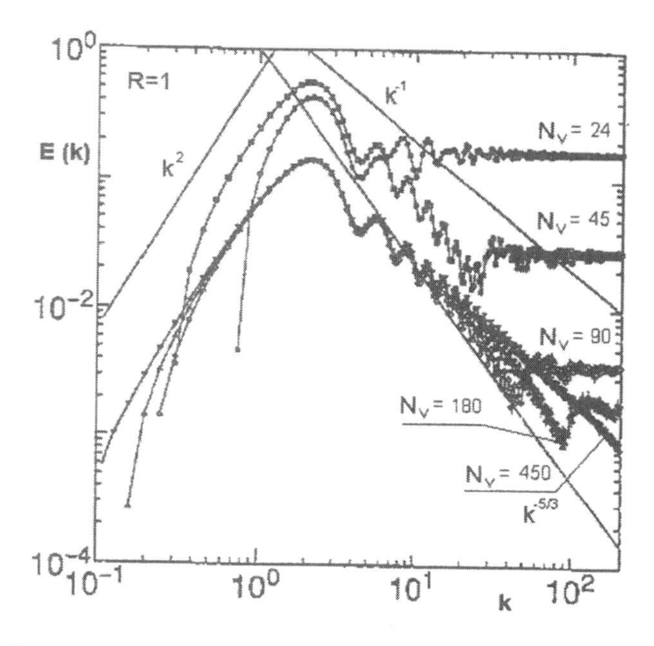

Fig. 2. Energy spectra of thin vortex rings with different amount of vortons

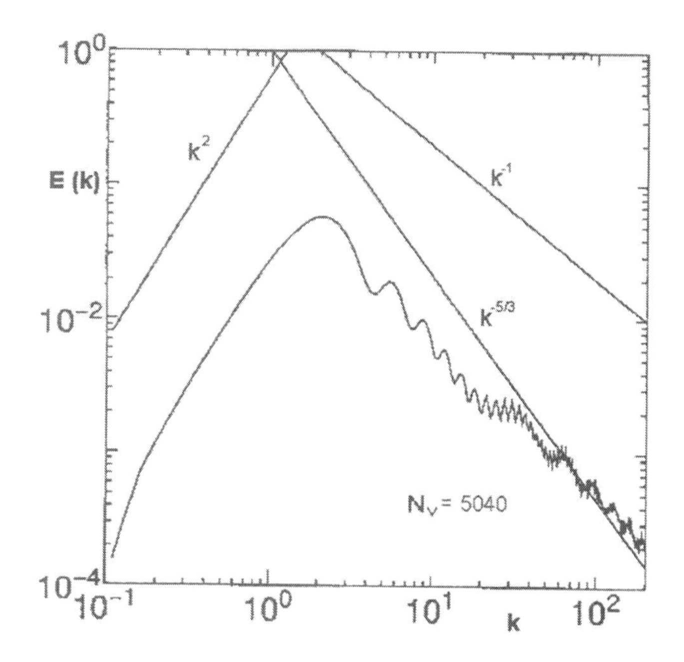

Fig. 3. Energy spectra of thick vortex ring with 5040 vortons

3.3 Interaction of the Two Vortex Rings Starting Side by Side

The interaction of two vorton rings starting in a side by side configuration has been studied extensively by many researchers. Kida et al. (1991) have provided a detailed analysis of the interaction problem. Our results show that, similar to the previous observations, the rings undergo two successive reconnection. The views of the instantaneous vorton distribution confirms most of the flow characteristics observed experimentally and numerically.

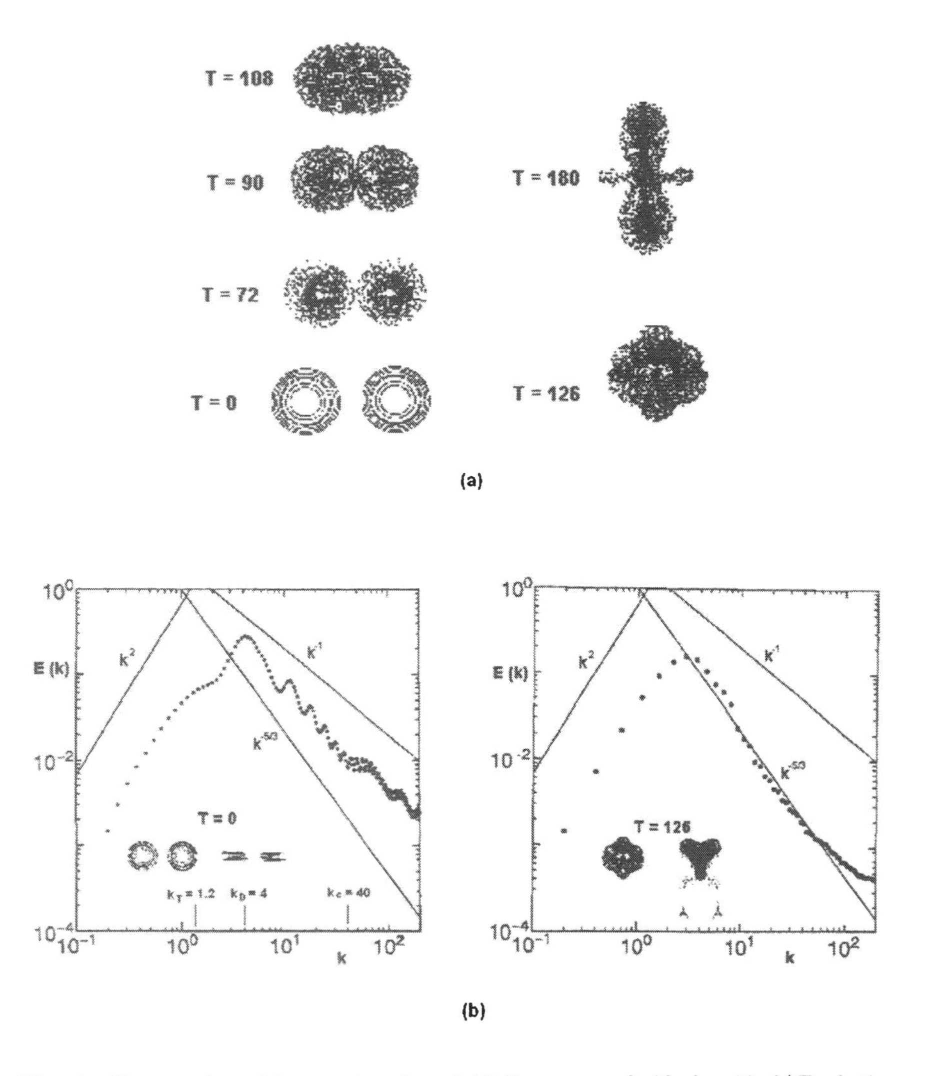

Fig. 4. a)Interaction of two vortex rings initially arranged side by side b)Evolution of energy spectra and realization of $-5/3$ power spectrum

Figure 4a shows the details of the interaction from top at selected instants of nondimensional time. Two vortex rings traveling in the same direction are drawn through one another in the region of closest approach. This results in a reconnection of the vortex tubes and the two rings become one ring. The vortex cores get pushed towards one another by the self-induction which squeezes the two vortex cores. The radial motion of the vortons leads to vortex stretching.

Figure 4b shows the evolution of energy spectra during the interaction. The energy is observed to cascade very efficiently to small scales. The energy spectrum of the system of two vorton rings at the initial moment has $1/k$ behavior for large wavenumbers and k^2 behavior for small wavenumbers. Energy spectrum has a peak and two important humps. The $k_D = 4$ corresponds to the ring diameter. The left hump at wavenumber $k_T = 1.2$ corresponds roughly to the total span of the vorton rings while the right hump at wavenumber $k_C = 40$ corresponds to the effective core size of the vorton rings. The minor fluctuations are due to the trigonometric characters of the functions representing the energy spectrum (Leonard, 1985). As the rings interact, the right hump disappears as the indication of the manifestation of the small scale structures inside the core developed by vortex stretching and artificial diffusion due to VIC method. The slope of the energy spectrum at large wavenumbers deviates from $1/k$ behavior and shows a trend which confirms the decay of energy in turbulent flow regimes. At the later stages of interaction the energy spectrum has a part of the approximate $-5/3$ power law range in about half a decade of wavenumbers. Although the overall structures of the vorton rings are not showing the signs of large scale fully developed turbulence, small scale motions in the fusion zones between two vorton rings are believed to be the cause of the turbulent behavior.

3.4 Interaction of Two Vortex Rings One Starting Behind the Other

The second process which shows the evidence of the $-5/3$ power law spectrum is due to the coaxial interaction of two vorton rings. It is well known that two identical vortex rings, placed one behind the other, show leapfrogging type motion (Yamada and Matsui, 1978). Three dimensional vorton method simulation of this historic problem is studied by two vorton rings system. For the computational time allowed to study the interaction process, one full period of rings' mutual interpenetration was demonstrated. Figure 5a shows the complex ring interaction. At a nondimensional time $T = 15$, the trailing ring vortex radius has already reduced while the leading edge vortex ring's radius starts to elongate. Due to the induced resultant velocity on upper vortex, the vortex structure bends and become wavy at the next time step $T = 30$. At a later instant of time full penetration of the trailing vortex is achieved.

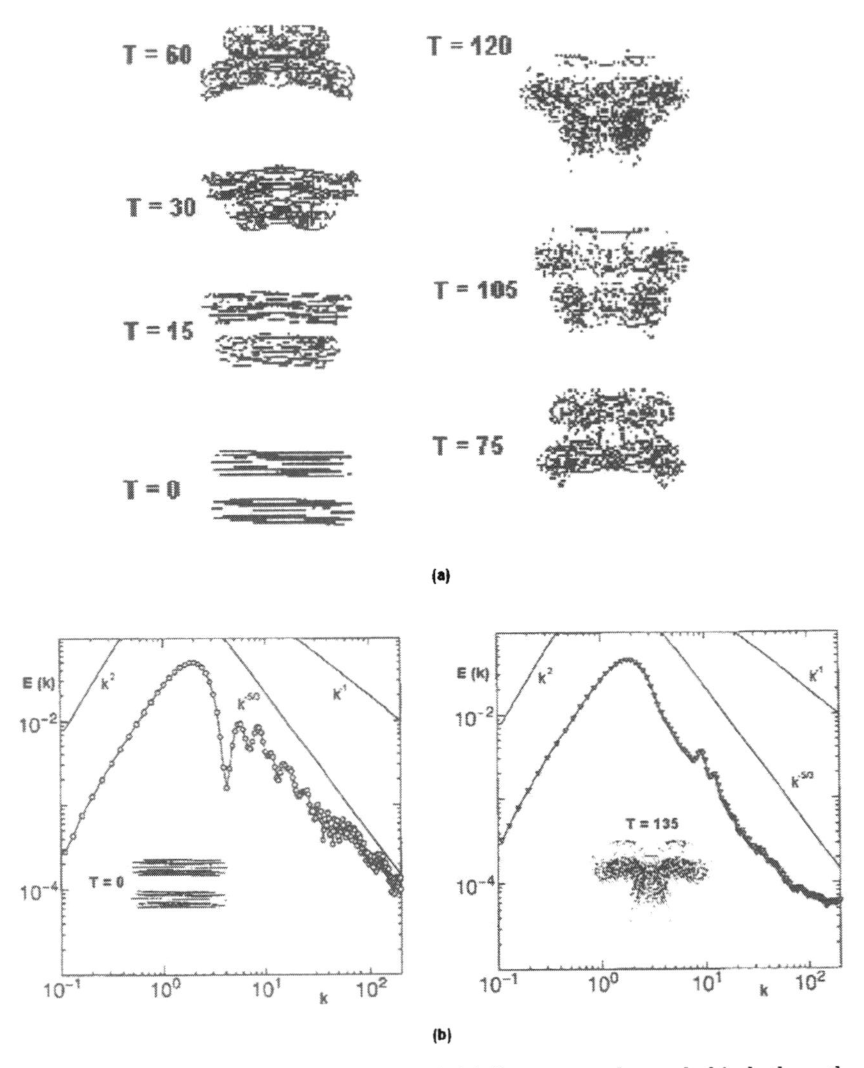

Fig. 5. a)Interaction of two vortex rings initially arranged one behind the other b)Evolution of energy spectra and realization of $-5/3$ power spectrum

The energy spectrum of the system is shown in Figure 5b at two instants of interaction .The energy spectrum is seen to have an approximate $-5/3$ power law over an approximately one decade at the nondimensional time of $T = 135$. The main peak of the energy spectrum correspond to the diameter of the ring.

3.5 Interaction of Four Vortex Rings Starting Side by Side

Results for the unsteady motion of four vortex rings along parallel axes are shown in Figure 6a. Similar to the two vortex ring interactions the rings turn toward each other and merge to form a single distorted ring. Figure 6b shows the behavior of the energy spectra for two instants of time. Again the presence of $-5/3$ power law region for short range of wavenumbers clearly supports the possibility of simulating turbulent flows with Vortex Element Methods.

4 Concluding Remarks

A three-dimensional Vortex-In-Cell code based on vorton elements was successfully used to study the dynamics of inviscid vortex ring interactions. The author has shown that the vorton method provides a reasonably accurate representation of the three important physical processes which are of great interest in simulating the interaction process in turbulent flows. The present computations have also demonstrated that the interaction of two vorton rings side by side or one behind the other can produce energy spectra which have parts of the approximate $-5/3$ law spectrum. Overall fluid topologies confirm previous experimental and computational observations reasonably well. Kolmogorov -5/3 similarity law are shown to be realized by the interaction of vorton rings over one decade for certain times of the interaction The Vortex Element Method based on vorton representation is shown to be a good candidate to simulate the turbulent behavior in the inertial range. The next phase of the research will concentrate on the quantitative character of the vorticity field resulting during multiple ring interactions.

Acknowledgment

Some parts of this study were carried out at Hokkaido University in Sapporo, Japan, during the visit of the author as a JSPS fellow. The author thanks Prof. M. Kiya for the helpful discussions.

References

Aksman, M.J., Novikov, E.A. and Orszag, S.A. (1985): Vorton Method in Three-Dimensional Hydrodynamics, Physical Review Letters, **54,22**, 2410–2413.

Ashurst, W.T. and Meiburg, E. (1988): Three Dimensional Shear Layers via Vortex Dynamics, J. Fluid Mech., **19**, 87

Brecht, S. and Ferrante, J.R. (1989): Vortex-In-Cell Simulations of Buoyant Bubbles in Three Dimensions, Physics of Fluids A, **1**, No 7, 1166–1191

Chorin, A. J. (1989): Vortex Filaments and Turbulence Theory, Proceedings of the IUTAM Symposium on Topological Fluid Mechanics, 607–616

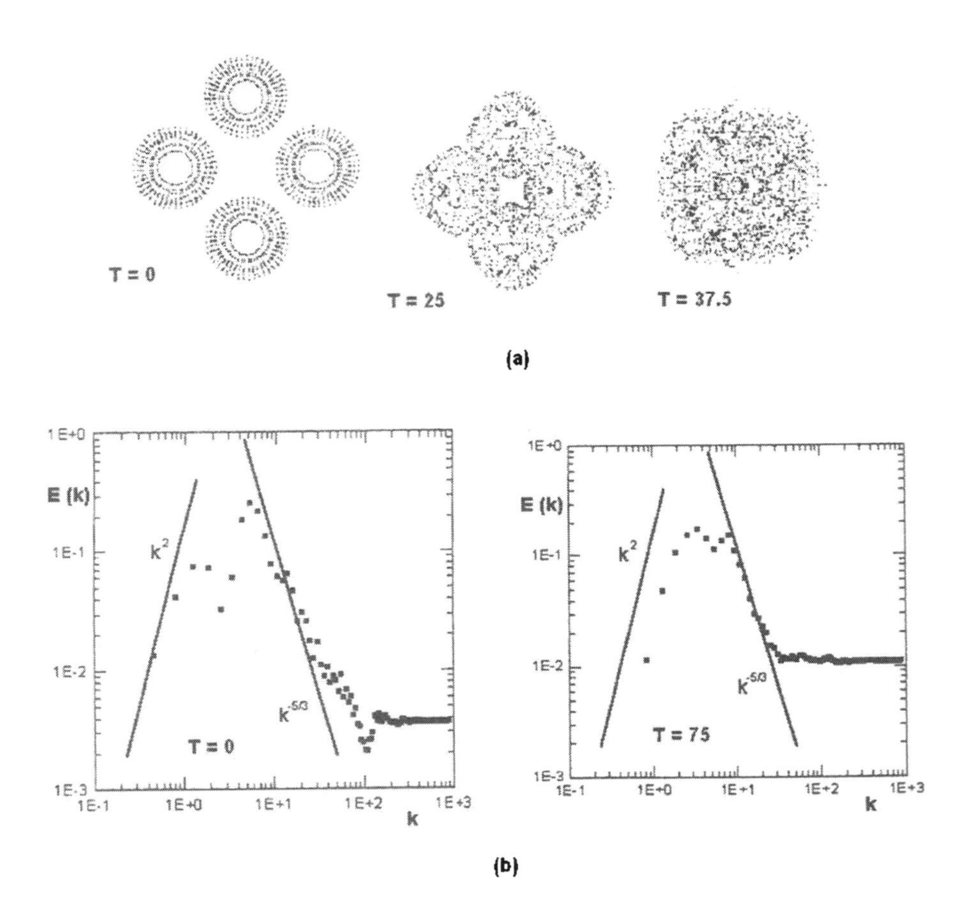

Fig. 6. a)Interaction of four vortex rings along a parallel axes b)Evolution of energy spectra and realization of $-5/3$ power spectrum

Chorin, A. J. (1994): *Vorticity and Turbulence* Springer Verlag, New York.

Christiansen, J.P. (1973): Numerical Simulation of Hydrodynamics by the Method of Point Vortices, J. of Comp. Phys., **13**, 363–379

Couet, B., Buneman, O, and Leonard, A., (1981): Simulation of Three-Dimensional, Incompressible Flows with Vortex–in–Cell Method, Journal of Computational Physics, **39**, 305–328.

Inoue, O. (1989): Vortex Simulation of Spatially Growing Three Dimensional Mixing Layers, AIAA Journal, **27**, 1517.

Kambe, T., Minota, T., Murakami, T. and Takaoka, M. (1989): Oblique collision of Two Vortex Rings and Acoustic Emission, Proceedings of the IUTAM Symposium on Topological Fluid Mechanics, 515–524

Kaykayoğlu, C.R., M. Kiya, M. and Ido, Y. (1995): Simulating Turbulence with the Interaction of Vorton Rings, Proceedings of the Int. Symposium on Mathematical Modeling of Turbulent Flows, Tokyo, 193–198

Kida, S., Takaoka, M. and Hussain, F. (1991): Collision of Two Vortex Rings, J. Fluid Mech., **230**, 583–646.

Kiya, M. and Ishii,H. (1991): Vortex Interaction and Kolmogorov Spectrum, Fluid Dynamics Research, **8**, 73–83.

Kiya, M., Ogura, Y. and Ido, Y. (1994): Axisymmetric, Helical and Multiple Forcing of the Round Jet, Proceedings of the Third JSME–KSME Fluids Engineering Conference, Sendai, Japan, 554–559

Knio, O.M. and Ghoniem, A.F. (1990): Numerical Study of a Three–Dimensional Vortex Method, Journal of Comp., Phys., **6**, 75–106

Leonard, A. (1980): Vortex Methods for Flow Simulation. Journal of Comp. Phys. **37**, 289–335

Leonard, A. (1985): Computing Three Dimensional Incompressible Flows With Vortex Elements. Ann. Rev. Fluid Mech., **17**, 523–559

Lesieur, M., Comte, P. and Metais, O. (1995): Applied Mechanics Review, **48** 121–149.

Lewis, R. I. (1991): *Vortex Element Methods for Fluid Dynamics Analysis of Engineering Systems* Cambridge University Press.

Meiburg, E. (1995): Three Dimensional Vortex Dynamics Simulations in *Fluid Vortices*, S. I. Green , Editor), Kluwer Academic Publishers, 651–685.

Novikov ,E.A. (1983): Generalized Dynamics of Three–Dimensional Vortical Singularities (vortons), Sov. Phys. JETP57, 566–569

Rehbach, C. (1978): Numerical Calculation of Unsteady Three–Dimensional Flows with Vortex Sheets, AIAA Paper No. 78–111

Saffman, P.G., (1993): *Vortex Dynamics*, Cambridge University Press.

Saffman, P.G. and Baker, G. R. (1979), Vortex Interactions. Ann. Rev. Fluid Mech., **11**, 95.

Saffman, P.G. and Meiron, D. I. (1986): Difficulties with Three Dimensional Weak Solutions for Inviscid Incompressible Flow, Physics of Fluids, **29**, No:8, 2373–2375

Sarpkaya, T. (1989): Computational Methods with Vortices–The 1988 Freeman Scholar Lecture, Journal of Fluids Engineering, **111**, 52.

Shariff, K. and Leonard, A. (1992): Vortex Rings. Ann. Rev. Fluid Mech., **24**, 235–279

Widnall, S. E. (1975): The Structure and Dynamics of Vortex Filaments, Ann. Rev. Fluid Mech., **7**, 141–153

Winckelmans, G. S. and Leonard, A. (1993): Contributions to Vortex Particle Methods for the Computation of Three-Dimensional Incompressible Unsteady Flows J. Comput. Phys.,, **109, 2**, 247–273

Winckelmans, G. S. and Leonard, A. (1988): Weak Solutions of the Three Dimensional Vorticity Equation With Vortex Singularities, Phys. of Fluids, **31, 7**, 1838–1839

Yamada, H. and Matsui, T. (1978): Preliminary Study of Mutual Slip–Through of a Pair of Vortices, Physics of Fluids, **21**, No.2, 292-294.

Zawadzki, I. and Aref, H. (1991): Mixing During Vortex Ring Collision, Phys. Fluids A, **3**, 1405–1410

Boundary-Layer Turbulence Modeling and Vorticity Dynamics: I. A Kangaroo-Process Mixing Model of Boundary-Layer Turbulence

H. Dekker[1,2], G. de Leeuw[1], and A.Maassen van den Brink[2]

[1] TNO Physics and Electronics Laboratory, P.O.Box 96864, 2509 JG Den Haag, The Netherlands
[2] Institute for Theoretical Physics, University of Amsterdam, Amsterdam, The Netherlands

Abstract. A nonlocal turbulence transport theory is presented by means of a novel analysis of the Reynolds stress, *inter alia* involving the construct of a sample path space and a stochastic hypothesis. An analytical sampling rate model (satisfying exchange) and a nonlinear scaling relation (mapping the path space onto the boundary layer) lead to an integro-differential equation for the mixing of scalar densities, which represents fully-developed boundary-layer turbulence as a nondiffusive (Kubo-Anderson or kangaroo) type stochastic process. The underlying near-wall behavior (i.e. for $y_+ \to 0$) of fluctuating velocities fully agrees with recent direct numerical simulations. The model involves a scaling exponent ε, with $\varepsilon \to \infty$ in the diffusion limit. For the (partly analytical) solution for the mean velocity profile, excellent agreement with the experimental data yields $\varepsilon \approx 0.58$. The significance of ε as a turbulence Cantor set dimension (in the logarithmic profile region, i.e. for $y_+ \to \infty$) is discussed.

1 Introduction

The logarithmic velocity profile is one of the few established results in the study of turbulent flows. It was first given by Von Kármán and, independently, by Prandtl. A purely dimensional derivation is due to Millikan (and to Landau; see Landau and Lifshitz (1987)) and clearly shows the universality (i.e. model independence) of the result. Namely, the dimensionless mean velocity $\bar{U}_+ = \bar{U}/u_*$ for Newtonian shear flow (in the x-direction) along an infinitely extended smooth surface can only be a function of the dimensionless distance $y_+ = u_* y/\nu$ to that surface, where ν is the kinematic viscosity and where the friction velocity u_* is defined by means of the total stress $\tau = \rho u_*^2$ in the boundary layer (ρ being the fluid mass density). However, in the inertial sublayer of the flow (i.e. the outer part), stress due to molecular viscosity will be unimportant. This is obviously possible only if $d\bar{U}_+/dy_+ = 1/\kappa y_+$, which implicitly defines the Von Kármán constant κ, so that

$$\bar{U}_+ = \frac{1}{\kappa}[\ln(y_+) + \gamma] \tag{1}$$

for $y_+ \to \infty$. Clearly, the integration constant γ can not be determined in this way. On the other hand, in the viscous sublayer (i.e. the inner part of the flow) one has $\bar{U}_+ = y_+$ as $y_+ \to 0$. Universality breaks down in the crossover region (where $y_+ \approx 10$), which therefore requires more specific turbulence transport modeling (see e.g. Launder and Spalding (1972)).

In the boundary layer one is particularly interested in the lateral transport of the longitudinal momentum density $P_x = \rho\bar{U}$. Under zero pressure gradient conditions (fully developed turbulence) the Reynolds-Navier-Stokes equation for this density reads

$$\frac{\partial \bar{U}}{\partial t} = \frac{\partial}{\partial y}(\tau_R/\rho) + \nu\frac{\partial^2 \bar{U}}{\partial y^2} \ , \tag{2}$$

τ_R being the Reynolds stress:

$$\tau_R/\rho = -\overline{uv} \ , \tag{3}$$

where $u = U - \bar{U}$ ($v = V - \bar{V}$) is the fluctuating velocity component in the x(y) direction. Eq.(3) reveals the hierarchy closure problem.

A unified modeling of the viscous and inertial sublayer (including the crossover region) should *ipso facto* yield a value for the integration constant γ in the logarithmic profile (1), for which the experimental data yield $\gamma/\kappa \approx 4 - 6$ (along with $\kappa \approx 0.39 \pm 0.02$). Both Landau and Lifshitz (1987) and Monin and Yaglom (1971) quote $\kappa \approx 0.4$ and $\gamma/\kappa \approx 5.0$ as an established result. So, any nontrivial boundary-layer model should -by smoothly connecting the viscous and inertial sublayers -predict a value of $\gamma \approx 2$. Such modeling should provide a realistic description of scalar transport, not only of $P_x = \rho\bar{U}$ but also of particles, temperature, humidity, etc. The crossover region is of considerable practical importance, especially if the surface is a source (or sink) of e.g. aerosol particles (Dekker and de Leeuw (1993)). Whether such particles are entrained in the flow or are readsorbed by the surface is effectively determined by the physics in this bufferlayer. For atmospheric flow ($\nu = 0.15\ cm^2/s$) with $u_* \approx 0.1$ m/s the relevant scale is a centimeter.

A widely used closure method models the Reynolds stress according to the ideas of molecular viscosity, i.e. as $\tau_R/\rho \equiv \nu_R \partial\bar{U}/\partial y$. In fact, a great variety of local gradient models has been proposed (see e.g. Launder and Spalding (1972)), such as local K- theory (where $\nu_R = \ell u_*$, with a Prandtl-type mixing length $\ell = \kappa y$). All models yield the logarithmic velocity profile (1) as a consequence of Eq.(2) in the steady state ($\partial\bar{U}/\partial t = 0$) outside the viscous sublayer, i.e. if τ_R is constant. Deep inside the viscous sublayer, Eq.(2) always implies constant viscous stress, i.e. the linear profile $\bar{U}_+ = y_+$. Local gradient closure models only account for turbulence mixing by infinitesimally small eddies. In contrast with molecular diffusion, however, such a local description (on the hydrodynamical scale) of turbulence transport is well-known to be inadequate (see e.g. Bernard and Handler (1990) and Dekker et al. (1994)).

Indeed, we will show that nonlocal effects (i.e. finite size eddies) are crucial for modeling boundary-layer turbulence.

The theory will be developed by means of a novel analysis of turbulence sample paths and a stochastic closure hypothesis, in Sec.2. The ensuing description will be seen to be intimately related to the general theory of continuous (but *not* necessarily *diffusive*) Markov processes (Bharucha-Reid (1960), van Kampen (1981)). The analysis gives rise to an integro-differential transport equation of the Chapman-Kolmogorov (or master) type, which describes the migration of an average density (e.g. $P_x = \rho\bar{U}$). If the transition rates in this master equation (which can be computed on the basis of the sampling rates) satisfy certain symmetry conditions, turbulence is mixing by "exchange".

In Sec.3 the analysis of Sec.2 will be applied to the (semi-infinite) boundary layer. This involves a scaling hypothesis for the sampling (hence, for the transition) rates. The scaling also defines a statistical mapping of trajectories in a fictitious path space (attached to each point in real space) onto physical space proper (i.e. of Eulerian onto Lagrangian sample paths). The boundary-layer scaling function is determined on the basis of a self-consistent analysis of the fully three-dimensional fluctuating velocity field. The consistency between the (analytical) predictions of the present theory and the data for near-wall fluctuations will be worth noticing.

Explicit results are obtained for the case of exponential Eulerian sampling rates, which transform into algebraic rates for Lagrangian trajectories. In Sec.4 the associated transition rates are shown to generate a nondiffusive stochastic process of the kangaroo (or strong collision) type (van Kampen (1981), Anderson (1954), Kubo (1962), Stenholm (1984)), which underscores the inadequacy of gradient (i.e. diffusive) type turbulence modeling. The model involves only two intrinsic parameters, viz. a novel exponent ε (local transport amounting to $\varepsilon \rightarrow \infty$) and the viscous sublayerlength a_+. Fortunately, the latter is fairly accurately known from data concerning the normal velocity fluctuations near the wall, both from experiments (see e.g. Johansson and Karlsson (1989)) and direct numerical simulations (e.g. Kim et al. (1987)); in addition, we will show its relation to a (experimentally) well-studied correlation function.

While our model applies to the mixing of any scalar tracer (with density P(y,t)), the solution of the transport problem in the case of $P_x = \rho\bar{U}$ allows for an easy comparison of the theory with a large body of existing data on the mean velocity profile (in particular for pipe and channel flow) thereby yielding a value for ε. Therefore, in Sec.5 the integro-differential master equation for the mean velocity (or rather: its gradient) is solved (partly analytically) in the steady state. Integrating the gradient from the surface (at y=0) through the crossover region, the asymptotic logarithmic behavior (as $y \rightarrow \infty$) will be found including the constant $\gamma(\varepsilon)$. This result describes the velocity profile in the boundary layer in a unified manner, i.e. in both the inertial and the viscous sublayer. With $\varepsilon \approx 0.58$ (for $a_+ \approx 16$ and $\kappa \approx 0.39$, corresponding to $\gamma \approx 1.8$; see (Dekker et al. (1995)) for details) the resulting mean velocity profile is in excellent agreement with the experimental data.

2 Nonlocal Stochastic Closure

The Reynolds stress (3) may be written as

$$\frac{\tau_R}{\rho} = -\frac{1}{T}\int_0^T u(y,t+\tau)v(y,t+\tau)d\tau \ , \tag{4}$$

where T is understood to be sufficiently–but not too–large (see App.A). A sample path $\eta(y,t)$ is defined on the basis of the fluctuation velocity v(y,t) through $v(t+\tau) = d\eta(\tau)/d\tau$, viz.

$$\eta(y,\tau) = \int_0^\tau v(y,t+\tau')d\tau' \ , \tag{5}$$

where we have let $\eta(y,0) = 0$. Eq.(4) thus becomes

$$\frac{\tau_R}{\rho} = -\frac{1}{T}\int_{R(y,T)} u[y,t+\tau(\eta)]d\eta \tag{6}$$

with $R(y,T)$ being the range of values covered by η during $\tau \in (0,T)$. Noticing that $\tau(\eta)$ is a multivalued function, Eq.(6) is rewritten as

$$\frac{\tau_R}{\rho} = -\frac{1}{T}\int_{-\infty}^{\infty} \sum_{n=1}^{N(\eta,y,T)} u[y,t+\tau_n(\eta)]d\eta \ , \tag{7}$$

where $R(y,T)$ is now accounted for by the number N of crossings (or visits) of the sample path $\eta(y,\tau)$ with a fixed value of the coordinate η during $\tau \in (0,T)$.

Now let $N_+(\eta,y,T)$ be the number of times the sample path visits the value η with $v_n > 0$. Similarly, let $N_-(\eta,y,T)$ be the number of times the path visits the value η with $v_n < 0$ (Note that N_+ and N_- always differ at most by one). Eq.(7) then reads

$$\frac{\tau_R}{\rho} = -\frac{1}{T}\int_{-\infty}^{\infty} \left[\sum_{n=1}^{N_+(\eta,y,T)} u[y,t+\tau_n(\eta|+)] - \sum_{n=1}^{N_-(\eta,y,T)} u[y,t+\tau_n(\eta|-)] \right] d\eta \ , \tag{8}$$

where the conditional functions $\tau_n(\eta|\pm)$ have been defined as the nth $\tau(\eta)$ for which $v > 0$ (upper sign) or $v < 0$ (lower sign).

We now introduce the mean visiting rates

$$\lambda_\pm(y,\eta) = \frac{1}{T} N_\pm(y,\eta,T) \quad , \tag{9}$$

as the limiting values of the right hand side (see App.A), so that upon defining the conditional averages

$$\overline{u}(y,\eta,t|\pm) = \frac{1}{N_\pm(y,\eta,T)} \sum_{n=1}^{N_\pm(y,\eta,T)} u[y, t+\tau_n(\eta|\pm)] \ , \tag{10}$$

Eq.(8) becomes

$$\frac{\tau_R}{\rho} = -\int_{-\infty}^{\infty} [\lambda_+(y,\eta)\overline{u}(y,\eta,t|+) - \lambda_-(y,\eta)\overline{u}(y,\eta,t|-)] \ . \tag{11}$$

Notice that Eq.(11) is merely a reorganized, but as yet exactly equivalent, version of Eq.(4). However, Eq.(11) is convenient for the implementation of the (time-reversal symmetry breaking) stochastic closure hypothesis. For the purpose of the present subsection, we assume that the statistical properties of the turbulence are spatially homogeneous, so that the coordinate η defined -for fixed y- by the sample path (5) can be mapped onto the real space coordinate y' without further ado. The inhomogeneity of boundary-layer turbulence is considered in Sec.3.

With $u = U - \overline{U}$ in Eq.(3), one has $\overline{u}(y,\eta,t|\pm) = \overline{U}(y,\eta,t|\pm) - U(y,t)$ in Eq.(11). Now let the transport (of momentum, in this case) over a distance $\Delta y = |\eta|$ be effectively instantaneous (within the averaging-time window T) and consider how it contributes to τ_R/ρ at the point y. Transport *to* y then amounts to events with $v_n < 0$ while $\eta < 0$ (i.e. downward) or with $v_n > 0$ while $\eta > 0$ (i.e. upward), which are therefore counted with $\overline{U}(y,\eta,t|\pm) = \overline{U}(y-\eta,t)$. On the other hand, events with $v_n < 0$ and $\eta > 0$ (downward) or $v_n > 0$ and $\eta > 0$ (upward), represent transport *from* y and thus amount to $\overline{U}(y,\eta,t|\pm) = \overline{U}(y,t)$. Compactly written, the model is defined by $\overline{U}(y,\eta,t|\pm) = U[y-\eta\theta(\pm\eta),t]$. Thus setting

$$\begin{aligned} \overline{U}(y,\eta,t|+) &= \overline{U}(y-\eta,t) \quad if \quad \eta > 0 \ , \\ \overline{U}(y,\eta,t|-) &= \overline{U}(y-\eta,t) \quad if \quad \eta < 0 \ , \end{aligned} \tag{12}$$

and $\overline{U}(y,\eta,t|\pm) = \overline{U}(y,t)$ in all other cases, Eq.(11) leads to

$$\frac{\tau_R}{\rho} = \int_{-\infty}^{\infty} \Lambda(y,\eta) \left[\overline{U}(y+\eta,t) - \overline{U}(y,t)\right] d\eta \ , \tag{13}$$

where $\Lambda(y,\eta<0) = -\lambda_+(y,-\eta)$ and $\Lambda(y,\eta>0) = \lambda_-(y,-\eta)$. This result is in line with a recent Lagrangian analysis of the Reynolds stress by Bernard and Handler (1990) (Sec.9).

If the visiting rates obey mirror symmetry (i.e. $\lambda_-(y,\eta) = \lambda_+(y,-\eta)$, so that $\Lambda = \pm\lambda_+(y,|\eta|)$ is odd in η), Eq.(13) implies Fiedler's (Fiedler (1984)) version of Stull's transilient turbulence theory (Stull (1988)). If in addition the sampling rates λ only depend on η, Eq.(13) further agrees with spectral diffusivity theory (Berkowicz (1984)). In the latter case the rates are modeled as $\lambda(\eta) = \int_{|\eta|}^{\infty} p(\ell)\omega(\ell)d\ell$ where $p(\ell)$ is the probability density of occurrence of an eddy of length ℓ with a typical transport frequency $\omega(\ell)$. Indeed, mixing over a distance $\Delta y = |\eta|$ can only be due to eddies of size $\ell \geq |\eta|$. Neither transilient theory nor spectral diffusivity has been developed under boundary-layer scaling.

3 Boundary-Layer Scaling

Inserting Eq.(13) for τ_R/ρ in the transport equation (2) –for the moment neglecting molecular diffusion–, differentiating the first term in the integrand with respect to y and partially integrating it, gives

$$\frac{\partial \overline{U}}{\partial t} + \frac{\partial}{\partial y}\left[\Lambda(y)\overline{U}(y,t)\right] = \int_{-\infty}^{\infty} \mathcal{W}(y,\eta)\overline{U}(y+\eta,t)d\eta \ , \tag{14}$$

with $\Lambda(y) = \int_{-\infty}^{\infty} \Lambda(y,\eta)d\eta$ and

$$\mathcal{W}(y,\eta) = \frac{\partial \Lambda(y,\eta)}{\partial y} - \frac{\partial \Lambda(y,\eta)}{\partial \eta} \ . \tag{15}$$

$\Lambda(y,\eta)$ is discontinuous at $\eta = 0$ with $\Lambda(y,+0) - \Lambda(y,-0) = 2\lambda(y,0)$. Hence, $\mathcal{W}(y,\eta)$ defines a transition rate $W(y,\eta)$ according to (see e.g. van Kampen (1981))

$$\mathcal{W}(y,\eta) = W(y,\eta) - 2\lambda(y,0)\delta(\eta) \ . \tag{16}$$

The inverse of Eq.(14) follows by considering it as an ordinary differential equation for the $\lambda(y,\eta)$. Putting $\eta = 0$ in the resulting line integral, one has

$$2\lambda(y,0) = \int_{-\infty}^{\infty} W(y-\eta,\eta)d\eta \ . \tag{17}$$

Using Eqs.(16) - (17), Eq.(14) becomes

$$\frac{\partial \overline{U}}{\partial t} + \frac{\partial}{\partial y}\left[\Lambda(y)\overline{U}(y,t)\right] = \int_{-\infty}^{\infty} \left[W(y|y')\overline{U}(y',t) - W(y'|y)\overline{U}(y,t)\right] dy' \tag{18}$$

where we have let $\eta = y' - y$. Eq.(18) describes a stochastic process with transition rates $W(y|y') = W(y, y'-y)$. The corresponding $\mathcal{W}(y|y')$ obviously satisfy $\int_{-\infty}^{\infty} \mathcal{W}(y|y')dy = 0$, as it should. Of course, $\mathcal{W}(y|y')$ has significance as a statistical matrix only if its offdiagonal elements are nonnegative, i.e. if $W(y|y') \geq 0$).

While the above direct mapping of the sample path coordinate η (constructed for fixed y) onto y' may be correct for statistically homogeneous turbulence, it does obviously not apply to the boundary layer (where typical eddies scale with the distance y from the surface). Let $\eta(y,t)$ be the sample path attached to the point y, and let $v(y,t)$ be the fluctuating velocity at y (as in Sec.2), so that $d\eta = v(y, t+\tau)d\tau$. Further, let $\eta_*(y,t)$ denote the actual trajectory of a (co-moving) fluid particle with velocity $v(y_*,t)$, where $y_* = y + \eta_*$. That is, $d\eta_* = v(y+\eta_*, t+\tau)d\tau$. The corresponding trajectory is given by

$$\eta_*(y,\tau) = \int_0^\tau v(y+\eta_*(\tau'), t+\tau')d\tau' \ . \tag{19}$$

Boundary-layer scaling then implies the existence of space-time scaling such that for the coordinate η, one has $d\eta = \mathcal{S}(y)d\varphi_o$, where $\varphi_o(\tau_o)$ is a dimensionless function of an invariant time scale τ_o (defined at some y_o). Similarly, $d\eta_* = \mathcal{S}(y+\eta_*)d\varphi_o$. Hence, one has

$$J(\eta,\eta_*) = \frac{\mathcal{S}(y)}{\mathcal{S}(y+\eta_*)} \tag{20}$$

for the Jacobian $J(\eta,\eta_*) = |d\eta/d\eta_*|$ of the mapping $\eta(\eta_*)$, so that

$$\frac{\eta}{\mathcal{S}(y)} = \int_0^{\eta_*} \frac{1}{\mathcal{S}(y+\eta_*')}d\eta_*' \ . \tag{21}$$

With $\mathcal{S}(y) \geq 0$ and $\mathcal{S}(0) = 0$, Eq.(21) maps the fictitious path space $\eta \in (-\infty,\infty)$ onto the actual path space $\eta_* \in (-y,\infty)$, as it should be for the boundary layer. For instance, under pure time scaling (see below), one has $\mathcal{S}(y) = y$, and Eq.(21) readily yields the inverse mapping $\eta_* = y[\exp(\eta/y)-1]$. The sampling rates at η_* are, of course, given by $\lambda_*(y,\eta_*) = J(\eta,\eta_*)\lambda(y,\eta)$.

The scaling function $\mathcal{S}(y)$ is the product of two functions, viz. $\mathcal{V}(y)$ for the normal velocity component and $\mathcal{T}(y)(= d\tau/d\tau_o)$ for the scaling of time, such that $\mathcal{S} = \mathcal{V}\mathcal{T}$. The basic scaling hypothesis for fully developed boundary-layer turbulence —i.e. $\mathcal{T}(y) = y$— implies that $\mathcal{V}(y) = 1$ in the inertial sublayer, since the logarithmic velocity profile (2) implies $\mathcal{S}(y) = y$ for $y \to \infty$. Indeed, there exists ample experimental evidence that $(\overline{v^2})^{1/2} \sim \mathcal{V}(y)$ does not depend on y in the logarithmic region. Since the flow in the viscous sublayer is fully turbulent (see e.g. Monin and Yaglom (1971)), linear time scaling is expected to hold down to molecular distances from the surface. In Appendix C of Dekker et al. (1995), it is shown that $\mathcal{T}(y) = y$ is consistent with a systematic calculation of the properties of the (fluctuating) flow adjacent to the surface. In fact, nonlinear time scaling is ruled out.

Near $y = 0$ the normal velocity goes to zero according to $\mathcal{V}(y) = (y/a)^2$ by virtue of continuity. The possibility of $a = \infty$ has been considered either ex- or implicitly via the Reynolds stress τ_R/ρ (see e.g. Monin and Yaglom (1971)). In that case one would have $\mathcal{V}(y) \sim y^3$ (if $y \to 0$) and mixing length scaling $\mathcal{S}(y) \sim y^4$. However, the latter is ruled out since it can be shown that $\mathcal{S}(y) \sim y^n$ (with $n > 1$) implies $\tau_R/\rho \sim y^n$ (if $y \to 0$) for the Reynolds stress (see Dekker et al. (1995)). On the other hand, it can also be shown that in the power series expansion of τ_R/ρ the term with $n = 4$ is always absent. Therefore, we take $\mathcal{V}(y) = (y/a)^2$ if $y \to 0$.

Appendix C of Dekker et al. (1995) also yields an expression for the length scale a, viz.

$$a = \left[\frac{2\nu}{\kappa u_*} \int_0^\infty R_{22}(x,0,0) dx \right]^{1/2} \tag{22}$$

where $R_{22}(x,0,0)$ represents the correlation of the normal velocity component close to the surface, with $R_{22}(0,0,0) = 1$ (see e.g. Tennekes and Lumley (1972)). The integral in Eq.(22) defines a correlation length $\overline{x}$ such that $a_+ = (2\overline{x}_+/\kappa)^{1/2}$. The experimental data indicate that $\overline{x}_+ \approx 30 - 60$ (with an average $\overline{x}_+ \approx 45$), which implies $a_+ \approx 12-18$ (with an average $a_+ \approx 15$). This value for a_+ is perfectly consistent with data from measurements concerning the root-mean-square normal velocity fluctuations $(\overline{v^2})^{1/2}$ (see below).

Consider now $\mathcal{S}^{-1} = 1/\mathcal{S}(y)$ as a function of y^{-1}. It has the property $\mathcal{S}^{-1} \to y^{-1}$ if $y^{-1} \to 0$, while $\mathcal{S}^{-1} \to y^{-n}$ if $y^{-1} \to \infty$. In addition, it should be a continuous and differentiable function on $y^{-1} \in (0, \infty)$. Therefore, it may have the polynomial representation $\mathcal{S}^{-1} = \sum_{k=1}^{n} c_k y^{-k}$. In view of the preceding, $n = 3$ while $c_1 = 1$ and $c_3 = a^2$. Since throughout $\mathcal{T}(y) = y$, the inverse $\mathcal{V}^{-1} = 1/\mathcal{V}(y)$ therefore reads $\mathcal{V}^{-1} = 1 + c_2 y^{-1} + (y/a)^{-2}$. Consequently, one has $\mathcal{V}(y) = (y/a)^2[1 - c_2 y/a^2 + ...]$ if $y \to 0$. However, it has been shown (see Dekker et al. (1995)) that in the Taylor expansion of $(\overline{v^2})^{1/2} \sim \mathcal{V}(y)$ the cubic term $\sim y^3$ is always absent. Hence, $c_2 = 0$ so that

$$\mathcal{V}(y) = \frac{(y/a)^2}{1 + (y/a)^2} \quad . \tag{23}$$

The ensuing result for the (inverse) mixing-length scaling function reads

$$\frac{1}{\mathcal{S}(y)} = \frac{1}{y} \left[1 + \left(\frac{a}{y} \right)^2 \right] \quad . \tag{24}$$

Comparison of $(\overline{v^2})^{1/2} \approx u_* \mathcal{V}(y)$ with the available data from both experiments and numerical simulations (see e.g. Monin and Yaglom (1971), Johansson and Karlsson (1989), Kim et al. (1987)) indeed yields $c_2 \approx 0$ and confirms that $a_+ \approx 10 - 20$ with an average $a_+ \approx 15$.

4 Kangaroo-Process Mixing

Using the Jacobian (20), one may rewrite Eq.(13) in terms of η_*. The closure hypothesis (12) now implies the mapping of η_* onto y'. Letting $\eta_* = y' - y$ and defining $\Lambda_*(y|y') = J(\eta, \eta_*)\Lambda(y, \eta)$, the Reynolds stress formula becomes

$$\tau_R/\rho = \int_0^\infty \Lambda_*(y|y')\, \overline{U}(y', t)\, dy' \quad , \tag{25}$$

where we have already accounted for exchange [$W(y|y') = W(y'|y)$; see Eq.(30)), so that $\Lambda_*(y) = 0$ (see Eq.(14)).

To proceed, we consider the following rate model (Dekker et al. (1994), Dekker et al. (1995)):

$$\lambda(y, \eta) = \frac{D}{\mathcal{S}(y)} \exp[-\varepsilon|\eta|/\mathcal{S}(y)] \ . \tag{26}$$

Using Eqs.(20)-(21), one obtains

$$\Lambda_*(y|y') = -\frac{D}{\varepsilon}\frac{\partial}{\partial y'}\left[\frac{\mathcal{K}(y)}{\mathcal{K}(y')}\right]^{\pm\varepsilon} \ , \tag{27}$$

with

$$\mathcal{K}(y) = \mathcal{K}_o \exp\left[\int_{y_o}^{y} \frac{dy'}{\mathcal{S}(y')}\right] \ , \tag{28}$$

the upper (lower) sign in the exponent applying if $y' > y$ ($y' < y$), and $\mathcal{K}_o$ and y_o being constants (of integration of $\partial\mathcal{K}/\partial y = \mathcal{K}/\mathcal{S}(y)$) which are immaterial for the rates. E.g., let $\mathcal{K}_o = \mathcal{K}(y_o)$ be such that $\mathcal{K}(y)/y \to 1$ if $y \to \infty$.

The transport equation (18) (i.e. Eq.(1)) now becomes

$$\frac{\partial \overline{U}}{\partial t} = \int_0^\infty \left[W_*(y|y')\,\overline{U}(y',t) - W_*(y'|y)\,\overline{U}(y,t)\right] dy' + \nu\frac{\partial^2 \overline{U}}{\partial y^2} \ , \tag{29}$$

with $W_*(y|y') = \partial\Lambda_*(y|y')/\partial y$. Eq. (27) yields

$$W_*(y|y') = \frac{\varepsilon D}{\mathcal{S}(y)\mathcal{S}(y')}\left[\frac{\mathcal{K}(y)}{\mathcal{K}(y')}\right]^{\pm\varepsilon} \ , \tag{30}$$

which is always nonnegative and, therefore, indeed generates a proper stochastic process. Note that (30) factorizes as $W_*(y|y') = \mathcal{P}(y)\mathcal{Q}(y')$. Such a stochastic process is known as a *kangaroo process* (see e.g. van Kampen (1981), Stenholm (1984)).

A kangaroo process (in particular, with $\mathcal{P}(y) = 1$ or $\mathcal{Q}(y) = 1$) is also known as a Kubo-Anderson process. It has *inter alia* been applied to motional narrowing in spin systems, Mössbauer spectroscopy and laser linewidth calculations, in the strong-collision limit. This limit is the opposite of diffusive motion. In fact, since boundary-layer scaling amounts to both $\mathcal{S}(y) \to y$ and $\mathcal{K}(y) \to y$ if $y \to \infty$, a gradient expansion in Eqs.(29)-(30) is not rigorously possible for any finite value of the scaling exponent $\varepsilon < \infty$ (see also Bernard and Handler (1990)).

5 Velocity Profile

Let us now consider Eqs.(29)-(30) in the steady state $\partial\overline{U}/\partial t = 0$. In that case it is more convenient to return to Eq.(2) and to invoke Eq.(25) for τ_R/ρ which yields

$$\nu\frac{\partial\overline{U}}{\partial y} + \int_0^\infty \Lambda_*(y|y')\,\overline{U}(y')dy' = u_*^2 \ . \tag{31}$$

With Eq.(27) for the rates and a partial integration, one obtains

$$\nu\frac{\partial\overline{U}}{\partial y} + \frac{D}{\varepsilon}\left[\int_y^\infty \left(\frac{\mathcal{K}(y)}{\mathcal{K}(y')}\right)^{\varepsilon}\frac{\partial\overline{U}}{\partial y'}\,dy' + \int_0^y \left(\frac{\mathcal{K}(y)}{\mathcal{K}(y')}\right)^{-\varepsilon}\frac{\partial\overline{U}}{\partial y'}\,dy'\right] = u_*^2 \ . \tag{32}$$

Eq.(32) can be mapped onto an equivalent differential equation. Namely, let $z = \ln[\mathcal{K}(y)/\mathcal{K}_o]$ so that

$$\nu\frac{\partial\overline{U}}{\partial y} + \frac{D}{\varepsilon}\int_{-\infty}^\infty \exp(-\varepsilon|z-z'|)\,\frac{\partial\overline{U}}{\partial z'}\,dz' = u_*^2 \ , \tag{33}$$

with $\partial/\partial z = \mathcal{S}(y)\partial/\partial y$. Note that if $z \to \infty$ (i.e. $y \to \infty$) one has $\partial\overline{U}/\partial z \to u_*/\kappa$ according to (1), so that $D = \frac{1}{2}\varepsilon^2\kappa u_*$. Now, with $K(z) = \exp(-\varepsilon|z|)$, considering $\Im = \int_{-\infty}^{\infty} dz'\ K(z-z')$ as a linear operator which maps the function $\varphi = \partial\overline{U}/\partial z$ onto another function, say $\psi = \Im\varphi$, one obtains $\Im^{-1} = \frac{1}{2}\varepsilon[1-\varepsilon^{-2}(\partial/\partial z)^2]$ for the inverse operation (i.e. $\Im^{-1}\Im = I$). Therefore, operating with $\Im^{-1}$ from the left on (5.3) reduces it to an inhomogeneous second order differential equation for the velocity gradient. In terms of $\mathcal{Y} = \kappa y_+$ and $\overline{\mathcal{U}} = \kappa\overline{U}_+$ it becomes a Sturm-Liouville equation for $\mathcal{F}(\mathcal{Y}) = \overline{\mathcal{U}}'$, viz.

$$-\varepsilon^{-2}\mathcal{S}(\mathcal{Y})[\mathcal{S}(\mathcal{Y})\mathcal{F}']' + [1+\mathcal{S}(\mathcal{Y})]\mathcal{F}(\mathcal{Y}) = 1 \ , \tag{34}$$

where a prime denotes differentiation with respect to $\mathcal{Y}$, where $\mathcal{S}(\mathcal{Y}) = \mathcal{Y}^3/\mathcal{A}^2 + \mathcal{Y}^2)$ with $\mathcal{A} = \kappa a_+$ so that $\mathcal{A} \approx 6$. The ensuing mean velocity profile $\overline{\mathcal{U}}(y)$ is universal in the sense that it is independent of the Von Kármán constant κ.

The new exponent ε may be determined by comparing the solution of Eq.(34) with measured mean velocity distribution data over a smooth surface. In fact, a rough estimate for ε can be made analytically since (34) allows for an exact solution in two limiting cases (see Dekker et al. (1995)). *First*: for $\varepsilon \to \infty$, Eq.(34) reduces to its local limit $\mathcal{F}(\mathcal{Y}) = 1/(1+\mathcal{S})$. In particular, for the profile (1) it leads to $\gamma = \gamma_a + \ln\kappa$, with $\gamma_a \approx (2\pi/3^{3/2})\mathcal{A}^{2/3} - \ln(\mathcal{A}^{2/3}) - \frac{1}{3}$ if $\mathcal{A} >> 1$. This yields $\gamma \approx 1.5$ for $\mathcal{A} \approx 6$. *Second*: for $a \to 0$, Eq.(34) reduces to its pure time-scaling limit where $\mathcal{S}(\mathcal{Y}) = \mathcal{Y}$, which allows for a solution in terms of Lommel functions. The resulting logarithmic profile constant is $\gamma = \gamma_\varepsilon + \ln\kappa$, with $\gamma_\varepsilon = \varepsilon^{-1} - 2[\psi_E(1+\varepsilon) - \ln\varepsilon]$ and ψ_E being Euler's psi-function. With $\gamma \approx \gamma_a + \gamma_\varepsilon + \ln\kappa$, the experimental value $\gamma \approx 2$ implies $\gamma_\varepsilon \approx 0.5$. This corresponds to $\varepsilon \approx 0.5$.

A more accurate value for ε can be obtained by a best-fit to the entire measured profile, i.e. including the crossover region between viscous and inertial sublayers. Eq.(34) has been integrated by means of a simple (matrix inversion) routine, starting at $\mathcal{Y} = 0$ with $\mathcal{F}(0) = 1$. A final integration then yields the universal profile $\overline{\mathcal{U}}(\mathcal{Y})$. An excellent *unconditional* fit (better than 2%) to the experimental data $\overline{U}_+ = \overline{\mathcal{U}}/\kappa$ as a function of $y_+ = \mathcal{Y}/\kappa$ for $1 < y_+ < 10^4$ (taken from standard references like Nikuradse (1932), and processed for their residual Reynolds number dependence as $Re \to \infty$) is obtained for $\kappa = 0.39$, $a_+ \approx 16$, and yields the value $\varepsilon \approx 0.58$. A detailed discussion can be found in Dekker et al. (1995).

6 Conclusion

We have shown that mixing in the turbulent boundary layer can be described as a kangaroo process. This nondiffusive stochastic process is found by means of (i) the construct of a local sample path space $\eta(y)$ at each point y in the fluid; (ii) a nonlinear scaling $\mathcal{S}(y)$ which maps η onto a global path space $\eta_*(\eta)$ with $\eta_* = y' - y$; (iii) a stochastic closure hypothesis; and (iv) an analytical sampling rate model.

In Appendix A the closure hypothesis will be shown to be exact for a Rayleigh particle in the Smoluchovski limit (i.e. for diffusion). The sampling rate model (26) is suggested by the nature of the mapping (21) of Eulerian onto Lagrangian sample paths in the inertial sublayer (i.e. where $\mathcal{S}(y) = y$). The general expression (24) for the scaling function in the semi-infinite boundary layer arises from an analysis of fully developed 3D-turbulence over an infinite flat plate, and an assumption concerning its simplest analytical form (see Dekker et al. (1995)).

Our model involves both an eddy viscosity (defining the Von Kármán constant $\kappa \approx 0.39$) and a novel exponent ε. The latter defines a Prandtl type mixing length ($l \approx y/\varepsilon$ if $\varepsilon \to \infty$; $l \approx y\varepsilon^{1/\varepsilon}$ if $\varepsilon \to 0$). In the limit $\varepsilon \to \infty$, the model reduces to local K-theory. While in the inertial sublayer turbulence mixing is self-similar by perfect time scaling $\mathcal{S}(y) \sim y$, it involves a characteristic length $a_+ \approx 16$ to yield viscous sublayer scaling $\mathcal{S}(y) \sim y^3$.

The steady state solution of the transport equation (29)-(30) is compared with the data for the mean velocity profile, which yields $\varepsilon \approx 0.58$ for the scaling exponent and shows the nonlocality of turbulent transport. As a consequence, our turbulence kangaroo process is suggested (in Appendix B) to have fractal features.

Appendix A: Markovian Sampling Rates

Let us write the mean sampling rates (9) as a time-average, viz.

$$\lambda_\pm(y,\eta) = \frac{1}{T}\int_0^T \sum_n \delta\left[\tau - \tau_n(\eta|\pm)\right] d\tau \quad , \tag{35}$$

the conditional functions $\tau_n(\eta|\pm)$ being defined in Eq.(8). Eq.(35) equals

$$\lambda_\pm(y,\eta) = \frac{1}{T}\int_0^T |v(\tau)|\delta[\eta - \eta(\tau)]\ \theta[\pm v(\tau)]d\tau \quad , \tag{36}$$

where $v(\tau) = d\eta/d\tau$ is the velocity at y and $\eta(\tau) = y_\xi(\tau|y',0)$ is the trajectory with $y = y'$ at $\tau = 0$ for the sample ξ. Using $f[v(\tau)] = \int_{-\infty}^{\infty} f(v)\delta[v - v(\tau)]dv$ with $f(v) = |v|\theta(\pm v)$ and recalling the definition of $\Lambda(y,\eta)$ in Eq.(13), leads to

$$\Lambda(y, y'-y) = \pm\frac{1}{T}\int_0^T d\tau \int_0^\infty dv \ \ vP_\xi(y, \mp v, \tau|y', 0) \quad , \tag{37}$$

with

$$P_\xi(y,\mp v,\tau|y',0) \equiv \delta\left[y - y_\xi(\tau|y',0)\right]\delta[v \pm v_\xi(\tau)] \quad , \tag{38}$$

where the upper (lower) sign applies if $y' > y$ ($y' < y$). The rate Λ in Eq.(37) will be a stochastic function if $\{y_\xi(\tau), v_\xi(\tau)\}$ is a stochastic process.

By imposing the initial condition $y_\xi = y'$ at time t (i.e. $\tau = 0$) on the sample functions of a stochastic process one selects a subensemble (see e.g. van Kampen (1981)). If the subensemble is such that during the time-lapse $T \to 0$ (on the relevant time scale) arrivals at y at time τ with $v > 0$ (< 0) almost surely come from the region $y' < y$ ($> y$), it is allowed to extend the velocity integration in Eq.(37) to $v = -\infty$, which yields

$$\Lambda(y, y'-y) = \lim_{T\to 0}\frac{1}{T}\int_0^T d\tau < -v_\xi(\tau)\ \delta\left[y - y_\xi(\tau|y',0)\right] > \quad , \tag{39}$$

where the brackets indicate the average with respect to $\{\xi\}$. The integrand in Eq.(39) is the time-derivative of the gap distribution (50), i.e.

$$\Lambda(y, y'-y) = -\int_y^\infty \frac{\mathrm{T}_\tau(y''|y') - \delta(y''-y')}{\tau} dy'' \quad , \tag{40}$$

where the short time propagator ($\tau \to 0$) is given by

$$\mathrm{T}_\tau(y|y') = < \delta\left[y - y_\xi(\tau|y',0)\right] > \quad . \tag{41}$$

For a stationary Markov process the subensemble defined by T_τ will be time-homogeneous and imply a time-independent transition rate $\mathcal{W} = \partial\mathrm{T}_\tau/\partial\tau$. In that case (see Eq.(49)) Eq.(40) readily yields

$$\mathcal{W}(y, y'-y) = \frac{\partial\Lambda(y, y'-y)}{\partial y} \quad , \tag{42}$$

which is identical to Eq.(15).

Brownian motion is also time-homogeneous, but its Green's function does not obey Eq.(49). However, Eqs.(40)-(41) do apply as Brownian motion has the properties implied in (39). Namely, consider Eqs.(37)-(38) and average over $\{\xi\}$. This gives

$$\Lambda(y, y'-y) = \pm\frac{1}{T}\int_0^T d\tau \int_0^\infty dv \;\; vP(y, \mp v, \tau|y', 0) \;. \tag{43}$$

For Brownian motion, $P(y, v, \tau|y', 0) =< P_\xi(y, v, \tau|y', 0) >$ obeys Kramers' Fokker-Planck equation for the Rayleigh particle, corresponding to the Langevin equation

$$dx/dt = v, \;\; dv/dt = -\gamma v + \xi(t) \;, \tag{44}$$

with

$$< \xi(t) >= 0, \;\; < \xi(t+\tau)\xi(t) >= 2\gamma\vartheta\delta(\tau) \;, \tag{45}$$

where $\vartheta = k_B\Theta/m$ (Θ being temperature, m the particle mass). In this case the averaging period in Eq.(43) should be very large on the Rayleigh scale $1/\gamma$. In this limit (i.e. $\gamma T \to \infty$) the particle loses all memory of its initial velocity and its propagator becomes (see e.g. Risken (1983))

$$P(y, v, \tau|y', 0) = N\exp\left[-\frac{1}{2\vartheta}(v - \frac{y-y'}{2\tau})^2 - \frac{\gamma}{4\vartheta\tau}(y-y')^2\right] \;, \tag{46}$$

where $N = (\gamma/2\tau)^{1/2}/2\pi\vartheta$. Using Eq.(46) in Eq.(43), and taking the limit $T \to 0$ (on the Smoluchovski scale), yields

$$\Lambda(y, y'-y) = \pm(2\nu/\ell^2)\; erfc(|y-y'|/\ell) \;, \tag{47}$$

with $\ell = 2(\nu T)^{1/2} \to 0$, and $\nu = \vartheta/\gamma$ being Einstein's viscosity. Upon comparing Eq.(47) with Eq.(26) for $y \approx y'$ one finds the relation $\varepsilon/y \lesssim (\pi\nu_R T)^{-1/2}$, where $\nu_R = \kappa u_* y$ and T is a so-called integral time scale. Taking for the latter $T \gtrsim y/u_*$ [see e.g. Tennekes and Lumley (1972)), one has $\varepsilon \lesssim (\pi\kappa)^{-1/2}$. Hence, $\varepsilon \lesssim 0.9$ in close agreement with the result $\varepsilon \approx 0.58$ (Sec.5).

Appendix B: Fractal Features

Let

$$Q(y') = \int_0^\infty W(y|y')dy \tag{48}$$

be the intensity function and let $\mathsf{T}_\tau(y|y')$ denote the transition probability over a short period of time τ (e.g. $\tau \approx T$, the integration time in the Reynolds stress; see also App.A). $\mathsf{T}_\tau(y|y')$ is known as short time propagator or Green's function. Then

$$\mathsf{T}_\tau(y|y') = [1 - \tau Q(y)]\,\delta(y-y') + \tau W(y|y') \quad . \tag{49}$$

Notice that $\mathcal{W}(y|y') = \partial \mathsf{T}_\tau / \partial \tau$ equals the transport kernel (15)-(16). Now let $Pr(Y > y)$ denote the probability that during τ the system (e.g. a tracer particle) jumps from y_o to beyond y. $Pr(Y > y)$ may also be called the gap distribution (Mandelbrot (1983)). In terms of Eq.(49) one has

$$Pr(Y > y) = \int_y^\infty \mathsf{T}_\tau(y''|y_o)dy'' \quad . \tag{50}$$

Using $W = W_*(y|y')$ from Eq.(30), with both y and y' in the inertial sublayer, Eq.(50) yields

$$Pr(Y > y) = C_o y^{-\varepsilon} \quad , \tag{51}$$

where the constant is given by $C_o = \tau D_o / y_o^{1-\varepsilon}$. Upon comparing the Paretian distribution (51) with the number of gaps $Nr(L > \ell) = N_o \ell^{-d}$ of length $L > \ell$ in a Cantor set $\mathcal{C}_\infty$ with fractal dimension $0 < d < 1$, one has $\varepsilon = d$.

References

Anderson P. W. (1954): J. Phys. Soc. Japan **9**, 316

Berkowicz R. (1984): in *Boundary-Layer Meteorology* (eds. H.Kaplan and N. Dinar, Reidel, Dordrecht), **30**

Bernard P. S., Handler R. A. (1990): J. Fluid. Mech. **220**, 99

Bharucha-Reid A. T. (1960): *Elements of the Theory of Markov Processes and Their Applications* (McGraw-Hill, New York)

Dekker H., de Leeuw G. (1993): J. Geophys. Res. **C98**, 10223

Dekker H., de Leeuw G., Maassen van den Brink A. (1994): Mod. Phys. Lett. **B8**, 1655 also see Phys. Rev. E, (1995) **52**, 2549

Dekker H., de Leeuw G., Maassen van den Brink A. (1995): Physica A **A218**, 335

Fiedler B. H. (1984): J. Atmos. Sci. **41**, 674

Johansson T. G., Karlsson R. I. (1989): in *Applications of Laser Anemometry to Fluid Mechanics* (ed. Adrian, Springer-Verlag, Berlin) p.3.

van Kampen N. G. (1981): *Stochastic Processes in Physics and Chemistry* (North-Holland, Amsterdam)

Kim J., Moin P., Moser R. (1987): J. Fluid. Mech. **177**, 133

Kubo R. (1962): in *Fluctuations, Relaxation and Resonance in Magnetic Systems* (ed. ter Haar, Oliver and Boyd, Edinburgh)

Landau L. D., Lifshitz E. M. (1987): *Fluid Mechanics* (Pergamon, Oxford)

Launder B. E., Spalding D. B. (1972): *Mathematical Models of Turbulence* (Academic, London)

Mandelbrot B. B. (1983): *The Fractal Geometry of Nature* (Freeman, New York)

Monin A. S., Yaglom A. M. (1971): *Statistical Fluid Mechanics* (MIT Press, Cambridge), Vol. 1-2

Nikuradse J. (1932): VDI-Forschungsheft, **356**
Risken H. (1983): *The Fokker-Planck Equation* (Springer-Verlag, Berlin)
Stenholm S. (1984): in *Essays in Theoretical Physics* (ed. Parry, Pergamon, Oxford), p.247.
Stull R. B. (1988): *An Introduction to Boundary Layer Meteorology* (Kluwer, Dordrecht)
Tennekes H. and Lumley J. L. (1972): *A First Course in Turbulence* (MIT Press, Cambridge, MA)
Townsend A. A. (1976): *The Structure of Turbulent Shear Flow* (Cambridge University Press, Cambridge, UK)

Boundary-Layer Turbulence Modeling and Vorticity Dynamics: II. Towards a Theory of Turbulent Shear Flow?

H. Dekker[1,2]

[1] TNO Physics and Electronics Laboratory, P.O.Box 96864, 2509 JG Den Haag, The Netherlands
[2] Institute for Theoretical Physics, University of Amsterdam, Amsterdam, The Netherlands

Abstract. I present a first, brief report on progress in the formulation of a novel systematic theory of fully-developed 3D boundary-layer turbulence.

1 Introduction

In Part I of this contribution to turbulence modeling and vortex dynamics, the emphasis has been on modeling, in particular of the sampling rates $\Lambda(y, \eta)$. The ensuing nonlocal Markovian mixing model was thus based on statistical properties of the underlying velocity fluctuations, such as the boundary-layer scaling function $\mathcal{S}(y)$. Presently, emphasis will be shifted to dynamics.

I will outline a theory of 3D turbulent shear flow as the irreducible, asymptotic part of fully-developed boundary-layer turbulence (i.e. for $Re = u_* y/\nu \to \infty$). Higher order corrections are typically of order $Re^{-1/2}$. While (in view of its asymptotic nature) the theory does not admit a low-order description of the viscous sublayer, it does allow a unified treatment of both inertial (vorticity) and thermal (buoyancy) features in the inertial sublayer.

Conceptually, thermal shear flow is a prototype turbulence. Namely, since turbulence production amounts to $\mathcal{P} = -(1 - Ri_f)\overline{u_i u_j}\overline{S_{ij}}$ (where $\overline{S_{ij}}$ is the mean strain rate and Ri_f is the flux-Richardson number) it is imperative to include shear in a full-fledged theory, especially in connection with the emergence of the spectral inertial subrange à la Kolmogorov (where $\mathcal{P} = \varepsilon$, with ε being the energy dissipation rate). Moreover, boundary-layer shear introduces an external length scale into the problem (which is a key to the description of e.g. bursts and sweeps) while the inclusion of thermal effects (i.e. $Ri \neq 0$) allows studying Monin-Obukhov universal functions (as background fields).

2 Thermal Vorticity Dynamics

Consider the 3D Navier-Stokes and heat transport equations (in the Boussinesq approximation) for incompressible turbulent flow over an infinite, flat plate (at $y = 0$). Gravity acts in the y-direction. Let the mean flow be steady, and in the x-direction, so that $\overline{U}_1 = \overline{U}(y)$, $\overline{U}_2 = \overline{U}_3 = 0$. Reynolds decomposition ($U_i = \overline{U}_i + u_i$, $T = \overline{T} + \theta$ for the temperature, and $P = \overline{P} + p$ for the pressure) then yields

$$-\overline{uv} + \nu \partial \overline{U}/\partial y = u_*^2 , \quad -\overline{v\theta} + \chi \partial \overline{T}/\partial y = u_* \theta_* , \quad \overline{vw} = 0 , \qquad (1)$$

while $\overline{P}$ is fixed by $\partial(\overline{P} + \rho\overline{v^2})/\partial y = \rho g(\overline{T} - T_o)/T_o$ (where T_o is the adiabatic background temperature). In order to eliminate the pressure (which is nonlocal in the velocities) from the equations, let me consider the vorticity fluctuations $\omega_i = \varepsilon_{ijk}\partial u_k/\partial x_j$. For the mean vorticity in the boundary layer one has $\overline{\Omega}_1 = \overline{\Omega}_2 = 0$ and $\overline{\Omega}_3 = -\partial\overline{U}/\partial y$.

I now focus on a line y_o , and expand the slowly varying mean fields about it by letting $y = y_o + \eta$. The irreducible part of the problem is then defined as the zeroth order set of equations which nontrivially includes the mean shear $\overline{S}_{12} = -\frac{1}{2}\overline{\Omega}_3$, and conserves transversality (i.e. $\partial\omega_i/\partial x_i = 0$). Upon introducing a 4D spectral decomposition (so that $\omega_i \Rightarrow \tilde{\omega}_i(k_i, \omega)$, and $\theta \Rightarrow \tilde{\theta}$), Taylor advection is eliminated by taking $\omega = \omega_o + f$ with $\omega_o = k_1\overline{U}_o$ (where $\overline{U}_o = \overline{U}(y_o)$). Putting $\tilde{\omega}_i = \hat{\omega}_i \exp(isf/\overline{\Omega}_o)$ with $\overline{\Omega}_o = \overline{\Omega}(y_o)$ and $s = k_2/k_1$ (and similarly $\tilde{\theta} \Rightarrow \hat{\theta}(k_i, f)$), one arrives at a set of coupled, nonlinear Bloch-type equations for $(\hat{\omega}_i, \hat{\theta})$ with s as fictitious time-variable, viz.

$$\begin{aligned} \kappa^2\partial\hat{\omega}_1/\partial s &= s\hat{\omega}_1 - \hat{\omega}_2 + \kappa^2\hat{\vartheta} + \hat{\mathcal{L}}_1(s,f) , \\ \kappa^2\partial\hat{\omega}_2/\partial s &= -(1+\alpha^2)\hat{\omega}_1 - s\hat{\omega}_2 + \hat{\mathcal{L}}_2(s,f) , \end{aligned} \qquad (2)$$

($\hat{w}_3$ following from $\hat{w}_1 + s\hat{w}_2 + \alpha\hat{\omega}_3 = 0$) where $\hat{\vartheta} = ig_ok_3\hat{\theta}$, $\kappa^2 = 1 + \alpha^2 + s^2$, $\alpha = k_3/k_1$, $g_o = -g/(\overline{\Omega}_oT_o)$; the viscous terms are included in the functionals $\hat{\mathcal{L}}$ of the nonlinear interactions. Further, the ϑ-equation implies that $\kappa^2\partial(\hat{\vartheta} - \hat{\omega}_2 Ri)/\partial s = \hat{\mathcal{F}}$ ($Ri = g_o d\overline{T}_o/d\overline{U}_o$ is the gradient-Richardson number; $\overline{T}_o = \overline{T}(y_o)$) is a functional similar to $\hat{\mathcal{L}}$. This leads to

$$\frac{\partial}{\partial s}\left[\kappa^2\frac{\partial\hat{\vartheta}}{\partial s}\right] = \mathcal{A}\hat{\vartheta} + \hat{\mathcal{J}}(s,f) , \qquad (3)$$

where $\mathcal{A} = -(1+\alpha^2)Ri$ and $\hat{\mathcal{J}}$ follows from $\hat{\mathcal{F}}$ and $\hat{\mathcal{L}}_i$. Eq. (3) determines the (thermal) response on the input field $\hat{G}$ as

$$\begin{aligned} \hat{\vartheta}(\tau) &= \left[A - \hat{\mathcal{I}}_P(\tau)\right] Q_\nu(i\tau) + \left[B + \hat{\mathcal{I}}_Q(\tau)\right] P_\nu(i\tau) , \\ \hat{\mathcal{I}}_P(\tau) &= i\int_0^\tau P_\nu(i\tau')\hat{\mathcal{J}}(\tau')d\tau' , \end{aligned} \qquad (4)$$

$\hat{\mathcal{I}}_Q = \hat{\mathcal{I}}_{P\Rightarrow Q}$; P_ν, Q_ν are Legendre functions with $\nu = \frac{1}{2}[(1+4\mathcal{A})^{1/2} - 1]$; $\tau = (1+\alpha^2)^{-1/2}s$; $A = -i(B^{(+)} - B^{(-)})/\pi$ and $B^{(\pm)} = -\hat{\mathcal{I}}_Q(\pm\infty)$ follow from boundary conditions at the cut $\tau = 0$ and at $\tau = \pm\infty$, respectively;

the vorticity $\hat{w}_2$ also follows from (3)-(4), *mutatis mutandis* for $\hat{\mathcal{J}}$. With $\hat{w}_2$ and $\hat{\phi} = (\partial\hat{\omega}_2/\partial s) - \hat{\mathcal{L}}_2/\kappa^2$, the velocity fluctuations read $\hat{u} = i\beta(s\hat{\phi} - \alpha^2\hat{\omega}_2)$, $\hat{v} = -i\gamma\hat{\phi}$, $\hat{\omega} = i\alpha\beta(s\hat{\phi} + \hat{\omega}_2)$ (where $\gamma = k_3^{-1}, \beta = \gamma/(1+\alpha^2)$).

3 Correlation Functions

The formal (rigorous) representation of the fluid dynamics in the form of (4) makes sense upon introducing systematic approximations for the input fields $\hat{\mathcal{J}}$. I define $\tau_o \equiv \ell/\overline{U}_o$, $\tau_S \equiv |\overline{S_{12}}^{-1}|$, $\tau_n \equiv |s_{ij}^{-1}|$ and $\tau_\nu \equiv \ell^2/\nu$ ($\nu =$ viscosity) with $\ell \approx k^{-1}$. Note that $\hat{\mathcal{J}}$ involves τ_n and τ_ν. One then shows (e.g. see Tennekes and Lumley (1972) and use $\tau_S \approx y/u_*$) that for $\ell \gtrsim \ell_K$ (Kolmogorov microscale) always $\tau_n, \tau_\nu >> \tau_o$ while $\tau_n \lesssim \tau_S$ only in the inertial subrange and beyond ($\ell << y$). Further, $\tau_\nu \lesssim \tau_S$ only in the viscous subrange ($\ell \approx \ell_K$). Hence, to leading order (in powers of $(\ell/y)^{2/3} \approx Re^{-1/2}$) $\hat{\mathcal{J}}$ in Eq. (3) is significant only for $\ell_K \lesssim \ell << y$, so that it can be treated statistically as isotropic turbulence input (using equilibrium theory, see Batchelor (1976) and Tennekes and Lumley (1972)).

Rewriting $\hat{\mathcal{J}}$ (and $\hat{\mathcal{L}}$, $\hat{\mathcal{F}}$) in terms of the original fields $\tilde{\omega}_i$ and $\tilde{\theta}$ (Sec.1), I use the dynamical output of Sec.2 for the second order correlation tensor (using local homogeneity and time translation invariance). The input tensor can be reduced rigorously to second order (as $\hat{\mathcal{J}}$, etc., contain *all* nonlinearities) and thus will only involve ε and ε_θ (the thermal fluctuations dissipation rate). Hence, I also calculate the output dissipation rates and self-consistently determine $\varepsilon(Ri)$, and $\varepsilon_\theta(Ri)$. Finally, since $Ri = \zeta\phi_h/\phi_m^2$ (with $\zeta = y/L$, $L = u_*^2 T_o/\kappa g\theta_*$), subjecting the output tensor to the constraints (1) will self-consistently fix the (so far unspecified) Monin-Obukhov functions $\phi_m(\zeta) \equiv \zeta\partial\overline{\mathcal{U}}/\partial\zeta$ and $\phi_h(\zeta) \equiv \zeta\partial\overline{\mathcal{T}}/\partial\zeta$ (with $\overline{\mathcal{U}} = \kappa\overline{U}/u_*$, $\overline{\mathcal{T}} = \kappa\overline{T}/\theta_*$).

4 Final Remarks

The above sketched analysis is akin to particle-on-a-string modeling of dissipation in mesoscopic systems (Dekker (1996)). Presently the particle and the environment can only be disentangled in Fourier space, the former (latter) being recognized in the low (high) k-region. 3D boundary-layer turbulence thus appears as the dynamics of a nonlinearly self-driven self-consistent stochastic field. This brief report (especially Sec.3) concerns work in progress. A full account will be given elsewhere.

References

Batchelor G. K. (1976): *The Theory of Homogeneous Turbulence* (Cambridge University Press, Cambridge, UK)

Dekker H. (1996): in *Tunneling and its Implications*, eds. D. Mugnai, A. Ranfagni and L. S. Schulman (World-Scientific, Singapore)

Tennekes H. and Lumley J. L. (1972): *A First Course in Turbulence* (MIT Press, Cambridge, MA)

Subject Index

Lecture Notes in Physics

For information about Vols. 1–455
please contact your bookseller or Springer-Verlag

Vol. 456: H. B. Geyer (Ed.), Field Theory, Topology and Condensed Matter Physics. Proceedings, 1994. XII, 206 pages. 1995.

Vol. 457: P. Garbaczewski, M. Wolf, A. Weron (Eds.), Chaos - The Interplay Between Stochastic and Deterministic Behaviour. Proceedings, 1995. XII, 573 pages. 1995.

Vol. 458: I. W. Roxburgh, J.-L. Masnou (Eds.), Physical Processes in Astrophysics. Proceedings, 1993. XII, 249 pages. 1995.

Vol. 459: G. Winnewisser, G. C. Pelz (Eds.), The Physics and Chemistry of Interstellar Molecular Clouds. Proceedings, 1993. XV, 393 pages. 1995.

Vol. 460: S. Cotsakis, G. W. Gibbons (Eds.), Global Structure and Evolution in General Relativity. Proceedings, 1994. IX, 173 pages. 1996.

Vol. 461: R. López-Pen˜a, R. Capovilla, R. Garcı´a-Pelayo, H. Waelbroeck, F. Zertuche (Eds.), Complex Systems and Binary Networks. Lectures, México 1995. X, 223 pages. 1995.

Vol. 462: M. Meneguzzi, A. Pouquet, P.-L. Sulem (Eds.), Small-Scale Structures in Three-Dimensional Hydrodynamic and Magnetohydrodynamic Turbulence. Proceedings, 1995. IX, 421 pages. 1995.

Vol. 463: H. Hippelein, K. Meisenheimer, H.-J. Röser (Eds.), Galaxies in the Young Universe. Proceedings, 1994. XV, 314 pages. 1995.

Vol. 464: L. Ratke, H. U. Walter, B. Feuerbach (Eds.), Materials and Fluids Under Low Gravity. Proceedings, 1994. XVIII, 424 pages, 1996.

Vol. 465: S. Beckwith, J. Staude, A. Quetz, A. Natta (Eds.), Disks and Outflows Around Young Stars. Proceedings, 1994. XII, 361 pages, 1996.

Vol. 466: H. Ebert, G. Schütz (Eds.), Spin - Orbit-Influenced Spectroscopies of Magnetic Solids. Proceedings, 1995. VII, 287 pages, 1996.

Vol. 467: A. Steinchen (Ed.), Dynamics of Multiphase Flows Across Interfaces. 1994/1995. XII, 267 pages. 1996.

Vol. 468: C. Chiuderi, G. Einaudi (Eds.), Plasma Astrophysics. 1994. VII, 326 pages. 1996.

Vol. 469: H. Grosse, L. Pittner (Eds.), Low-Dimensional Models in Statistical Physics and Quantum Field Theory. Proceedings, 1995. XVII, 339 pages. 1996.

Vol. 470: E. Martı´nez-González, J. L. Sanz (Eds.), The Universe at High-z, Large-Scale Structure and the Cosmic Microwave Background. Proceedings, 1995. VIII, 254 pages. 1996.

Vol. 471: W. Kundt (Ed.), Jets from Stars and Galactic Nuclei. Proceedings, 1995. X, 290 pages. 1996.

Vol. 472: J. Greiner (Ed.), Supersoft X-Ray Sources. Proceedings, 1996. XIII, 350 pages. 1996.

Vol. 473: P. Weingartner, G. Schurz (Eds.), Law and Prediction in the Light of Chaos Research. X, 291 pages. 1996.

Vol. 474: Aa. Sandqvist, P. O. Lindblad (Eds.), Barred Galaxies and Circumnuclear Activity. Proceedings of the Nobel Symposium 98, 1995. XI, 306 pages. 1996.

Vol. 475: J. Klamut, B. W. Veal, B. M. Dabrowski, P. W. Klamut, M. Kazimierski (Eds.), Recent Developments in High Temperature Superconductivity. Proceedings, 1995. XIII, 362 pages. 1996.

Vol. 476: J. Parisi, S. C. Müller, W. Zimmermann (Eds.), Nonlinear Physics of Complex Systems. Current Status and Future Trends. XIII, 388 pages. 1996.

Vol. 477: Z. Petru, J. Przystawa, K. Rapcewicz (Eds.), From Quantum Mechanics to Technology. Proceedings, 1996. IX, 379 pages. 1996.

Vol. 479: H. Latal, W. Schweiger (Eds.), Perturbative and Nonperturbative Aspects of Quantum Field Theory. Proceedings, 1996. X, 430 pages. 1997.

Vol. 480: H. Flyvbjerg, J. Hertz, M. H. Jensen, O. G. Mouritsen, K. Sneppen (Eds.), Physics of Biological Systems. From Molecules to Species. X, 364 pages. 1997.

Vol. 481: F. Lenz, H. Grießhammer, D. Stoll (Eds.), Lectures on QCD. VII, 276 pages. 1997.

Vol. 482: X.-W. Pan, D. H. Feng, M. Vallières (Eds.), Contemporary Nuclear Shell Models. Proceedings, 1996. XII, 309 pages. 1997.

Vol. 483: G. Trottet (Ed.), Coronal Physics from Radio and Space Observations. XVII, 226 pages. 1997.

Vol. 484: L. Schimansky-Geier, T. Pöschel (Eds.), Stochastic Dynamics. XVIII, 386 pages. 1997.

Vol. 486: G. Chavent, P. C. Sabatier (Eds.), Inverse Problems of Wave Propagation and Diffraction. XV, 379 pages. 1997.

Vol. 488: B. Apagyi, G. Endrédi, P. Lévay (Eds.), Inverse and Algebraic Quantum Scattering Theory. XV, 385 pages. 1997.

Vol. 489: G. M. Simnett, C. E. Alissandrakis, L. Vlahos (Eds.), Solar and Heliospheric Plasma Physics. VIII, 278 pages, 1997.

Vol. 491: O. Boratav, A. Eden, A. Erzan (Eds.), Turbulence Modeling and Vortex Dynamics. XII, 245 pages, 1997.

New Series m: Monographs

Vol. m 1: H. Hora, Plasmas at High Temperature and Density. VIII, 442 pages. 1991.

Vol. m 2: P. Busch, P. J. Lahti, P. Mittelstaedt, The Quantum Theory of Measurement. XIII, 165 pages. 1991. Second Revised Edition: XIII, 181 pages. 1996.

Vol. m 3: A. Heck, J. M. Perdang (Eds.), Applying Fractals in Astronomy. IX, 210 pages. 1991.

Vol. m 4: R. K. Zeytounian, Mécanique des fluides fondamentale. XV, 615 pages, 1991.

Vol. m 5: R. K. Zeytounian, Meteorological Fluid Dynamics. XI, 346 pages. 1991.

Vol. m 6: N. M. J. Woodhouse, Special Relativity. VIII, 86 pages. 1992.

Vol. m 7: G. Morandi, The Role of Topology in Classical and Quantum Physics. XIII, 239 pages. 1992.

Vol. m 8: D. Funaro, Polynomial Approximation of Differential Equations. X, 305 pages. 1992.

Vol. m 9: M. Namiki, Stochastic Quantization. X, 217 pages. 1992.

Vol. m 10: J. Hoppe, Lectures on Integrable Systems. VII, 111 pages. 1992.

Vol. m 11: A. D. Yaghjian, Relativistic Dynamics of a Charged Sphere. XII, 115 pages. 1992.

Vol. m 12: G. Esposito, Quantum Gravity, Quantum Cosmology and Lorentzian Geometries. Second Corrected and Enlarged Edition. XVIII, 349 pages. 1994.

Vol. m 13: M. Klein, A. Knauf, Classical Planar Scattering by Coulombic Potentials. V, 142 pages. 1992.

Vol. m 14: A. Lerda, Anyons. XI, 138 pages. 1992.

Vol. m 15: N. Peters, B. Rogg (Eds.), Reduced Kinetic Mechanisms for Applications in Combustion Systems. X, 360 pages. 1993.

Vol. m 16: P. Christe, M. Henkel, Introduction to Conformal Invariance and Its Applications to Critical Phenomena. XV, 260 pages. 1993.

Vol. m 17: M. Schoen, Computer Simulation of Condensed Phases in Complex Geometries. X, 136 pages. 1993.

Vol. m 18: H. Carmichael, An Open Systems Approach to Quantum Optics. X, 179 pages. 1993.

Vol. m 19: S. D. Bogan, M. K. Hinders, Interface Effects in Elastic Wave Scattering. XII, 182 pages. 1994.

Vol. m 20: E. Abdalla, M. C. B. Abdalla, D. Dalmazi, A. Zadra, 2D-Gravity in Non-Critical Strings. IX, 319 pages. 1994.

Vol. m 21: G. P. Berman, E. N. Bulgakov, D. D. Holm, Crossover-Time in Quantum Boson and Spin Systems. XI, 268 pages. 1994.

Vol. m 22: M.-O. Hongler, Chaotic and Stochastic Behaviour in Automatic Production Lines. V, 85 pages. 1994.

Vol. m 23: V. S. Viswanath, G. Müller, The Recursion Method. X, 259 pages. 1994.

Vol. m 24: A. Ern, V. Giovangigli, Multicomponent Transport Algorithms. XIV, 427 pages. 1994.

Vol. m 25: A. V. Bogdanov, G. V. Dubrovskiy, M. P. Krutikov, D. V. Kulginov, V. M. Strelchenya, Interaction of Gases with Surfaces. XIV, 132 pages. 1995.

Vol. m 26: M. Dineykhan, G. V. Efimov, G. Ganbold, S. N. Nedelko, Oscillator Representation in Quantum Physics. IX, 279 pages. 1995.

Vol. m 27: J. T. Ottesen, Infinite Dimensional Groups and Algebras in Quantum Physics. IX, 218 pages. 1995.

Vol. m 28: O. Piguet, S. P. Sorella, Algebraic Renormalization. IX, 134 pages. 1995.

Vol. m 29: C. Bendjaballah, Introduction to Photon Communication. VII, 193 pages. 1995.

Vol. m 30: A. J. Greer, W. J. Kossler, Low Magnetic Fields in Anisotropic Superconductors. VII, 161 pages. 1995.

Vol. m 31: P. Busch, M. Grabowski, P. J. Lahti, Operational Quantum Physics. XI, 230 pages. 1995.

Vol. m 32: L. de Broglie, Diverses questions de mécanique et de thermodynamique classiques et relativistes. XII, 198 pages. 1995.

Vol. m 33: R. Alkofer, H. Reinhardt, Chiral Quark Dynamics. VIII, 115 pages. 1995.

Vol. m 34: R. Jost, Das Märchen vom Elfenbeinernen Turm. VIII, 286 pages. 1995.

Vol. m 35: E. Elizalde, Ten Physical Applications of Spectral Zeta Functions. XIV, 228 pages. 1995.

Vol. m 36: G. Dunne, Self-Dual Chern-Simons Theories. X, 217 pages. 1995.

Vol. m 37: S. Childress, A.D. Gilbert, Stretch, Twist, Fold: The Fast Dynamo. XI, 410 pages. 1995.

Vol. m 38: J. González, M. A. Martín-Delgado, G. Sierra, A. H. Vozmediano, Quantum Electron Liquids and High-T_c Superconductivity. X, 299 pages. 1995.

Vol. m 39: L. Pittner, Algebraic Foundations of Non-Commutative Differential Geometry and Quantum Groups. XII, 469 pages. 1996.

Vol. m 40: H.-J. Borchers, Translation Group and Particle Representations in Quantum Field Theory. VII, 131 pages. 1996.

Vol. m 41: B. K. Chakrabarti, A. Dutta, P. Sen, Quantum Ising Phases and Transitions in Transverse Ising Models. X, 204 pages. 1996.

Vol. m 42: P. Bouwknegt, J. McCarthy, K. Pilch, The W_3 Algebra. Modules, Semi-infinite Cohomology and BV Algebras. XI, 204 pages. 1996.

Vol. m 43: M. Schottenloher, A Mathematical Introduction to Conformal Field Theory. VIII, 142 pages. 1997.

Vol. m 44: A. Bach, Indistinguishable Classical Particles. VIII, 157 pages. 1997.

Vol. m 45: M. Ferrari, V. T. Granik, A. Imam, J. C. Nadeau (Eds.), Advances in Doublet Mechanics. XVI, 214 pages. 1997.

Vol. m 46: M. Camenzind, Les noyaux actifs de galaxies. XVIII, 218 pages. 1997.

Vol. m 48: P. Kopietz, Bosonization of Interacting Fermions in Arbitrary Dimensions. XII, 259 pages. 1997.

Printed by Publishers' Graphics LLC